Understanding Complexity in the Prehistoric Southwest

UNDERSTANDING COMPLEXITY IN THE PREHISTORIC SOUTHWEST

Editors

George J. Gumerman
Santa Fe Institute
Santa Fe, New Mexico

Murray Gell-Mann
Santa Fe Institute
Santa Fe, New Mexico

Proceedings Volume XVI

Santa Fe Institute
Studies in the Sciences of Complexity

Addison-Wesley Publishing Company
The Advanced Book Program
Reading, Massachusetts Menlo Park, California New York
Don Mills, Ontario Wokingham, England Amsterdam Bonn
Sydney Singapore Tokyo Madrid San Juan
Paris Seoul Milan Mexico City Taipei

Publisher: *David Goehring*
Editor-in-Chief: *Jack Repcheck*
Production Manager: *Michael Cirone*
Production Supervisor: *Lynne Reed*

Director of Publications, Santa Fe Institute: *Ronda K. Butler-Villa*
Publications Assistant, Santa Fe Institute: *Della L. Ulibarri*

This volume was typeset using TEXtures on a Macintosh II computer. Camera-ready output from a Hewlett Packard Laser Jet 4M Printer.

ISBN 0-201-52763-4 (Hardcover)
ISBN 0-201-52766-9 (Paperback)

1 2 3 4 5 6 7 8 9 10–MA–97969594
First printing, March 1994

About the Santa Fe Institute

The *Santa Fe Institute* (SFI) is a multidisciplinary graduate research and teaching institution formed to nurture research on complex systems and their simpler elements. A private, independent institution, SFI was founded in 1984. Its primary concern is to focus the tools of traditional scientific disciplines and emerging new computer resources on the problems and opportunities that are involved in the multidisciplinary study of complex systems—those fundamental processes that shape almost every aspect of human life. Understanding complex systems is critical to realizing the full potential of science, and may be expected to yield enormous intellectual and practical benefits.

All titles from the *Santa Fe Institute Studies in the Sciences of Complexity* series will carry this imprint which is based on a Mimbres pottery design (circa A.D. 950–1150), drawn by Betsy Jones. The design was selected because the radiating feathers are evocative of the outreach of the Santa Fe Institute Program to many disciplines and institutions.

Santa Fe Institute
Studies in the Sciences of Complexity

Lectures Volumes

Vol.	Editor	Title
I	D. L. Stein	Lectures in the Sciences of Complexity, 1989
II	E. Jen	1989 Lectures in Complex Systems, 1990
III	L. Nadel & D. L. Stein	1990 Lectures in Complex Systems, 1991
IV	L. Nadel & D. L. Stein	1991 Lectures in Complex Systems, 1992
V	L. Nadel & D. L. Stein	1992 Lectures in Complex Systems, 1993

Lecture Notes Volumes

Vol.	Author	Title
I	J. Hertz, A. Krogh, & R. Palmer	Introduction to the Theory of Neural Computation, 1990
II	G. Weisbuch	Complex Systems Dynamics, 1990
III	W. D. Stein & F. J. Varela	Thinking About Biology, 1993

Reference Volumes

Vol.	Author	Title
I	A. Wuensche & M. Lesser	The Global Dynamics of Cellular Automata: Attraction Fields of One-Dimensional Cellular Automata, 1992

Proceedings Volumes

Vol.	Editor	Title
I	D. Pines	Emerging Syntheses in Science, 1987
II	A. S. Perelson	Theoretical Immunology, Part One, 1988
III	A. S. Perelson	Theoretical Immunology, Part Two, 1988
IV	G. D. Doolen et al.	Lattice Gas Methods for Partial Differential Equations, 1989
V	P. W. Anderson, K. Arrow, D. Pines	The Economy as an Evolving Complex System, 1988
VI	C. G. Langton	Artificial Life: Proceedings of an Interdisciplinary Workshop on the Synthesis and Simulation of Living Systems, 1988
VII	G. I. Bell & T. G. Marr	Computers and DNA, 1989
VIII	W. H. Zurek	Complexity, Entropy, and the Physics of Information, 1990
IX	A. S. Perelson & S. A. Kauffman	Molecular Evolution on Rugged Landscapes: Proteins, RNA and the Immune System, 1990
X	C. G. Langton et al.	Artificial Life II, 1991
XI	J. A. Hawkins & M. Gell-Mann	The Evolution of Human Languages, 1992
XII	M. Casdagli & S. Eubank	Nonlinear Modeling and Forecasting, 1992
XIII	J. E. Mittenthal & A. B. Baskin	Principles of Organization in Organisms, 1992
XIV	D. Friedman & J. Rust	The Double Auction Market: Institutions, Theories, and Evidence, 1993
XV	A. S. Weigend & N. A. Gershenfeld	Time Series Prediction: Forecasting the Future and Understanding the Past
XVI	G. Gumerman & M. Gell-Mann	Understanding Complexity in the Prehistoric Southwest
XVII	C. G. Langton	Artificial Life III

Contributors to This Volume

George J. Armelagos, Emory University
Linda Cordell, University of Colorado
Patricia L. Crown, Arizona State University
Jeffrey S. Dean, University of Arizona
Paul R. Fish, University of Arizona
Murray Gell-Mann, Santa Fe Institute
George Gumerman, Santa Fe Institute
Christopher G. Langton, Los Alamos National Laboratory and Santa Fe Institute
Jonathan Haas, Field Museum of Natural History
Edmund J. Ladd
Stephen Lekson, Crow Canyon Archaeological Center and Museum of New Mexico
Jane H. Kelley, University of Calgary
Keith W. Kintigh, Arizona State University
Timothy A. Kohler, Washington State University
Jerrold E. Levy, University of Arizona
Debra L. Martin, Hampshire College
Randall McGuire, State University of New York at Bringhamton
Ben A. Nelson, State University of New York at Buffalo
Fred Plog, New Mexico State University
Robert Reynolds, Wayne State University
Rolph M. Sinclair, National Science Foundation
Alan C. Swedlund, University of Massachusetts
Joseph Tainter, USDA Forest Service
Steadman Upham, University of Oregon
Wirt H. Wills, University of New Mexico
Norman Yoffee, University of Michigan

Contents

The Problem 1

Introduction
George J. Gumerman, Murray Gell-Mann, and Linda Cordell 3

Approaches to Understanding Southwestern Prehistory
Stephen Lekson, Linda Cordell, and George J. Gumerman 15

Southwestern Contributions to the Understanding of Core-Periphery Relations
Joseph Tainter 25

Human Biology and Health 37

Issues in Demography and Health
Alan C. Swedlund 39

Studies in Disruption: Demography and Health in the Prehistoric Southwest
Ben A. Nelson, Debra L. Martin, Alan C. Swedlund, Paul R. Fish, and George J. Armelagos 59

Demographic Alternatives: Consequences for Current Models of Southwestern Prehistory
Ben A. Nelson, Timothy A. Kohler, and Keith W. Kintigh 113

Offering Explanations 147

The Nature of Explanation in Archaeology: A Position Paper
Linda S. Cordell 149

Toward Increasing Our Knowledge of the Past: A Discussion
Linda S. Cordell, Jane H. Kelley, Keith W. Kintigh, Stephen H. Lekson, and Rolf M. Sinclair 163

Historical Processes and Southwestern Prehistory: A Position Paper
Randall H. McGuire 193

Historical Processes in the Prehistoric Southwest
Jonathan Haas, Edmund J. Ladd, Jerrold E. Levy, Randall H. McGuire, and Norman Yoffee 203

Ethnographic Analogs: Strategies for Reconstructioning Archaeological Cultures
Jerrold E. Levy 233

Systems Modeling and Political Evolution: A Position Paper
Steadman Upham 245

A Perspective on Systems Modeling in Southwest Archaeology
Steadman Upham, Fred Plog, Robert Reynolds, and Joseph Tainter 265

Evolutionary and Ecological Modeling in Southwestern Archaeology
Wirt H. Wills 287

Complex Adaptive Systems and Southwestern Prehistory
Wirt H. Wills, Patricia L. Crown, Jeffrey S. Dean, and Christopher G. Langton 297

Memorandum to Murray Gell-Mann Concerning: The Complications of Complexity in the Prehistoric Southwest
Norman Yoffee 341

Index 359

The Problem

George J. Gumerman,† Murray Gell-Mann,† and Linda Cordell‡
†Santa Fe Institute, 1660 Old Pecos Trail, Suite A, Santa Fe, NM 87501
‡University Museum, University of Colorado, Boulder, CO 80309

Introduction

The Santa Fe Institute (SFI) is a multidisciplinary center dedicated to the understanding of simplicity and complexity. A large part of its work is devoted to the study of the similarities and differences among many kinds of complex adaptive systems (CAS). A complex adaptive system is a system that learns or evolves by utilizing acquired information. It identifies perceived regularities in the data stream reaching it and compresses those regularities into concise packages that are often called schemata. A schema, supplemented by further information, leads to a description of the outside world or to predictions about it or to behavior of the complex adaptive system in its interaction with that world.

In biological evolution, the genome of an organism is a schema. In the scientific enterprise, a theory is a schema. In the evolution of a society, such things as laws, traditions, kinship rules, and myths constitute schemata—they are sometimes described as cultural DNA. Schemata are subject to variation and the different variants are in competition with one another. The descriptions or predictions provided by a schema can be tested by comparison with further inputs of data. Similarly, behavior prescribed by a schema is tested by its consequences. Thus selection pressures are created that feed back to the competition among schemata. Some schemata are rewarded by survival or at least by promotion, while others are suppressed or demoted. The selection pressures in biological evolution favor a genome that leads

Understanding Complexity in the Prehistoric Southwest,
Eds. G. Gumerman and M. Gell-Mann, SFI Studies in the Sciences of
Complexity, Proc. Vol. XVI, Addison-Wesley, 1994

to survival and to successful reproduction on the part of the organism or its close relatives. Scientific selection pressures favor a consistent body of theory that leads to explanation of observations and successful predictions of further observations. In cultural evolution, schemata will be favored that, among other things, offer advantages in responding to changing social and environmental conditions.

Adaptation to changing circumstances, say on the part of a society, can take place on at least three levels, with three different time scales: (1) The society can adapt to certain changes by following the rules of a particular schema for behavior. In response to a drought, a village may be moved to a higher elevation where more moisture is available. Or the villagers may perform a ceremony believed to enhance the probability of rain. Or both. (2) A schema for the behavior of the society can be exchanged for a different one. A new religion might be introduced with a different way of praying for rain. New agricultural methods might be adopted. (3) The society may disappear, carrying its unsuccessful schemata into extinction with it, while other societies, with other schemata, may survive.

In many cases a CAS is composed of smaller complex adaptive systems interacting with one another. Such a composite CAS is said to consist of adaptive agents. Those agents are typically engaged in constructing schemata for describing and predicting one another's behavior. A society is a composite CAS in which the agents can be taken to be individuals or else households or larger social units. Other examples are ecological communities, composed of individual organisms of many species, and financial markets, composed of individuals or firms as investors.

A particularly instructive kind of complex adaptive system is a computer programmed to evolve strategies for playing a game. If playing the game optimally were a solved problem, as in tic-tac-toe, then such a computer would soon reach a kind of equilibrium by finding the optimal strategy. However, in a game like chess, where the optimal strategy has not yet been discovered, the computer operates far from any equilibrium, in a continuing search among possible strategies. Complex adaptive systems usually find themselves far from perfect adaptation in this sense. That is true to an even greater extent when the external conditions keep changing. Thus a CAS is often highly dynamic.

SFI operates by means of multidisciplinary research teams of affiliated researchers drawn from a diverse group of disciplines, academic institutions, research centers, and research-oriented corporations. The research problem is the cohesive element binding these scientists together; they do not share a particular research method or ensemble of theories. The external faculty along with additional invited scholars interact largely through workshops focused on specific topics, by means of a graduate summer school in the sciences of complexity, and by research networks linked through electronic communication.

The form of organization at SFI, together with the method of carrying out research, is nearly unique. So too is the manner in which questions are addressed. SFI researchers believe that different methods of inquiry are most effective if they are used in combination with one another, rather than in opposition. For instance, while it is common to give lip service to the concept that critical and creative thinking are complementary methods of inquiry, these approaches are often treated, in carrying out actual research, as if they were in conflict. Critical thinking ability is a major factor in the advancement of knowledge. It is the capacity to evaluate ideas, to probe weaknesses in logic, to judge concepts, and to test ideas. Creative thinking, on the other hand, is crucial for the development of new hypotheses and the ability to find and understand relationships among phenomena. Some researchers remain in the critical thinking stage, and there are those creative thinkers whose grand schemes cannot stand up to critical scrutiny. SFI scientists try to blend elements of critical and creative thinking in the way they address their problems. The researchers understand that a single individual need not excel in both types of thinking skills but that the two abilities must nevertheless be used in order to examine successfully large questions in any scientific domain.

The SFI philosophy similarly holds that generalizing and specializing traditions in science are complementary and not contradictory, and that it is important to adopt a generalizing approach while paying careful attention to detail. Since the behavior of a nonlinear complex system cannot be described in terms of the sum of the behaviors of its parts, an attempt must be made to comprehend the whole situation, even crudely if necessary, while also examining the individual elements with care. The interplay between the specialist who understands the details and the generalist (usually but not always a different person) who can view the whole picture is a vital part of understanding complex systems.

The characteristics that epitomize SFI research can profitably be applied to archaeological problems because cultures are complex adaptive systems. When referring to culture, archaeologists traditionally use both the terms *system*[1,4] and *complexity*.[20,23] Despite criticisms of the appropriateness of the use of the systems concept for culture[3,24] and problems involving use of the term *complex* in specific situations,[10,18] recent advances in understanding how systems are organized and evolve and a broader definition of complexity enhance the usefulness of the terms for understanding cultural systems.[11,25,26]

Most archaeologists and other anthropologists use the word "complex" in two different ways. It can be used in the general sense of "complicated" or "intricate" or else in a more technical sense, referring to societies that are hierarchically organized, usually with features of organization that are not found in egalitarian societies such as bands or tribes.[27,26] In the second sense, societies have been described as complex when there are many different roles, often associated with different status, that distinguish individuals and when criteria other than age, sex, and ability are applied in filling positions of power or authority.

Two of us have commented as follows on the relation between "effective complexity" as used in some of the scientific work at SFI and the technical use of the term "complexity" or "social complexity" in archaeology:

Complexity is not easy to define. In order to capture the intuitive ideas that most of us have of what complexity means, several different quantities would probably have to be introduced. It turns out, however, that in most scientific usage and in much ordinary discourse what is meant by the complexity of a system being observed is, more or less, the length of the description given by the observer of the regularities of the system.[11] This is obviously a subjective or context-dependent—even a behavioral-definition, but no comparable objective definition has ever been found. Probably there is none. Granted the subjectivity of the definition, it is still necessary to provide a number of comments and qualifications before the concept can be used.

For one thing, the length of an ostensive description is no good; it is just as easy to point to a complicated system as to a simple one. Likewise, since a short nickname could be given to any system, the description should be in a language previously agreed upon with a correspondent (and a distant correspondent at that, to eliminate the possibility of pointing!).

Not only the language, but also the knowledge and understanding of the world that are shared by the observer and the correspondent may significantly affect the length of message required for the description. Just as important is the level of detail achieved in the description—what is known in physical science as the "coarse graining."

Given the coarse graining, the language, and the assumed level of knowledge and understanding, the description should be as concise as possible. The length could, of course, be artificially inflated by the use of unnecessary verbiage or just by repeating things that could be said only once.

So far, then, we are discussing a definition of complexity based on (1) the length of a concise description, (2) to a distant correspondent, (3) using language previously agreed upon, (4) of the regularities of an observed system, (5) given the coarse graining that is applied, and (6) the knowledge and understanding of the world shared by the observer and the correspondent. But this definition still leaves the notion of regularities to be examined. What does it mean to separate regularities from random details?

For a finite stream of data, there is no rigorous way to distinguish a system's regular features from those that are attributed to chance. For an infinite stream, the situation is more hopeful. If possible, we should therefore be dealing with a very large body of data. As the amount of coarse-grained information increases, so does the meaningfulness of extracting regularities, since a regular pattern will be more likely to recur frequently enough to set it off from incidental features.

We can see how the foregoing observations apply in the description of a prehistoric society, with emphasis on its social structure. Evidently, complexity does not depend on the length of a message that merely designates sites or names a branch or phase of a particular ruin. We also see that the concentration on social complexity means that the coarse graining will ignore features of the remains that do not appear to bear importantly on social structure. The language for discussing that structure is fairly standard these days and is heavily influenced by certain theories, of various degrees of plausibility and usefulness, of what is considered to be a typical sequence of stages of societal developemnt. The regularities that are actually identified are similarly constrained by the limitations of current theories. Finally, it is clear that the likelihood of recognizing patterns of social structure is increased if there is an abundance of material.

In the light of these rather obvious general remarks, we can see that the social complexity assigned by an archaeologist to a prehistoric culture really does typically relate to the length of the concise description of regularities—the various social roles, the patterns of residence, the distribution and architecture of public buildings, the arts and crafts, the technology, the utilization of plant and animal species, the relations with other cultures, and so forth—as well as to the "variety of mechanisms for organizing these into a coherent, functioning whole."[23]

Sometimes, societies are described as more or less complex merely according to how they are thought to fit into a particular presumed sequence of evolutionary stages. Such an interpretation of the extent that it agrees with the more general interpretation given here. Fortunately, it often does agree, and the concept of complexity as a measure of the evolutionary status of a society continues to have heuristic potential despite recent criticism.[20]

There is no question that after the adoption of agriculture, socieities tend to evolve into more complex entities. The intertwined relationships among agricultural intensification, population growth, political integration, and social role diversity are major aspects of the processes of cultural evolution. Usually those changes are associated with an increase in social inequality, and it is an interestng question whether that effect was less blatant in parts of the Southwest than in societies at comparable stages of evolution elsewhere in the world, as has sometimes been claimed.

—from Gumerman and Gell-Mann[15]
(references within this quoted material have been changed to reflect the reference list for this chapter)

Many archaeologists today find that the use of traditional categories such as simple vs. complex, does not promote understanding of how cultures work or develop and evolve over time.[2,8,17] Rather than trying to maintain the categories, recent approaches examine the trajectories of organizational dimensions such as community or site size, numbers of levels in site size hierarchies, the area encompassed by exchange networks, evidence of craft specialization, etc. Much of this research is concerned with organizational properties.[16,17,19]

The distinction between specialization and generalization in research orientation, in the more common meanings of detailed versus broad, has also been a major theme in archaeology, although it is not much discussed.[14] While emphasis in southwestern archaeology in the last few decades has been on specialization, a number of generalizing studies have attempted to bring the detail of more specialized efforts into broader interpretations for larger sections of the Southwest, even the region as a whole. But how can archaeologists use the richness of detail and uniqueness of the culture history of scores of specific locales in the Southwest to understand general notions about human societies?

One answer lies in enhancing the kind of structure for investigating research problems advocated by SFI and traditionally used by archaeologists—bringing groups of people together to work on a problem. Archaeologists have always used group efforts. Until the 1960s, however, while field work, analysis, and write up were accomplished by groups of people with diverse specialties, usually only one individual was responsible for obtaining funding, directing the intellectual thrust of the project, and supervising excavation, analysis, and interpretation—a structure called the "lone scholar" approach.[13] Archaeologists operating under this model of scientific enterprise successfully laid the foundations for the understanding of the past that we have today. In the 1960s and 1970s, however, it became necessary to test other organizational models because the size and number of projects in contract archaeology required more participatory team efforts.

The change, however, was not all due to contract work. One of the first problem-oriented group efforts was the Southwest Anthropological Research Group (SARG). In part, SARG was a response to Streuver's call for research that would transcend the "one scholar" approach that was not adequate to address problems of cultural interaction and culture change. SARG was a concerted group effort to collect data in a comparable way, to try to understand why people distributed themselves across the landscape the way they did.[12] The exercise was notable for two reasons. First, it was a coordinated group effort in what was then an atmosphere of divisiveness in southwestern archaeology. SARG engendered tremendous intellectual excitement because it involved many scholars of different intellectual persuasions who were willing to share unpublished data, and engage in open intellectual debate. Second, SARG was distinctive in that it met to pose and address problems, rather than to present results. It is unfortunately characteristic of archaeologists, worldwide, that we seldom get together to discuss problems and to formulate and prioritize research concerns.[6] Archaeologists tend towards individualism in conducting and reporting research. Even the newly founded biannual Southwest Symposium and advanced seminars at the School of American Research usually consist of pre-prepared single-author papers with introductory and synthesizing commentary. These formats hinder insights that can be cooperatively derived. SARG was conceived as an antidote to individualism.

The increasing sense of the commonality of cultural processes that operated regionally, cutting across the major southwestern traditions was heightened by an advanced seminar, "Dynamics of Southwest Prehistory," held at the School of American Research in 1983.[5] While the structure of the seminar emphasized subregions, such as the San Juan and Upper Rio Grande, the seminar participants named a number of "hinge points" that were characterized by common processes. At about the same time, in 1984, the founding of the Santa Fe Institute and the involvement of Murray Gell-Mann, Robert McCormick Adams, and Douglas W. Schwartz in both SAR and SFI led to the fortuitous coincidence of the renewed interest in viewing the prehistoric Southwest as a single entity while recognizing the great diversity within the region.

A few years thereafter two seminars were held. Jonathan Haas, then of the School of American Research (now at the Field Museum of Natural History), along with Gell-Mann and George Gumerman, planned the long-term effort and then, with the advice of many archaeologists, a structure for the meetings gradually evolved that resulted in the two meetings. The first meeting was an advanced seminar titled "The Organization and Evolution of Prehistoric Southwestern Society" convened in September, 1989, at the School of American Research.[14] Its goals were to examine the pan-southwestern themes and processes (some that resulted in increasing complexity) to attempt to understand the linkages that integrated the various themes and processes, and to lay a basic foundation of knowledge about southwestern archaeology. The advanced seminar was planned to set the stage for the more interdisciplinary workshop at SFI.

For five days in October, 1990, 27 archaeologists, ethnologists, biological anthropologists, and complex system theorists in biology, computer science, and physics met at the Santa Fe Institute to consider the archaeology of the Southwest. Supported by SFI, SAR, and private donors, the workshop, "Organization and Evolution of Southwestern Prehistoric Societies," was devoted to studying and attempting to explain the evolution of complexity in the Southwest. This volume is a result of that workshop.

Since the nature of archaeological data and reasoning is not generally understood outside the discipline, a day-long field trip prior to the conference was organized by Rolf Sinclair of the National Science Foundation. Stephen Lekson, then of the Museum of New Mexico (and now at Crow Canyon Research Center), led a group of workshop participants to archaeological sites near Santa Fe that represent various prehistoric eras and exemplify interpretive quandaries. General overviews of the history and goals of the workshop were given by Gumerman, complex adaptive systems by Gell-Mann, southwestern prehistory by Lekson, cultural evolution by archaeologist Joseph Tainter (U.S. Forest Service), and problems in modeling prehistoric population dynamics by archaeologists Ben Nelson (SUNY Buffalo), Keith Kintigh (Arizona State University), and Timothy Kohler (Washington State University).

Participants were divided into working sessions on a variety of topics of concern in contemporary archaeology such as the nature of explanation in archaeology, historical processes, environmental and evolutionary modeling, systems modeling, and environment, demography, and health. For each topic, an anthropologist was asked to write a 10- to 15-page draft platform statement. These are very general papers, without specific reference to the Southwest, meant to introduce the general subject to nonanthropologists. They are published in this volume as position papers. The position papers were distributed to all the participants a month before the workshop. In addition, each participant was given a copy of the draft papers presented at the SAR seminar and the nonarchaeologists a reading list of general overviews of Southwest prehistory.

The goal of the Explanation in Archaeology group was to examine how archaeologists in the Southwest structure their research and to suggest ways in which we might do better. The group affirmed the method of science as appropriate for archaeological inquiry, noting the general success of ecological approaches but also recognizing some of its shortcomings. The group examined both the process of archaeological research and the sociology of archaeology including how ideas are rewarded. For example, novelty is encouraged in tenure evaluations while testing existing ideas and using extant data sets are discouraged. The group offers a number of suggestions. These include ways in which more attention might be focused on selecting problems for research, working toward more robust methods of confirmation, promoting what they call a comparative archaeology that would develop a large cross-cultural data base of archaeological information for comparative research, rewarding joint-authorship of publications and encouraging journals to permit footnotes for more complete documentation.

The Historical Processes working group attempted to take historical factors into consideration in understanding culture change and stability. They were concerned with how the existing form of a society structures the manner in which it can adapt, or not adapt, to changing conditions. They focused on characteristics intrinsic to the society which shaped change and stability and to the effect of interaction with other cultures. They argue that all histories are constructed and that there are many possible constructions. They do present a history of development of societies in the Southwest, and in acknowledgment of the multiplicity of histories and voices, they encouraged other workshop participants to comment on their history in footnotes that they include in their chapter.

The Environmental and Evolutionary Modeling group used rugged fitness landscape models which are based on the concept of coupling, so that a change in the fitness of one system affects the fitness of other systems on the landscape. This concept may help explain the overall trajectory of increasing population despite the series of abandonments. A pattern common in the prehistoric Southwest is that of sequential subregional abandonment. For example, the Chaco Canyon and Virgin River regions were abandoned in the late A.D. 1100s, Mesa Verde in the late 1200s, and the Middle Little Colorado River area in the 1400s. Despite these, the overall temporal trajectory of population growth was positive. The working group suggests that some prehistoric settlement relocations may have been to increasingly stable environmental contexts where deformation of the fitness landscape was more difficult.

Another group examined the topic of demography and health, looking at the close articulation between political centralization, cooperative networks, demography, and health. They were able to evaluate several different models that have been proposed to explain various instances of southwestern cultural development in relation to population growth or decline and general health characteristics. The models are truly systemic including such components as birth spacing, maternal health, and so on.

Another working group examined how increases or decreases in the complexity of regional systems of exchange and sociopolitical alliances might be modeled using Holland's Genetic Algorithm, as modified by workshop participant Robert Reynolds (Wayne State University). The Genetic Algorithm is based on a Darwinian paradigm of fitness modified to include learning. Using surrogate measures to monitor dimensions of demographic, economic, social, and political behavior, the working group investigated the potential of a model generated from the Genetic Algorithm to reproduce systems of comparable complexity to those known archaeologically. Efforts such as this one will eventually allow us to specify the ways in which systems of human behavior mirror or depart from the behavior of other complex adaptive systems.

Archaeologists have long been consumers of ideas (e.g., uniformitarianism, gradualist evolution, optimum foraging models, taphonomic processes) and techniques (e.g., radiocarbon dating, playnology) from other disciplines. Unfortunately, sometimes we may adopt theories, concepts, and methods without fully understanding them in their original context, therefore running the risk of using them inappropriately or as simple heuristic devices (e.g., positive and negative feedback). Yet, when we ask how successful are the approaches advocated by SFI and whether or not we have done any more than show trends in Southwest prehistory, we see the integration of different perspectives as beneficial for everyone. The key seems to have been to bring together, in small groups, those with different talents and different data since those scholars who see relationships more often than boundaries.

Sometimes, archaeologists may adopt theories and concepts far too rigidly, making them more static than they should. It is for this reason, for example, that some archaeologists have made formal pronouncements about evolution that seem misguided to evolutionary biologists. At SFI, theoreticians in biological evolution and demography offer suggestions and alternative hypotheses to understand processes they are unsure about. Discovering that things we thought we knew about evolutionary biology are imperfectly understood by evolutionary biologists was a revelation. The nonarchaeologists provided the exceedingly rich and variegated detail about their disciplines that is essential for archaeologists to understand if we are to draw meaningful lessons from their data and useful metaphors from their theories when they are appropriately applied to past cultural behavior.

An important lesson from the experience is the futility of the proponents of rigid points of view, those advocates who do not allow for any flexibility. Cordell[6] in her position paper for the workshop listed a dozen or so philosophical flags under which some archaeologists march today. If these theoretical perspectives provide flexible guides for research direction, that is one thing. If they are adopted and applied with little flexibility, they are almost certainly wrong. Understanding something as fragmentary and as complex as past cultural systems requires the integration of different points of view—a harmony created by cooperative efforts, the cultivation of creative as well as critical thinking skills, and the application of new analytical and modeling techniques.

We are fully aware that these suggestions for restructuring archaeological research will not resolve all of our difficulties. But one of the participants in the SFI workshop, a nonarchaeologist, remarked on our tentativeness and lack of willingness to espouse partially formulated ideas and to take risks. What have we got to lose?

ACKNOWLEDGMENTS

This volume is the result of a multi-year effort on the part of many researchers—archaeologists and non-archaeologists alike. Both the workshop and the volume is truly a cooperative effort as the large number of multi-authored papers attest. The synergism resulting from the interaction of enthusiastic and knowledgeable researchers can be appreciated by the range and the high quality of the papers in this volume. We are grateful to all of the participants.

The staff of SFI, especially Ronda Butler-Villa, Ginger Richardson, Andi Sutherland, and Della Ulibarri provided not only the necessary support for the success of endeavors of this kind, but did so with an efficiency and friendliness that greatly enhanced our experience. All of the participants express their gratitude.

REFERENCES

1. Binford, Lewis R. "Archaeology as Anthropology." *Amer. Antiquity* **28** (1962): 217–225.
2. Braun, David P., and S. Plog. "Evolution of 'Tribal' Social Networks: Theory and Prehistoric North American Evidence." *Amer. Antiquity* **47(3)** (1982): 504–525.
3. Braun, David P. "Are There Cross-Cultural Regularities in Tribal Social Practices?" In *Between Bands and States*, edited by Susan A. Gregg, 423–444, Occasional Paper No. 9. Center for Archaeological Investigations, Southern Illinois University, Carbondale, 1991.

4. Clarke, David L. *Analytical Archaeology.* New York: Columbia University Press, 1968.
5. Cordell, Linda, and George J. Gumerman. *Dynamics of Southwest Prehistory.* Washington, DC: Smithsonian, 1989.
6. Cordell, Linda. "The Nature of Explanation in Archaeology: A Position Paper." This volume.
7. Flannery, Kent V. "The Cultural Evolution of Civilizations." *Ann. Rev. Ecol. Sys.* **3** (1972): 399–426.
8. Feinman, G., and J. Nietzel. "Too Many Types: An Overview of Sedentary Prestate Societies in the Americas." In *Advances in Archaeological Method and Theory*, edited by M. Schiffer, Vol. 7, 39–102. New York: Academic Press, 1984.
9. Fried, Morton H. *The Evolution of Political Society: An Essay in Political Anthropology.* New York: Random House, 1967.
10. Johnson, Gregory A. "Dynamics of Southwestern Prehistory." In *Far Outside—Looking In*, edited by Linda S. Cordell and George J. Gumerman, 371–389. Washington D.C.: Smithsonian Institution, 1989.
11. Gell-Mann, Murray. "Complexity and Complex Adaptive Systems." In *The Evolution of Human Languages*, edited by J. Hawkins and M. Gell-Mann. Santa Fe Institute Studies in the Sciences of Complexity, Proc. Vol. XI, Redwood City, CA: Addison Wesley, 1992.
12. Gumerman, George J., ed. *The Distribution of Prehistoric Population Aggregates. Proceedings of the First Annual Meeting of the Southwestern Anthropological Research Group.* Anthropological Papers, No.1. Prescott College, 1971
13. Gumerman, George J., ed. *Exploring the Hohokam: Prehistoric Desert Dwellers of the Southwest.* Albuquerque, NM: University of New Mexico Press, 1991
14. Gumerman, George J., ed. *Themes in Southwestern Prehistory.* Santa Fe, NM: School of American Research Press, 1993.
15. Gumerman, George J., and Murray Gell-Mann. "Cultural Evolution in the Prehistoric Southwest." In *Themes in Southwestern Prehistory.* Santa Fe, NM: School of American Research Press, 1993.
16. Kowalewski, Stephen A. "The Evolution of Complexity in the Valley of Oaxaca." In *Annual Review of Anthropology*, 39–58. Palo Alto, CA: Annual Reviews Inc., 1990.
17. Lightfoot, Kent G. *Prehistoric Political Dynamics: A Case Study from the American Southwest.* DeKalb: NIU Press, 1984.
18. Lightfoot, Kent G., and Steadman Upham, "Complex Societies in the Prehistoric American Southwest." In *The Sociopolitical Structure of Prehistoric Southwestern Societies*, edited by S. Upham, K. Lightfoot, and R. A. Jewett. Boulder, CO: Westview Press: 1989.

19. Neitzel, Jill. "Hohokam Material Culture and Behavior: The Dimensions of Organizational Change." In *Exploring the Hohokam Prehistoric Desert Peoples of the American Southwest*, edited by G. J. Gumerman, 177–230. Albuquerque, NM: University of New Mexico Press, 1991.
20. Plog, Fred. "Studying Complexity." In *The Sociopolitical Structure of Prehistoric Southwestern Societies*, edited by Steadman Upham, Kent G. Lightfoot, and Roberta A. Jewett, chapter 5, 103–125. Boulder, CO: Westview Press, 1989.
21. Service, Elman R. *Origins of the State and Civilization: The Process of Cultural Evolution.* New York: Norton, 1975.
22. Struever, Stuart. "Problems, Methods, and Organization: A Disparity in the Growth of Archaeology." In *Anthropological Archaeology in the Americas*, 131–133. Washington, D.C.: The Anthropological Society of Washington, 1968.
23. Tainter, Joseph A. *The Collapse of Complex Societies.* New York: Cambridge University Press, 1988.
24. Trigger, Bruce G. "Archaeology at the Crossroads: What's New?" *Ann. Rev. Anthropology* **13** (1984): 275–300.
25. Wills, W. H. "Evolution and Ecological Modeling in Southwest Archaeology." This volume.
26. Yoffee, N. "A Memo to Murray Gell-Mann Concerning: The Complications of Complexity in the Prehistoric Southwest." This volume.
27. Yoffee, Norman "Too Many Chiefs? (or, Safe Texts for the 1990s)." In *Archaeological Theory: Who Sets the Agenda?*, edited by Norman Yoffee and Andrew Sherratt, 60–78. NBW Directions in Archaeology. Cambridge: Cambridge University Press, 1993.

Stephen Lekson,* Linda Cordell,† and George J. Gumerman‡
*Crow Canyon Archaeological Center, 1777 S. Harrison Street, Denver, CO 80210 and Laboratory of Anthropology, Museum of New Mexico, P.O. Box 2087, Santa Fe, NM 87504
†University Museum, University of Colorado, Boulder, CO 80309
‡Santa Fe Institute, 1660 Old Pecos Trail, Suite A, Santa Fe, NM 87501

Approaches to Understanding Southwestern Prehistory

As archaeologists who are interested in understanding how prehistoric southwestern society was organized and how it evolved into more complex forms, we believe that valuable insights can be gained by listening to a variety of approaches and points of view. We share a baseline resource that is the land itself, together with its archaeological materials that are tangible remains of its ancient inhabitants. With these data in common, we are confident that it is productive to examine them by bringing together those who espouse different perspectives within archaeology, as well as those who represent diverse disciplinary backgrounds. The Santa Fe Institute (SFI) encourages discourse among people with different points of view and backgrounds. The diverse intellectual views people bring to SFI can be reconciled, at least in part, by moving to higher levels of abstraction. This is how scientific knowledge progresses. The SFI, by standing outside of specific campuses and traditional research institutions, provides an appropriate structure for people to discuss the Southwest without the intellectual baggage of institutional agendas.

Interdisciplinary approaches should be pursued, especially in this case, because by their very nature they "de-provincialize" the Southwest by providing a perspective that is more abstract and scientific. Using a variety of approaches and working at higher level of abstraction should offer new ways of viewing issues. We think it is important to remind our readers that archaeology has never been a "self-contained"

Understanding Complexity in the Prehistoric Southwest,
Eds. G. Gumerman and M. Gell-Mann, SFI Studies in the Sciences of
Complexity, Proc. Vol. XVI, Addison-Wesley, 1994

endeavor. It has always incorporated techniques, methods, and models from other fields. While occasionally borrowing models from elsewhere has resulted in interpretations that have been considered naive, archaeology as a whole is consistently enriched and advanced by its intellectual eclecticism.

For those interested in the behavior of complex adaptive systems (CAS), the archaeological record of prehistoric southwestern societies offers a unique and challenging resource. This record includes detailed information on 2,000 years of uninterrupted human occupation during which there were patterned technological, social, and ideological changes constrained by a harsh and unpredictable natural environment. The record of a variable climate and changing environment, documented with unusual precision in tree-rings, pollen, and alluvial sediments, forms one dimension along which to view social responses that included changes in subsistence practices, economy, and settlement strategy.

During the prehistoric period of interest, transitions included accepting domestic plants and developing farming economies, and, at times, intensifying agricultural production. At other times, some societies shifted more effort to foraging wild plant and animal food. Economic behaviors included periods during which there were strong inter-regional networks of exchange, but there were also intervals of subregional isolation. Settlement strategies involved communities that were dispersed and those that were aggregated. Some settlements reflect planning at the suprahousehold level but others do not.

Despite local variation, broad regional patterns reveal similarities in the timing of particular economic and social changes as reflected in archaeological sites and artifact assemblages. The trajectories of the patterned changes may be viewed as the result of CAS evolving by similar processes or as affected by similar historical contingencies. Viewing patterned changes in these ways may permit us to take advantage of recent advances in the mathematics of modeling physical and biological CAS in order to distinguish statistically universal processes from unique historical events in cultural CAS.

Most CAS have a number of similar features. They all consist of many relatively independent and interactive components that also interact with their environments and that, in doing so, modify their total behavior (that is, they "learn" in the broadest sense). Initially, at least, they are both self-organizing and self-reproducing; i.e., there is an emergence of order deriving from the dynamics of the system itself, and this order is capable of reproducing itself. In part, these characteristics are derived from the CAS ability to generate, acquire, and process information. This information is often stored in compressed form. Examples of compression include the information stored in DNA and in some computer programs. In the case of human culture traditions, myths, and ritual appear to store information in similarly compressed form.[3,10] CAS usually are far from equilibrium, and many typically evolve to a critical state, i.e., self-organized criticality.[1] While small changes in critical states, may or may not cause major transformations in the systems, they are like the straw, no heavier than any of the others, that broke the proverbial camel's back. It has also been shown that many dynamic systems move toward the boundary between chaotic or random states and more ordered states, and that it is at this boundary that major transformations occur.[7,12]

All CAS appear to be terribly complicated to those who study and seek to understand them and their histories over time. Many of these systems are studied by entire disciplines of specialists who define and describe their intricacies and changes in the systems' states over time. The essential complexity of these systems virtually insures that, within the specialist disciplines, there will be different (often conflicting or competing) theoretical models to "explain" or predict their behavior. For example, "laws" or rules of microeconomic theory are very different from macroeconomic models, yet both may be used to describe the same national recession. The micro- and macroeconomic perspectives will also have vastly different, and conflicting, implications for the enactment of policy that might end the recession. Similarly, the biological evolution of some classes of organisms may be described as examples of Darwinian competition or, by another group of biological theorists, as examples of synergistic mutualism. The novel advantage of viewing all complex systems and their trajectories first or primarily as complex adaptive systems is that they should share certain characteristics that can be modeled mathematically. As Wills[20] and others so clearly point out, while each historical process is unique from the perspective of a particular discipline, when viewed across disciplines it becomes another in a series of examples of historical processes that are similar in specifiable ways.

Anthropologists and archaeologists are not alone in considering their particular subject matter to be unique and uniquely confounding. Beyond that, in the context of trying to model our data as CAS, it is our responsibility to describe the systems we are studying in as much detail as possible. In charting the histories of these systems, the more precisely elaborate we can be the more accurate the model. The more texture and context we can provide in describing the history of the systems we study, the more accurate will be the eventual derived "Canonical Ensemble" (Wills et al.[21]).

It may well be the potential richness and the quantitative nature of some of the data of southwestern archaeology that makes the field particularly interesting to students of CAS, a point we return to later. First though, it is important to explain the effort the workshop participants made in the context of anthropological archaeology in general.

Although this observation is irrelevant to the system theorists, all but the most extreme post-processual and deconstructionist positions in anthropology are concerned with looking for cross-cultural regularities and patterns. The focus on comparison appears as early as Tylor's classic 1889 paper on kinship[15] and continues through the pages of all our modern journal articles and monographs. While anthropologists acknowledge that each society (like each archaeological site) is unique, aspects of societies (and sites) are comparable. There would be no recognizable anthropology as a field of study if anthropology were concerned with describing a universe of unique phenomena. Anthropologists wish to generalize about kinship, descent systems, household economies, techniques of farming, the nature of ideological systems, language acquisition and use, and the myriad other aspects of cultural behavior. For the system theorist each culture or culture history may be used as a "case" in the Canonical Ensemble. In that sense, each case might well be viewed as unique and still be useful for analysis.

Within anthropology, archaeology has its own perspective and constraints. On the positive side and in contrast to ethnology, archaeology has data that represent periods of time that are long enough to encompass multiple instances of culture change. Those of us in archaeology who work in the Southwest are additionally fortunate in having long and detailed records of paleoenvironmental conditions and change. On the negative side, although most archaeologists are interested in the cultures and culture histories they learn about through excavation, archaeology obviously cannot access these cultures or culture histories directly. Archaeologists must rely on theory to interpret the materials they excavate in terms that are meaningful with respect to human cultural behavior. Although archaeologists do routinely draw upon a body of theoretical constructs for this purpose, they are also keenly aware of the unevenness with which this theory is developed and the inadequate ways in which it has been evaluated.[2,14] Archaeologists who wish to provide system theorists with a richly detailed portrait of a past society and its various changes over time are also acutely aware of how little they really know and how fundamentally inaccurate their reconstruction may be.

As noted above, from the physical scientist's perspective, some archaeological data are wonderfully quantifiable. Southwestern archaeologists have millions of pottery shards, thousands of stone tools, and millions of pieces of the waste debris from producing them, and hundreds of radiocarbon dates. Most important in the Southwest, archaeologists have thousands of tree-ring dates and detailed, seasonal and annual, paleoenvironmental records that go back for fourteen centuries. The archaeologist, however, understands that there is often an ambiguous relationship between the artifacts available for study and the human behavior of interest. For example, the presence in Chaco Canyon of thousands of shards of pottery made of nonlocal clays may indicate the operation of an elaborate exchange network. Alternatively, the thousands of shards of nonlocal clays and the well-documented paucity of wood in the canyon may suggest that the ancient Chacoans produced their own pottery in remote locations near the firewood they needed for fuel. The admittedly impressive quantity of shards is far less important than accurate inferences about the underlying processes involved. Archaeologists of all theoretical proclivities have long looked upon artifacts of exotic style or materials as potentially informative about exchange. This is the underpinning of some of our most basic inferences, such as cross-dating, as well as for less secure interpretations of diffusion. Yet, it is apparent that exchange as a process must be better understood in order to be unambiguously documented and to be comprehended in relation to other processes.[19]

Perhaps, from the point of view of physical science, archaeologists are too concerned with causal explanations. We are vitally interested in the paleoenvironment of Mesa Verde, for example, because we want to know what caused its ancient inhabitants to abandon the mesa for other regions. Discovering a great drought in the tree-ring record is therefore a reason for celebration among archaeologists. In addition to the interest in cause, however, the indirect way in which archaeologists measure and "see" their subjects of inquiry (the societies of the past) dictate that we establish confidence in our bases of inference. We do not know how many people lived in eleventh century Chaco Canyon or at Mesa Verde. Although we do have some idea, we do not know precisely how climatic variables affected the habitats, life cycles, productivity, or behavior of all the plants and animals used for food by ancient southwestern peoples. We may never understand these things to the degree that we might wish in order to clarify the causes of past cultural change. Equally relevant though is that we must also understand these things in order to begin to describe—with any accuracy—the societies we want to model and to understand. There is a great deal of difference between a model of Chaco Canyon in the eleventh century that indicates a total human population in the hundreds, perhaps with small task groups moving seasonally to higher elevations to make pottery to bring back, and a model of the same area and time indicating thousands of residents involved in intraregional exchange. Which description do we offer to the system theorist as part of the "Canonical Ensemble?"

Archaeology is still a relatively new discipline. It is an area of study in which there is a high incidence of novel findings and in which accidental discoveries can have major impacts on interpretations. For example, although the Indus Valley civilizations have been known since the 1920s, the way in which they developed has only been documented since 1979/1980. At that time, excavations at accidentally discovered sites began to yield data of a very long sequence of sedentism and village farming that preceded the appearance of Harappan Civilization. The new information indicates that Harappan Civilization resulted from a very brief period of transformation, "a veritable revolution within the urbanization process," rather than from a long, slow period of cultural growth and change that would be predicted from other cases of the appearance of urban society.[1] In the ancient Southwest, we are well aware that interpretations of Chaco Canyon that were offered before the 1970s and 1980s are by and large inadequate because they did not take account of the size of the region integrated by the prehistoric roads and outliers that were documented at that time.[8,16] Again, descriptions of past cultural systems as cases for system modeling are not as secure as most archaeologists would wish.

On a world-wide basis, it seems that most of archaeology is moving ahead on two fronts. One is the still essential exploratory quest to learn the basics, in broad outline, of what happened in the past. There are many parts of the world where even quite general sequences simply do not exist. The prehistoric Southwest is obviously beyond this primary stage. The second direction in which most of archaeology seems to be going is toward developing a better understanding of the cultural and behavioral processes reflected in the material record. For example, much of the

[1]See Possehl,[13] pp. 261–262.

current literature is concerned with defining past systems of trade and exchange, or kinds of craft production or methods of resource procurement. The literature of the Southwest is currently dominated by these studies, and that is entirely appropriate given the state of the art of archaeology.

The contributors to the seminar have gone beyond both the stage of describing basic sequences and defining the kinds of behavioral systems (i.e., trade, types of agriculture, etc.) that were most likely operating in the past. They have, in admirable fashion, attempted to look at the complex processes that seem to have operated in various parts of the Southwest in terms that are amenable to more general system analysis. For example, one pattern common in the prehistoric Southwest is that of sequential subregional abandonment. The Chaco Canyon and Virgin River regions were abandoned in the late A.D. 1100s, Mesa Verde in the late 1200s, and the Middle Little Colorado River area in the 1400s. Despite these patterns of abandonment, the overall temporal trajectory of population growth was positive. One workshop group investigated the utility of viewing the sequential abandonments in terms of rugged fitness landscape models. The working group suggests that some prehistoric settlement relocations may have been to increasingly stable environmental contexts where deformation of fitness landscape was more difficult.

Very few archaeologists seem to be looking at their data at the same scale.[2] The issue of scale is fundamental to most, if not all, comparative questions. Archaeological entities, as complex adaptive systems, reflect social units larger than the individual or family. How much larger? What are the boundaries of the relevant archaeological phenomenon? Questions of regional scale have received considerable archaeological attention but remain largely unresolved. Again, Chaco illustrates some of the historical trends and conceptual quandaries in regional thinking.

Initially, Chaco Canyon was perceived as the setting of a series of independent, self-sufficient, agrarian villages: Pueblo Bonito, Chetro Ketl, and each of the other large structures was considered to be separate entities, discrete "complex adaptive systems" in today's terms. Later, the canyon was conceptualized as a single large settlement, containing a variety of "neighborhoods" (represented by individual buildings) of ethnically and socially distinct residence groups. More recently, Chaco was recognized as a "center" of a much larger region, marked by the geographic distribution of Chaco-like architecture. The precise nature of Chaco's "centrality" is unclear. Arguments were offered that Chaco's centrality was political, economic, ceremonial, or all three simultaneously. Even as the nature of the Chacoan center was debated, the size of its attendant region (the "Chacoan regional system") grew by leaps and bounds. First thought to be an area of 53,1002 km^2, some archaeologists now suggest that the Chacoan region incorporated over 300,000 km^2 of the Colorado Plateau.[9] Historically, the scales considered appropriate for understanding Chaco have changed dramatically. The ruins, artifacts, and remains are substantially the same; it is the archaeological understanding of those remains that has changed.

[2]For an exception, see Kowalewski,[11] pp. 39–58.

The relation of size and scale to complexity is not necessarily direct. Human formations, however, do tend to become more elaborate, to incorporate more diversity, to develop social and political complexity (in the anthropological usage), and to become more complex (sensu SFI) with increasing scale. Research by Johnson,[5] Kosse,[6] and Lekson[9] suggests scalar "thresholds" of organizational transformations, which appear to be related to inherent human capacities for processing information. Thus scale and size probably have direct implications for comparative studies of human complex adaptive systems, such as those in the ancient Southwest.

If we wish to study Chaco as a complex adaptive system, we must first determine the scales that are appropriate and the structure of the system at various scales. Definition of scale, so basic to the task at hand, is by no means straightforward. There has been and continues to be considerable debate about the size of the Chacoan region and the role of the Chacoan center within it. Indeed, the historical trend toward larger and larger scales in archaeological thinking may not yet have reached its limits. Chaco was one of two major "regional systems" in the tenth, eleventh, and twelfth century Southwest, the other being the very impressive Hohokam developments of the southern Arizona dessert. Archaeologists are only beginning to research and to understand the relationships of these two systems. It is likely that the dynamic between Chaco and Hohokam was neither minor nor trivial, but reflect the fundamental structure of the larger regional ecology of the Southwest. While there is little evidence of direct interaction between the two systems, their large scale suggests that they altered and redefined the world around them, including the larger environments in which both operated.

On the very largest scale, the Southwest itself was not an isolated human "laboratory." If we were to draw a series of increasingly inclusive geographic distributions, levels only one or two steps above that of Chaco and Hohokam would incorporate the Southwest into the much larger Mesoamerican sphere. A series of empires and states occupied the central Mexican region before Chaco took form; states such as Tula were at their height while Chaco grew and declined. The effects of these developments on Chaco, Hohokam, and the larger Southwest have not yet been comprehensively evaluated in other than purely historical terms. Nevertheless, the coexistence of these systems and preliminary work[17] suggest the larger continental context must have been critically important to the evolution of complex adaptive systems in the American Southwest.

In sum, the issue of scale is basic to our endeavor, but scale is as yet an unresolved issue in most southwestern archaeological inquiries. The theater for the evolution of human complex adaptive systems is biogeographic; and it is likely that on the largest scales, comparative studies of complex adaptive systems will have to embrace the concept of the New World as a mega-island, with all the constraints and considerations of island biogeography, to explain the evolutionary disparities between the human complex adaptive systems of the Old and New Worlds. Local studies, such as the southwestern case, are invaluable; but ultimately our search for realistic scales will necessarily lead us to considerations of the Southwest as part of the much larger whole.

Several of the nonanthropologist workshop participants commented that the archaeologists seemed reluctant to take the scholarly risks necessary for the exploration of new concepts about the past. This, they argue, is especially true considering the relatively high quality of some of the data of southwestern prehistory, particularly the paleoenvironmental and chronological data. The risks of being wholly inaccurate in regard to inferences about the past are less than for those working in other parts of the world. Yet, given the difficulties archaeology suffers because of its youth, it is understandable that the contributors to this seminar felt as though they were quite possibly out on various limbs with respect to interpretation. They nevertheless clearly believe that the potential for greatly enriching their understanding of the past that is offered through the perspective of complex adaptive system analyses is worth the risks. We urge that readers recall a quotation from Darwin before reading the rest of the book.

> Anyone whose disposition leads him to attach more weight to unexplained difficulties than to the explanation of a certain number of facts will certainly reject my theories.[3]

[3]See Darwin,[4] p. 453.

REFERENCES

1. Bak, Per, and Kan Chen. "Self-Organized Criticality." *Sci. Am.* **264** (1991): 46–53.
2. Binford, Lewis R. "Objectivity, Explanation and Archaeology 1980." In *Theory and Explanation in Archaeology*, edited by C. Renfrew, M. J. Rowlands, and B. A. Segraves, 125–138. New York: Academic Press, 1983.
3. Connerton, Paul. *How Societies Remember.* Cambridge, MA: Cambridge University Press, 1989.
4. Darwin, Charles. *The Origin of Species by Means of Natural Selection or the Preservation of Favored Races in the Struggle of Life.* Harmondsworth: Penguine Classics edition, Penguin Ltd., 1859 (1968).
5. Johnson, Gregory, A. "Organizational Structure and Scalar Stress." In *Theory and Explanation in Archaeology*, edited by C. Renfrew, M. J. Rowlands, and B. A. Segraves, 389–423. New York: Academic Press, 1982.
6. Kosse, Krisztina. "Group Size and Societal Complexity: Thresholds in the Long-Term Memory." *J. Anthro. Arch.* **9** (1990): 275–303.
7. Langton, Christopher. "Studying Artifical Life with Celluar Automata." *Physica* **22D** (1986): 120–149.
8. Lekson, Stephen, H., Thomas C. Windes, John R. Stein, and W. James Judge. "The Chaco Canyon Community." *Sci. Am.* **256(7)** (1988): 100-127.
9. Lekson, Stephen. "The Community in Anasazi Archaeology." In *Households and Community: Proceedings of the 21st Annual Chacmool Conference*, edited by S. MacEachern, D. Archer, and R. Garvin, 181–185. Calgary, 1989.
10. Gell-Mann, Murray. "Complexity and Complex Adaptive Systems." In *The Evolution of Human Languages*, edited by J. A. Hawkins and M. Gell-Mann. Santa Fe Institute Studies in the Sciences of Complexity, Proc. Vol X, 3–13. Reading, MA: Addison-Wesley, 1992
11. Kowalewski, Stephen A. "The Evolution of Complexity in the Valley of Oaxaca." In *Annual Review of Anthropology*, 39–58. Palo Alto., CA: Annual Reviews, 1990.
12. Packhard, Norman. "Adaptation Towards the Edge of Chaos." Technical Report CCSR-88-5, Center for Complex Systems Research, University of Illinois, 1988
13. Possehl, Gregory L. "Revolution in the Urban Revolution: The Emergence of Indus Urbanization." In *Annual Review of Anthropology*, 261–395. Palo Alto, CA: Annual Reviews, 1990.
14. Schiffer, Michael B. *Formation Processes of the Archaeological Record.* Albuquerque, NM: University of New Mexico Press, 1987.
15. Tylor, E. G. "On a Method of Investigating the Development of Institutions, Applied to Laws of Marriage and Descent." *J. Roy Anthro. Inst.* **18** (1989): 245–269.

16. Vivian R. Gwinn. *The Chacoan Prehistory of the San Juan Basin.* San Diego, CA: Academic Press, 1990.
17. Upham, Steadman, Gary M. Feinman, and Linda M. Nicholas. "New Perspectives on the Southwest and Highland Mesoamerica, a Macroregional Approach." *Review* **XV (3)** (1992): 427–451.
18. Upham, Steadman, Fred Plog, Robert Reynolds, and Joseph Tainter. "A Perspective on Systems Modeling in Southwestern Archaeology." This volume.
19. Webb, Malcom C. "Exchange Networks in Prehistory. Annual Review of Anthropology." *Ann. Rev.* (1974): 357–385.
20. Wills, W. H. "Evolutionary and Ecological Modeling in Southwestern Archaeology." This volume.
21. Wills, W. H., P. L. Crown, J. S. Dean, and C. G. Langton. "Complex Adaptive Systems and Southwestern Prehistory." This Volume.

Joseph A. Tainter
USDA Forest Service, 2113 Osuna NE, Albuquerque, New Mexico 87113

Southwestern Contributions to the Understanding of Core-Periphery Relations

Several years ago I found great significance in David Clarke's characterization of archaeology as "...an undisciplined empirical discipline...,"[1] whose members turn out such a volume of descriptive information that we can scarcely hope to assimilate it. Today we produce, if anything, an even greater volume of detailed and seemingly endless descriptions of archaeological sites, their contents, and their settings. Yet partly in reaction to the situation Clarke described, archaeology has changed greatly in recent decades. It is now respectable to generalize, to theorize, and to interpret at least a few of our gigabytes of data. Perhaps it is too respectable. For the revolution in archaeology has opened Pandora's box and unleased on us a plague of theoretical archaeologies. We have become an undisciplined theoretical discipline.

I now find great significance in the list of theoretical archaeologies that Linda Cordell compiled in her position paper for this workshop.[5] She listed the following recognized types of archaeology: "...ahistorical, behavioral, cognitive, critical, ecological, empirical, evolutionary, feminist, functional, historical, Marxist, processual, post-processual, and structuralist." To this we must add the new field of hermeneutical archaeology, which was recently brought to our attention by Ian Hodder.[13] For those who, like me, need to look the word up, it means the methodology of

[1] Clarke,[4] xiii.

interpreting texts, and referred originally to the study of scriptures. A hermeneutical archaeology may be appropriate, for of course we all know certain colleagues who apparently do feel that their writings should be treated as scriptures. The *Oxford English Dictionary* characterizes hermeneutics as the opposite of exegesis, so perhaps we may anticipate the day when we are presented with an exegetical archaeology.

Philosophers like Lester Embree are finding us such a fertile field that they are devoting significant parts of their careers to determining what this efflorescence means. Yet despite the Sisyphean efforts we have devoted to theoretical archaeology, we are vulnerable to a criticism raised some years ago by Thomas King. Despite unparalleled amounts of work, King asserted, archaeology in recent years has not succeeded in answering a single, major research question.[17] That brings us to the point of these remarks.

Southwestern archaeology is fascinating in its own right and has been the subject of uncountable symposia, seminars, conferences, and workshops. Yet the ultimate value of the archaeology of any area is what we can learn about the evolution of our species. This should be, but rarely is, our ultimate purpose: to use the data of southwestern archaeology to discuss and evaluate general ideas about the evolution of complexity in human societies. This is the kind of research that King challenged us to undertake.

However much we like to quibble over just how complex southwestern societies were, we can, I trust, agree that they were more complex when the Spanish arrived than when the PaleoIndians did. In between these crucial events, southwestern societies evolved more complex institutions and behaviors. They did so at rates that varied over time and space, but they did not do so unilineally. Instead what we see are societies that developed greater complexity, collapsed, abandoned territories, and began somewhere else to develop complexity again. In undertaking to study this matter, we need not concern ourselves with finding the right evolutionary label to apply to southwestern societies, nor with fruitless debates over whether they were more or less complex than anyone else. The important matter is the *process* by which societies *become* more complex.

At one point in the development of our discipline, we were content to assume that human societies have a latent tendency toward complexity. Complexity was thought to be a desirable thing, and the inevitable result of surplus food, leisure time, and human creativity. Regrettably this view took hold among the public, where it remains deeply entrenched. There are many reasons why this scenario never was correct. One of the most important is simply that complexity costs. More complex societies are more expensive to maintain than simpler ones and require greater support levels per capita.[2] In a society that is more complex, there are more groups and social roles, more networks among individuals and groups, more controls (both horizontal and hierarchical), more processed information, more centralization of information, and more specialists who are not engaged directly in

[2] Tainter,[30] pp. 91–92.

producing resources.[14,20,30,31] If we as a discipline once believed that all this came about simply by virtue of human creativity, then we must also have believed that there really is such a thing as a free lunch.

In the days before fossil fuel subsidies, the higher cost of supporting a more complex society meant that *people* worked harder. To account then for complexity, we must assume one of two alternatives. The first is that complex societies emerged because people acted irrationally. That is, they acted irrationally in an economic sense: undertaking costly actions not in their own best interests. Various archaeologists do indeed make such an argument, some implicitly (e.g., Judge[16]; discussed in Tainter and Plog[33]), others quite unashamedly. Since the development of ever-more-complex societies is nearly a constant in human history, those who make this assumption bear an unenviable burden: to explain how so much of our history, and our success as a species, can be explained as a function of persistent irrationality.

If that sounds too abstract, then picture an overworked Chinese or Mayan or Mesopotamian peasant, or a Roman peasant selling his children into slavery because of the burden of taxes. Participating in a complex society is, for such people, a benefit/cost equation. They may not have the option of *not* participating, but their participation is not irrational. If supporting complex institutions is, for such people, too high in cost or too low in benefits, they have historically responded with degrees of resistance that range from withdrawal and apathy to outright rebellion (see Eisenstadt[3] and Wolf[38]).

It seems farfetched to believe, then, that complex institutions could, time and again, have arisen from economic irrationality, but precisely that assumption underlies much past and present archaeological thinking. We have, for example, recently been offered a model which sees the San Juan Basin Anasazi building vacant Chacoan Great Houses, and receiving for their effort religious inspiration and a few trinkets of turquoise.[16] It is hard to understand how a people could have mastered such a marginal environment and thrived in it for centuries, if they were so unable to make the most rudimentary calculations of return on effort.

If perpetual irrationality does not seem credible as the driving force of history, then we are forced to the alternative assumption. This is that complexity is a response to circumstances and comes about under conditions of compelling need or perceived benefit.[30] Complexity evolves under the force of pressures within a society; or of pressures from without; or because a simpler society emulates a more complex neighbor, seeking therein a path to wealth and power, or perhaps merely self-preservation. This assumption seems far more defensible than the alternative, and I will rely on it as a basis for further discussion.

As they experienced the development of greater complexity, southwestern societies underwent an evolutionary course alike in many respects to that of other societies. This makes southwestern archaeology a suitable database for generalizing studies. It is easy to point out some of the things that southwestern and other societies have in common: the development and intensification of agriculture; the

[3] Eisenstadt,[10] p. 209

establishment of large towns; the emergence of hierarchies; mobilization of labor for public works; the establishment of craft industries; the formation of regional systems; and episodes of sociopolitical collapse. These generalities are superficially obvious, but in their simplicity is a universe of subtler and richer observations. For the remainder of this discussion, I will concentrate on a topic that I have increasingly found to be essential in understanding much of southwestern prehistory. This is the relationship between core complex societies and their peripheries.

Randall McGuire and his colleagues have explored this topic in depth, in their revised paper from the School of American Research Advanced Seminar.[21] The approach I will take is not that of a Wallersteinian world-system (e.g., Wallerstein[34]), for the relations I see between core and periphery in the Southwest were not primarily cases of cores dominating peripheries and suppressing their development.

We know much about how complex societies interact with the peoples on their peripheries, and we know that this interaction has been a recurrent historic process. To list a few of the more notable examples:

- the north China periphery, where the interaction between nomads and settled farmers affected political evolution for millennia[18];
- the Mesopotamian alluvium, where early dominant states in the Tigris and Euphrates valleys contended with peoples like Gutians, Amorites, and Elamites, and came later to be ruled by what had been a peripheral hill people: the Persians[8,22,39];
- the Egyptian Middle Kingdom, which came to an end in 1668 B.C. with the Hyksos invasions (e.g., Butzer[3]);
- the Hittites of Anatolia, whose empire collapsed during a time of conflict with less complex peripheral peoples (e.g., Barnett[1] and Goetze[11]); and
- the interaction between central Mexican civilizations and northern barbarians, as recounted in Mesoamerican legends (e.g., Davies[7]).

No one argues that any southwestern society was as complex as those listed, but there is much to be learned of a general nature of how more and less complex societies interact. A comparative study of this process would clarify much of southwestern prehistory. It is to be expected also that southwesternists studying this phenomenon will someday produce results helpful in understanding the process in Europe, Mesopotamia, or China.

A few comments will emphasize some of the recurrent aspects of core-periphery relations. When a complex society dominates a region, it becomes the focus of peripheral societies. It is a locus of wealth, power, and prestige, and peripheral societies find it irresistible as a source of both trade and plunder. Probably the best-known consequence of core-periphery interaction is the process of secondary-state formation. This is an adaptive process in which societies emulate the organizational characteristics of a more complex neighbor. What is important is that this process does not depend on any specific level of complexity, nor on any specific form of organization, such as statehood. It is at base a *process* of the less complex emulating the more complex. Where secondary states do not form, typical courses of action for peripheral people are to try to join the wealthier and more stable center, or barring that, to obtain some of its wealth through raiding, or by accepting subsidies aimed at forestalling raids.

Political activity *among* peripheral peoples often leads to the formation of supra-tribal political aggregates that can militarily challenge the center. Such supra-tribal formations emerged at many points along the frontiers of Old World complex societies. These would include the Picts of Scotland, the Franks of northwest Germany, and the Alemanni to the south.[4] Elsewhere confederations emerged that had a lasting influence on world history: in desert Arabia, in central Asia, and to the north of China.[12] It is a regular and predictable process.

There are several cases in southwestern prehistory where scholars agree that more complex systems developed. These include Chacoan society of the San Juan Basin, the Mimbres, the Hohokam, and many of the large, aggregated Pueblo III and IV communities. In discussing this matter I will confine myself primarily to the Chacoan system, which I know better than the Mimbres or the Hohokam. I will also restrict the discussion to the southern and eastern edges of the San Juan Basin, partly again because it is those areas that I know best, and partly to avoid, for now, the issue of recent proposals to extend the Chacoan system far into eastern Arizona.[19]

Certainly within the San Juan Basin, the evolution of Chacoan communities was at least partially a function of participation in the Chacoan system. That is almost a matter of definition. A question that is of equal interest is to what extent societies peripheral to the Basin, and presumably outside the Chacoan system, evolved toward complexity under local pressures, or as a response to the presence of more complex neighbors.

One of the implicit premises of the Pecos Classification is that Anasazi societies evolved in a lockstep fashion. We now know, as Linda Cordell and Fred Plog pointed out, that the Pecos Classification is greatly overgeneralized.[6,24] Nevertheless, it is clear that many, if not all, Anasazi communities *did* evolve in roughly a lockstep manner. An experienced archaeologist can, for example, readily identify a Pueblo I jacal village in either the central Rio Puerco Valley or in southwestern Colorado. Across vast areas there are broad similarities. Is this because each community was responding to local circumstances and coincidentally devising similar solutions? Or did it possibly have something to do with the relationship between the core Chacoan system and the peripheral societies for whom Chaco may have been, metaphorically, an 800-pound gorilla that simply could not be ignored?

The Cebolleta Mesa area is an instructive example. It is located in west-central New Mexico, west of Acoma and southeast of the town of Grants. Ruppé[25] and Dittert[9] have traced what they see as divergent patterns of cultural evolution in this area. They see northern Cebolleta Mesa populations as having interacted with, and been influenced primarily by, people to the north. They see southern Cebolleta Mesa populations as having simultaneously interacted mainly with people to the

[4]see Brauer,[2] p. 75.

south. Northern Cebolleta Mesa is near the southern edge of the San Juan Basin, and includes the possibly Chacoan site of Las Ventanas. Southern Cebolleta Mesa was decidedly not part of Chaco. Yet there does not seem to have been, between northern and southern Cebolleta Mesa, any major or lasting differences in the rates at which greater complexity evolved among these populations. Both the northern population, which may have been involved heavily with Chaco, and the southern population, which apparently was not, display broadly similar patterns of cultural evolution. Indeed, in the period called the Pilares phase, ca. A.D. 1100–1200, it appears that the people of southern Cebolleta Mesa lived in communities that were, on average, over twice as large as those to the north.[26] Certainly in this period, at least, the more peripheral communities seem not to have been any less complex than their neighbors to the north.

Did the peoples of northern and southern Cebolleta Mesa develop complexity at similar rates because they were adapting to similar local environments? Or did they evolve at similar rates because they were adapting to each other, and to the decidedly more complex society to the north? Similar questions could no doubt be asked in regard to prehistoric societies all around the periphery of the San Juan Basin.

There is an interesting test case for evaluating this question. This is the Albuquerque area, a region where, for much of prehistory, populations *did not* evolve in a lockstep fashion with the rest of northwestern New Mexico. Albuquerque has always been a puzzle, in part because its archaeological assemblages consistently show a mixture of northern, western, and southern characteristics, and in part because, prior to the mid-thirteenth century, it did not parallel the cultural evolution seen in other areas peripheral to the San Juan Basin. While much of northwestern New Mexico was experiencing the development of large communities, great pueblos, political and economic hierarchies, and extensive trade systems, the middle Rio Grande was characterized by small populations occupying pithouse villages and showing little evidence of social differentiation or intercommunity hierarchies.[5] It is only with the abandonment of much of northwestern New Mexico and the Four Corners region that Albuquerque area populations began to show noticeable complexity.

What we see around Albuquerque are societies that did not evolve as, say, those of Cebolleta Mesa did. Over the past few years I have tried to develop an understanding of why this was so. Some of the results of a recently completed project have given us a significant clue.

In discussing elsewhere the mixed assemblages of Albuquerque-area archeology, I pointed out that Albuquerque populations participated minimally in the San Juan Basin exchange network. This was apparently because their territory was redundant to the San Juan Basin economic system. San Juan Basin people looking to trade with people possessing a riverine resource base would have found a closer candidate in the Rio Puerco, where Guadalupe Ruin is one of the earliest Chacoan Outliers.[23]

[5] See Tainter,[29] p. 19.

Extending such a trade network beyond the Rio Puerco to the Rio Grande would have been superfluous and costly. Hence Albuquerque-area populations never participated fully in the San Juan Basin exchange system.[28,33]

South of Guadalupe Ruin, in the Atrisco section of the Puerco (which lies in the area where the river crosses Interstate Highway 40), surveys were conducted in the early 1980s for a federal land exchange. Here there appears to have been a pattern of primarily seasonal use, with occasional longer term settlements. The most striking aspect of the archaeological record here is the pottery: virtually all the ceramics were imported, and petrological studies show that about 95% came from the west, primarily the higher elevation terrain from the vicinity of Acoma south to the upper Rio Salado drainage basin. What is most interesting is that this pattern persisted for *centuries*, from about A.D. 500 to about A.D. 1300.[31,35] This is the kind of strong, persistent patterning that gladdens the heart of any archaeologist, the kind of patterning we always hope to find but rarely do.

I have argued elsewhere that two factors account for this pattern.[31] The first was seasonal transhumance. Many of the seasonal sites at Atrisco (or at least many of the seasonal sites with pottery) must have been used by people who at other times of the year lived in the higher elevations to the west and southwest. Secondly, at times when Atrisco was occupied for longer periods, the people residing here apparently had their major economic ties with people to the west and southwest. In all likelihood their main social and cultural ties lay in these directions as well. The reason for these ties would have been to increase subsistence security by economic linkages between areas potentially experiencing different productivity cycles.

It is likely, but not yet certain, that this east-west trend in transhumance and economic ties stopped at the Rio Puerco. It did not extend the few miles further east to the Rio Grande. Unfortunately, few ceramic temper studies comparable to that done at Atrisco have yet been done with Albuquerque area ceramics. The lack of ceramic temper studies in the Albuquerque area makes it difficult to compare directly Rio Grande and Rio Puerco ceramics. We can say, however, that there are presently no indications that the Albuquerque-area was so dominated by western pottery imports as was Atrisco, and there are definite indications to the contrary.

If this statement can be provisionally accepted, two very important points can be suggested. The first is that the Atrisco area was at the eastern end of an east-west network of transhumance and economic exchange. The second, and more important, is that there was an economic boundary between Atrisco and Albuquerque, a boundary which persisted for centuries. The existence of such a boundary would imply that there was also between Atrisco and Albuquerque a social boundary and a cultural boundary. This would be the inevitable result of a situation where people using or occupying the Atrisco area interacted primarily with people to the west.

We can account for this boundary by the same factors that appear to explain why Albuquerque-area populations did not participate strongly in the San Juan Basin exchange system. If people resident in the higher elevation terrain of the Acoma area, or the upper Rio Salado drainage, sought to include a lower elevation river valley in their territorial round, or to trade with people living in such an area, the logical candidate to the east would be the Rio Puerco. In such a case the Rio Grande would be redundant, and its use would entail extra costs in travel and transport. Hence between Atrisco and Albuquerque we find a boundary: economic certainly, and probably social and cultural as well.

With these findings and inferences, we can begin to sketch why Albuquerque prehistory was so unsynchronized with the rest of northwestern New Mexico. The earlier evolution of complexity in and around the San Juan Basin was prompted by high population densities and a risky environment. The response was, in some places, the development of regional economic and political hierarchies, and exchange systems. These served to moderate subsistence uncertainties in a fluctuating environment.[6] By being incorporated into these exchange systems, and by emulating organizationally more complex communities, much of the Anasazi population of northwestern New Mexico became more complex in roughly a lockstep fashion.

The Anasazi of the Albuquerque area, in contrast, participated minimally in these exchange systems. To the people of northwestern New Mexico, the middle Rio Grande was redundant and, therefore, unimportant. Thus populations in what is now the Albuquerque area experienced little stimulus toward greater complexity. They did not need to aggregate around trade centers such as Guadalupe Ruin; they did not need to adopt the social conventions of more complex trading partners; and they did not need to intensify production to have surpluses for exchange. By living in an area that was superfluous to the economies of more complex communities, they were able to maintain low complexity and cultural stability until the more complex societies to the west collapsed, and their territories were abandoned. This changed forever the world of the Rio Grande Anasazi.

There are lessons to be learned from Albuquerque-area prehistory. Certainly we see here no intrinsic or latent tendency toward complexity. Societies here didn't become especially complex until the mid-thirteenth century, and then only after the San Juan Basin was abandoned, and much of its population apparently resettled in the Rio Grande Valley.[36,37] Complexity is a costly strategy: a society will not become complex unless it needs to. Albuquerque-area populations maintained relatively simple societies until they were *forced* to be otherwise.

Thus I argue that the reason for low complexity among the Albuquerque-area Anasazi is because, prior to the mid-thirteenth century, they did not interact on a persistent, intense level with the more complex societies to the west and northwest. This point leads us back to the question raised in regard to Cebolleta Mesa: did societies on the Chacoan periphery evolve complexity because they adapted to local circumstances, or because they adapted to more complex neighbors? In the Albuquerque area we see societies that did not need to adapt to more complex neighbors and that did *not* become more complex. If the *converse* holds, then it may be that the evolution of complexity in much of northwestern New Mexico

[6]See Tainter and Gillio,[32] pp. 100–113; Schelberg[27]; Tainter,[30] pp. 178–187; and Tainter and Plog[33].

was a function of core-periphery relations. The implication, which I raise only as a conjecture, for that is all it can be, is that if the Chacoan system had not existed, complex communities either would not have evolved in northwestern New Mexico, or would have emerged later than they did.

If this reasoning and these suggestions have value, they also have implications for cultural evolution on the peripheries of other complex southwestern societies, such as the Mimbres, the Hohokam, and of course Casas Grandes. They have implications as well for the evolution of peripheral societies everywhere. Did the North China frontier, for example, or the high country north of Mesopotamia, support societies as divergent in their evolutionary trajectories as we have seen in Cebolleta Mesa and Albuquerque? No doubt the answer is yes, and so the next question is: to what extent could such diversity be understood if we were to look at differences in the intensity of core-periphery relations? Definitive anwswers to such questions cannot be given at this point. What is clear is that southwestern archaeology has relevance to such questions, and that it can and should contribute to generalizing studies about cultural evolution.

REFERENCES

1. Barnett, R. D. "The Sea Peoples." In *The Cambridge Ancient History II(2)*, edited by I. E. S. Edwards, C. J. Gadd, N. G. L. Hammond, and E. Sollberger, 359–378. 3rd ed. Cambridge: Cambridge University Press, 1975.
2. Brauer, George C., Jr. *The Age of the Soldier Emperors: Imperial Rome,* A.D. *244–284*. Park Ridge: Noyes Press, 1975.
3. Butzer, Karl W. "Civilizations: Organisms or Systems?" *Amer. Sci.* **68** (1980): 517–523.
4. Clarke, David L. *Analytical Archaeology.* London: Methuen, 1968.
5. Cordell, Linda S. "The Nature of Explanation in Archaeology" A Position Paper." This Volume.
6. Cordell, Linda S., and Fred Plog. "Escaping the Confines of Normative Thought: a Reevaluation of Puebloan Prehistory." *Amer. Antiquity* **44** (1979): 405–429.
7. Davies, Nigel. *The Toltecs Until the Fall of Tula.* Norman: University of Oklahoma Press, 1977.
8. Diakonoff, I. M. "The Rise of the Despotic State in Ancient Mesopotamia." In *Ancient Mesopotamia: Socio-Economic History*, edited by I. M. Diakonoff, 173–203. Moscow: Nauka, 1969.
9. Dittert, Alfred E. "Culture Change in the Cebolleta Mesa Region, Central Western New Mexico." Ph.D. dissertation, University of Arizona, 1959. University Microfilms, University of Michigan, Ann Arbor.

10. Eisenstadt, S. N. *The Political Systems of Empires.* Glencoe: Free Press, 1963.
11. Goetze, A. "The Hittites and Syria (1300–1200 B.C.)." In *The Cambridge Ancient History II(2)*, edited by I. E. S. Edwards, C. J. Gadd, N. G. L. Hammond, and E. Sollberger, 252–273. 3rd ed. Cambridge: Cambridge University Press, 1975.
12. Grousset, René. *The Empire of the Steps: A History of Central Asia*, translated by Naomi Walford. New Brunswick: Rutgers University Press, 1970.
13. Hodder, Ian. "Intepretive Archaeology and its Role." *Amer. Antiquity* **56** (1991): 7–18.
14. Johnson, Gregory J. "Information Sources and the Development of Decision-Making Organizations." In *Social Archeology: Beyond Subsistence and Dating*, edited by Charles L. Redman, Mary Jane Berman, Edward V. Curtin, William T. Langhorne, Jr., Nina M. Versaggi, and Jeffrey C. Wansner, 87–112. New York: Academic Press, 1978.
15. Johnson, Gregory J. "Organization Structure and Scalar Stress." In *Theory and Explanation in Archaeology: The Southampton Conference*, edited by C. Renfrew, M. J. Rowlands, and B. A. Segraves, 389–421. New York: Academic Press, 1982.
16. Judge, W. James. "Chaco Canyon-San Juan Basin." In *Dynamics of Southwest Prehistory*, edited by Linda S. Cordell and George J. Gumerman. Washington, DC: Smithsonian Institution Press, 1989.
17. King, Thomas F. "The NART: A Plan to Direct Archeology Toward More Relevant Goals in Modern Life." *Early Man* **3(4)** (1981): 35–37.
18. Lattimore, Owen. *Inner Asian Frontiers of China.* Boston: Beacon Press, 1940.
19. Lekson, Stephen H., Thomas C. Windes, John R. Stein, and W. James Judge. "The Chaco Canyon Community." *Sci. Amer.* **256(7)** (1988): 100–109.
20. McGuire, Randall H. "Breaking Down Cultural Complexity: Inequality and Heterogeneity." In *Advances in Archaeological Method and Theory*, Volume 6, edited by Michael B. Schiffer, 91–142. New York: Academic Press, 1983.
21. McGuire, Randall H., E. Charles Adams, Ben A. Nelson, and Katherine Spielmann. "Drawing the Southwest to Scale: Perspectives on Macroregional Relations." In *Themes in Southwestern Prehistory: Grand Patterns and Local Variations in Culture Change*, edited by George J. Gumerman and Murray Gell-Mann. Santa Fe, NM: School of American Researc Press, in press.
22. Oates, Joan. *Babylon.* London: Thames and Hudson, 1979.
23. Pippen, Lonnie C. "Prehistory and Paleoecology of Guadalupe Ruin, New Mexico." *University of Utah Anthropological Papers*, Vol. 107. Salt Lake City, UT: University of Utah, 1987.
24. Plog, Fred. "Political and Economic Alliances on the Colorado Plateaus, A.D. 400–1450." In *Advances in World Archaeology*, Vol. 2, edited by Fred Wendorf and Angela E. Close, 289–330. New York: Academic Press, 1983.

25. Ruppé, Reynold J. "The Acoma Culture Province: An Archaeological Concept." Ph.D. dissertation, Department of Anthropology, Harvard University, 1953. Unpublished.
26. Ruppé, Reynold J. "The Archaeological Survey: A Defense." *Amer. Antiquity* **31** (1966): 313–333.
27. Schelberg, John D. "Economic and Social Development as an Adaptation to a Marginal Environment in Chaco Canyon, New Mexico." Ph.D. dissertation, Northwestern University, 1982. University Microfilms, University of Michigan, Ann Arbor.
28. Tainter, Joseph A. "Perspectives on the Northern Mogollon Boundary Phenomenon." In *Recent Research in Mogollon Prehistory*, edited by Steadman Upham, Fred Plog, David G. Batcho, and Barbara E. Kauffman, 45–58. New Mexico State University Museum Occasional Paper, vol. 10. Las Cruces, NM. New Mexico State University, 1984.
29. Tainter, Joseph A. "The Development and Collapse of Prehistoric Societies: A Perspective From Central New Mexico." In *Secrets of a City: Papers on Albuquerque Area Archaeology in Honor of Richard A. Bice*, edited by Anne V. Poore and John Montgomery, 18–24. Papers of the Archaeological Society of New Mexico, vol. 13. Albuquerque, NM: Archaeological Society of New Mexico, 1987.
30. Tainter, Joseph A. *The Collapse of Complex Societies.* Cambridge: Cambridge University Press, 1988.
31. Tainter, Joseph A. "Synthesis of the Prehistoric Occupation and Evaluation of the Research Design." In *Interim Final Report of Investigations: The Albuquerque Sector of the Elena Gallegos Project*, edited by Joseph A. Tainter, 988–1051. Manuscript on file, USDA Forest Service, Cibola National Forest, Albuquerque, New Mexico, 1991.
32. Tainter, Joseph A., and David A. Gillio. *Cultural Resources Overview, Mt. Taylor Area, New Mexico.* Albuquerque and Santa Fe: USDA Forest Service, Southwestern Regional Office and USDI Bureau of Land Management, New Mexico State Office, 1980.
33. Tainter, Joseph A., and Fred Plog. "Structure and Patterning: The Formation of Puebloan Archaeology." In *Themes in Southwestern Prehistory: Grand Patterns and Local Variations in Culture Change*, edited by George J. Gumerman and Murray Gell-Mann. Santa Fe, NM: School of American Researc Press, in press.
34. Wallerstein, I. *The Modern World System.* New York: Academic Press, 1974.
35. Warren, A. H., and Dan Warren. "Ceramics of the Albuquerque Project Areas." In *Interim Final Report of Investigations: The Albuquerque Sector of the Elena Gallegos Project*, edited by Joseph A. Tainter, 764–914. Manuscript on file, USDA Forest Service, Cibola National Forest, Albuquerque, New Mexico, 1991.
36. Wendorf, Fred. "A Reconstruction of Northern Rio Grande Prehistory." *Amer. Anthropologist* **56** (1954): 200–227.

37. Wendorf, Fred, and Erik Reed. "An Alternative Reconstruction of Northern Rio Grande Prehistory." *El Palacio* **62** (1955): 131–173.
38. Wolf, Eric R. *Peasant Wars of the Twentieth Century.* New York: Harper & Row, 1969.
39. Yoffee, Norman. "The Collapse of Ancient Mesopotamian States and Civilizations." In *The Collapse of Ancient States and Civilizations*, edited by Norman Yoffee and George L. Cowgill, 44–68. Tucson, AZ: University of Arizona Press, 1988.

Human Biology and Health

Alan C. Swedlund
Department of Anthropology, University of Massachusetts, Machmer Hall, Amherst, MA 01003

Issues in Demography and Health

INTRODUCTION

> It's complicated. Actually it involves food economics, nutrition, tolerances, population sizes, things like that, and you spend a lot more time working on programming statistical projections on a computer than you do digging in the field. Really dull stuff, unless you're weird enough to be into it (p. 111).
>
> —Maxie Davis to Leaphorn
> in *Thief of Time* by Tony Hillerman

Conventional wisdom has it that the human population maintained a growth rate close to zero for the first 3 million years, but in the last 10,000, since the advent of agriculture, the human population has grown at an ever increasing, exponential rate. While this certainly may be true in the global sense, it is quite misleading in a regional context, because local population rates have probably always fluctuated

Understanding Complexity in the Prehistoric Southwest,
Eds. G. Gumerman and M. Gell-Mann, SFI Studies in the Sciences of
Complexity, Proc. Vol. XVI, Addison-Wesley, 1994

and varied with respect to a number of density-dependent and independent factors. Some events will be random statistical departures from an underlying set of demographic parameters, while others are direct responses to real changes (demographic, environmental, political-economic). In attempting to understand prehistoric population change in the Southwest, we must tread very cautiously between the Grand Canyon of overgeneralization and the arroyo of particularism. Moreover, we must abandon simple, deterministic models that fail to capture the major variables and their interactions while at the same time we must not focus too much on the noise, meanwhile missing the signal.

In this brief overview I will: elucidate some of the demographic concepts that are important to the study of population and health in a large regional context; discuss some of the dimensions within which we might focus our questions; describe some of the lines of inference that are available; and use one case in point to illustrate how these have been and/or might be brought together. In no way can I hope to be comprehensive or exhaustive of the relevant issues or previous research, but, as the charge implies, I will try to give us a base upon which to build. The role of demography and health in cultural stability and change is not unidirectional, so care must be taken to ascertain effects including interactions.

A brief history of demographic anthropology would reveal at least two common themes. First, there has been a compelling interest in identifying closed systems under stable conditions and with a high level of internal integration, so that whatever variables we were studying would behave in highly predictable ways. In this we need not apologize, because closed systems make tidy theories, and much of demographic theory has been developed around assumptions of such systems, with little or no variation. Secondly, there has been a tradition of interest in issues of population pressure and carrying capacity as the driving forces in population history.[13] Work in the 1960s and 1970s on these issues gradually gave way to the critique that carrying capacity was a difficult if not impossible concept for human populations because of our behavioral and ecological complexity, and because we moved around all the time. In animal population studies, a similar "crisis" in theory developed, in which many population biologists discovered that "k" meant something quite interpretable when studying bacteria or phage in a chemostat, but something far less clear with mammals in a wildlife habitat.

Nevertheless, the related concepts of homeostasis and equilibrium are still important theoretical underpinnings to much of demography and population biology, and it has always been my belief that these are very important concepts for understanding population change in the Southwest. Obviously, this is because there are clear limiting factors to population growth in a semiarid environment. Those that are often invoked include water, food and storage, arable land, and wood for fuel and construction.

I would summarize the broad understandings we have about population history in the Southwest by stating the obvious—it has proven to be a very dynamic region—by that, I mean there is much activity and high variability over time and space, much more than previously thought to be the case. The factors which account for this change are no doubt numerous and complex, but we should be able to track density-dependent ones in a reasonable fashion. These would include the limiting resources noted above as well as intrinsic limits to growth (fertility and mortality), as well as boundary conditions (e.g., neighboring groups or resource limits). Also, since we are investigating a population that is part of the last 10,000 years of agricultural development, we would expect that the predominant pattern should be population growth.

WHAT IS THE QUESTION?

> All species, in order to succeed, must be capable of reproduction equal to or in excess of mortality. Survival of the species depends on it. The inherent, underlying strategy of populations should then be persistence; and biological and behavioral mechanisms evolve to help insure this persistence.[1]

If the general finding is that agricultural innovation during the Neolithic created an unprecedented pattern of positive growth for the human population, and that the underlying tendency for those populations that survived should be slightly positive growth rates or at least persistence, then growth is not the question, it is an assumption—the Malthusian assumption. Relevant hypotheses are more likely to be founded on issues regarding deviations from positive growth or by responses to that growth. Indeed, a lot of attention has been focused in Southwestern archaeology on the phenomenon everyone once referred to as abandonment—a phenomenon which is now probably better understood as a more complex process varying in time and space, and not necessarily preceded by constant, gradual growth.

An empirical generalization that most of us no doubt bring with us to this conference is a progressive one—that from Basketmaker times to Late Pueblo times, most regions did witness population growth, aggregation, cultural/technological elaboration, agricultural intensification, and probably some environmental degradation. Refinements on that generalization point to the fact that these are not strictly linear trends and there are perturbations. There is regional diversity in growth and decline and there are costs to pay—sometimes in the form of deteriorating health, or of migration, of disaggregation, and of modification of cultural and subsistence strategies.

A second assumption, and one much less easily stated in formal terms, is that population variables both produce social change and respond to it. Positive growth

[1]See Swedlund,[36] 141; paraphrasing Slobodkin and Rapaport.[34]

places demands on institutions to produce, allocate, or reallocate resources; as institutional changes occur, forms of population structure and composition may occur, or growth rates may be modified.

It appears to me that the nonlinear trends and deviations should form the questions, and often they do. To the extent that the explanatory variables are external, natural forces like hydrologic cycles and climatic change, researchers feel a sense of confidence, order, and even contentment that things are as they should be. But when these environmental causalities do not achieve the strength of association desired to account for the nonlinearities, the questions get more interesting, more problematic, and the discontentment level rises. This is good.

PARTITIONING THE VARIANCE

How do we proceed from here? Most of us probably agree that this is not a simple, multivariate statistical problem; otherwise, it would have been resolved some time ago and we would not be here. Yet, some of the approaches to multivariate problems are likely still useful in teasing out the more substantive questions regarding demography and health and cultural change in the Southwest.

First, perhaps, we might ask how much of the population variability can we attribute to stochastic and sampling processes? What do the residuals look like? And so forth. Although this has no doubt been done many times before, it may be time to do it again. Secondly, it seems logical to account for as much of the variation as possible through hydrologic cycles and climatic change models, and see what is left. I realize that this is so easy to say and yet so difficult to do. Time-series models usually require much more precise data points and estimates than may be available, but Jorde's[16] use of dendrochronology series to infer cultural buffering and aggregation stands out as an example. Also, much progress might be made through non-parametric or even qualitative assessments (or maybe even by voice vote?).

Can this exercise reveal pivotal points in both regional and subregional trajectories that warrant increased attention? Will those points share similarities?

ARE THE QUESTIONS ABOUT "SUCCESS" OR "FAILURE"?

There seems implicit in the Southwestern literature of which I am familiar a tone of "what went wrong?" Things were good and then they got bad. The issue of the role of demography and health in cultural stability and change has a much more appealing and neutral tone, and yet I still find myself frequently thinking about what went wrong. This is an issue that may require some of our attention. After all, 1100 years of positive growth, on average, can hardly be considered failure. But we must remember, too, that biological success as measured by growth can ultimately lead to "failure" by overexploitation of available resources.

Stress is a concept and reality with which we are all familiar and which appears more and more frequently in the literature on the Southwest. It is a response variable that is amenable to measurement by health and population variables and by elements of material culture given certain assumptions. Are the cycles of aggregation/disaggregation one of the more important questions? In the form of stress models?

ARE DEMOGRAPHY AND HEALTH TO BE PREDICTOR OR RESPONSE VARIABLES?

As noted, population can be modeled as both cause and consequence, and I mean this in a larger sense than that implied by Dean et al.[7] when they refer to the fact that population size can be adjusted by infanticide or starvation. Rather, changes in population can occur by intrinsic processes (like fertility), but also we need to be very certain about those cases in which we are attempting to predict a population response (e.g., abandonment) by other factors (e.g., political disintegration). Likewise, we are sometimes just as apt to explain a social phenomenon, like political disintegration, by population pressure. Frankly, I think there has been a lot of circularity in models by not making the lines of causality explicit. If we consider stress in the equation, we can note that stress has been used both to explain the response *of* aggregation,[23] and as response *to* aggregation (e.g., Cohen and Armelagos[5]). However, in these examples the lines of causality have been explicit in both cases and the implications provocative.

MODELS

The models available to us for the analysis of change in prehistoric societies may be limited by the kinds and qualities of data, but several approaches are possible. While it may be somewhat arbitrary, I will briefly consider a few descriptive and explanatory examples.

DESCRIPTIVE

A. DIMENSIONS OF ANALYSIS I have suggested a fairly simplistic model for the dimensions of analysis in anthropological demography (Figure 1).

There are no answers here, and I do not wish to argue some unilineal point, but it may help us to focus on levels of the problem, especially in dimensions (a) and (b). If stress or adaptation models are to be invoked it is very important to identify the level in which we anticipate responses. For example, changes in fertility have population-size consequences, but are sometimes best understood, if not measured,

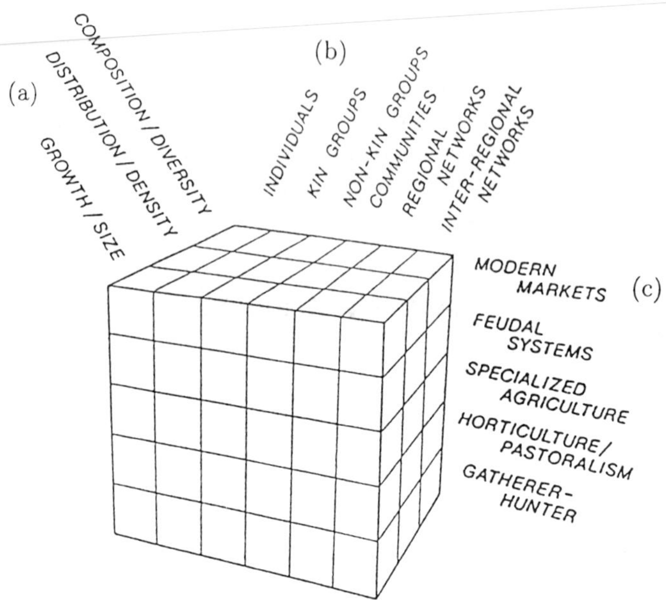

FIGURE 1 A three-dimensional matrix of variables used in human population ecology. (a) Population characteristics; (b) units of analysis; and (c) levels of economic integration.

at the household level. Some of the broad modeling done in the past might gain resolution by further consideration of other units of analysis in the future.

In this same vein we can observe that there has been a tradition in paleodemography that has almost exclusively looked at issues of population growth with little regard to issues of population distribution and composition. Fortunately, this has not tended to be the case in the Southwest, but there have not been, to date, the levels of integration between growth processes and structural processes that may be desired.

My admittedly superficial reading of the genetic and biodistance literature on the Anasazi suggests a fairly homogeneous population biologically. Thus, most growth processes may be primarily the result of intrinsic processes (fertility and mortality) and most population movement a result of internal migration until the protohistoric period.

B. AGE-SEX DISTRIBUTIONS, MORTALITY/SURVIVORSHIP, AND LIFE TABLES. There is an ever-growing body of population composition data for the prehistoric Southwest. If asked the question of whether life tables are appropriate for these cases, I would have to answer with a definitive "I dunno," followed by an unequivocal, "it depends."

One of the attractive features of skeletal analysis is that the aging process falls under a number of genetic and "natural" limitations which reduce the potential variability in aging and age structure. The recognition of this fact coupled with the observation of regional and national differences in life expectancy has led to the development of model life tables.[3,17,40] It has also led to the *Uniformitarian* assumption in paleodemography.[15] In its benign form, uniformitarian theory argues that the life expectancies and survival rates we observe in prehistoric populations should mirror those observed in model life tables in present-day populations since we share this common biological substrate. In its more corrupt forms, it can lead to arguments that all crossover events—where two survivorship curves or mortality profiles cross each other—are the result of defective data rather than real events.[4] *A priori* use of uniformitarian assumptions can potentially mask the results of real exogenous sources of variation.

I am not at all prepared, as yet, to argue that prehistoric populations had age structures and survivorships that closely mirror those of modern or recent historical populations. The deviations should not be extraordinarily different and the curves should be the same approximate shape, but they may be significantly different in many cases.[10]

Most sites do not offer the number and quality of burials that permit the construction of realistic abridged life tables, yet there are literally thousands of skeletons from a wide variety of sites for which sex and age data are known.[24,35] However, in so many of these cases, the base population, growth rates, and other parameter estimates are not known in sufficient detail to hope for any reliability from these tables.

I have spoken before with a forked tongue about the general problem of life table reconstructions in paleodemography[26] and this debate continues.[2,6,18,19,28,41] The problem of estimating stability and growth rates in the Southwest is somewhat attenuated by the very good information we occasionally have from independent sources on site size, population growth, and provenance of burials, but there are still many potential pitfalls.

Another problem with life tables has been not only in their construction but also obviously their application. There has been too much attention paid to life expectancy without enough attention being paid to the inferences that may be made from the relationships in the q_x and d_x columns themselves. Life expectancy requires the most assumptions about the population and is probably most misleading because of the accumulated error estimation problems in sexing and aging, but other estimates of the basic mortality structure, given some more elastic assumptions, are potentially quite reliable. Suffice to say that this is a very important domain for further consideration.

Survivorship curves are also subject to many of the same problems as life tables and in some cases are worse since they often involve cumulative rates that compound errors in estimation among younger age sets even more than life tables will. However, our ability to estimate population structure and growth are dependent on knowing something about survivorship in these sites. Models based on a variety of living populations can be helpful in this regard.

Better use of the data is perhaps being made in the paleoepidemiological models that are increasingly being employed.[24] Models using a combination of pathological diagnoses with age structure and sex composition approach the methodologies used in survival and hazards analysis, can use nonparametric designs and are already providing many insights into prehistoric health. We do know, for example, that nutritional and disease indicators imply a deteriorating profile of health in many of the major sites and across subregions in the nineth and especially in the later centuries A.D. Also, since nutritional stress is directly linked to fecundity and hence to fertility rates and growth, we may be able to infer stress as a precursor to population decline.

C. LOCATIONAL ANALYSIS. The ecological and adaptational frameworks usually describe resources in terms of their (1) diversity, (2) density, and (3) predictability over space and time. In the study area we have much confirming evidence for the patchiness of critical resources, especially arable land and water, but also of many other plant and animal populations. Guided by this framework and by the accumulated knowledge of human habitation patterns in the Southwest, archaeologists have been able to predict quite well the principal patterns of occupation over the early Basketmaker to Later Pueblo periods. The more fine grained analyses of site abandonment and reoccupation, between uplands and lowlands and between areas with alternative resources, has greatly enriched our understanding of population distribution and density. The formal locational analyses that flourished in the 1970s, largely as a result of the integrative SARG projects, seems to have waned somewhat. What additional considerations need to be addressed or amplified here? The presence/absence/distribution of central places? Their relation to alliances?[30,31]

D. GROWTH AND CHANGE. As noted above, previous work has been driven by deterministic models of growth and change in the Southwest. The Positivist Paradigm and processual approach very much affected the nature of the questions and probably the results. As a person grounded somewhat in population biology, it is normal for me to think about persistence and growth, as implied by my assumptions above. However, I am hardly wedded to concepts like "success" or to "persistence," nor, though, am I wedded to just discoursing about the Anasazi "being." Positivists have to acknowledge that human groups do stupid things (in hindsight, at least) and can behave nonadaptively at fairly high costs.[12,27] What role will growth have in our descriptive or explanatory models? Growth can be an assumption, a variable, and a constant. Is it a driving variable? Probably.

EXPLANATORY MODELS

A. HOMEOSTASIS/EQUILIBRIUM/STRESS All of us know what these mean, I suspect, but there is still a lot of imprecision and inconsistency in their use. *Homeostasis*, in its conventional context, refers to an individual organism's physiological ability to maintain a steady state, or a population's ability to do so demographically. At the population level it almost necessarily goes hand-in-hand with the assumption that there are *exogenous* limiting (density-dependent) factors that are responsible for maintaining population equilibrium. That is, in the individual case the forces are *endogenous* to the organism; in the population case we assume outside forces are inhibiting and equilibrating. *Dynamic equilibrium* has come to refer to conditions where populations are not in a pure steady state, are in fact growing, but at a rate that is constant and unchanging, and homeostatic behaviors are those that tend to maintain the same trajectory when and if slight deviations occur. *Stress* refers to those cases of disruption, on the individual or population level, that are met with either adequate or inadequate adaptive responses (see Thomas et al.[39] and Goodman et al.[11] for lengthy discussions on the anthropological use of these concepts). Harking back to my introductory comments, it appears that in the Southwest these concepts have had their fair share of use (and abuse) in explaining demographic-social change. How do we synthesize and/or improve on that work?

The Malthusian assumption implies that populations will not self-regulate until they are forced to by limiting factors. The predominant pattern in the Southwest for several centuries is one of positive growth (see Euler,[8] Plog,[32] and many other sources). Are homeostatic models appropriate prior to the ninth century? Where? In many respects, with the advent of agriculture, subgroups are like frontier, colonizing populations. There is a lot of expansion and a lot of opportunity. There are a lot of experiments, and some of them, at least, will be marginal and unsuccessful. Even in the comparatively rich environment of Colonial New England, several communities failed, sometimes due to clear resource limitations (e.g., upland farming in marginal areas; even timber becomes a scarce resource in some communities from over exploitation).

If positive growth is the pattern, then negative, density-dependent regulation is not the predominant explanation, at least not until fairly late. Population biologists note that in the absence of density dependence, the population size varies independently, primarily as a product of the underlying rates in the past—when density variables kick in, population size and growth respond dependently.

Positive growth at its basic level is the result of a NRR (net reproductive rate) above zero. Looking beyond the general growth curves, we must recognize this as the fertility events of individual women, articulated with family and household units, situated in moieties and clans (or whatever structures), and clustered in communities. On which level(s) are the homeostatic mechanisms operating most influentially? Also, how strong are these effects?

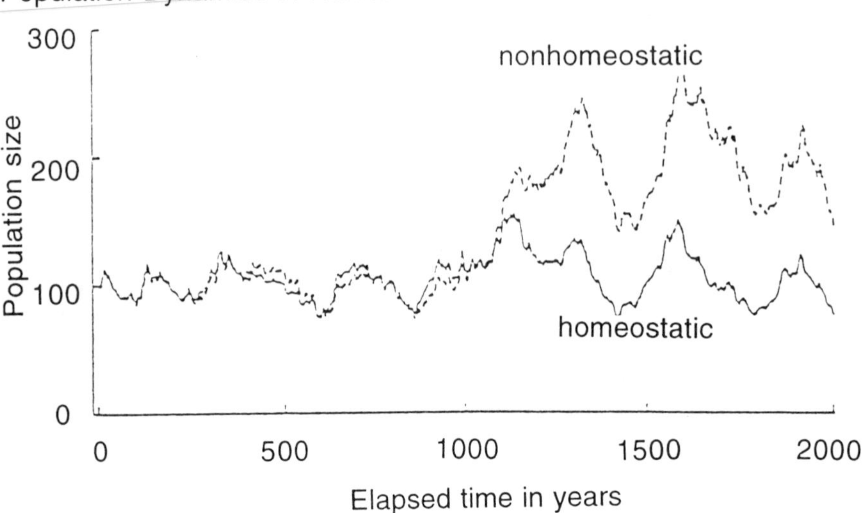

FIGURE 2 Stochastic simulation of population size with and without homeostasis (2,000 years). From Lee.[21]

Ron Lee[21] has considered homeostatic mechanisms over time by examining historical European populations. Using populations in which underlying fertility rates and mortality rates can be estimated with some accuracy, and incorporating information on wages and food prices, he has been able to characterize some fairly important relationships between density-dependent mechanisms and demographic rates. One such result is a stochastic simulation of a small population over several centuries with assumptions of homeostatic and nonhomeostatic conditions. Using feedback elasticities that are admittedly fairly low, but empirically valid, he shows that it takes several centuries for the two populations to achieve any significant differences (Figure 2). Moreover, he argues that the feedback elasticities would have to be much higher to seriously affect the results. Lee also discusses the fact that *density-independent* historical influences can have marked effects and recovery to equilibrium conditions can be quite slow.

Do these simulation approaches have application to the prehistoric Southwest?

Do the homeostasis/equilibrium/stress models have broad application to the issues of stability and change? I think they still do, but for the episodes of change, not for the long-term periods of positive growth. Stress is a very useful concept for some of these purposes and I will return to this briefly below.

B. SOCIO/POLITICAL/ECONOMIC. Quantifying or specifying these variables in the prehistorical context is something we can acknowledge as a problem without spending a lot of time or space justifying here. Yet this may be the most important objective. What I am often struck with is how much is known about the Southwest, not how little; but how does one capture the variability in a meaningful, operational fashion? It seems to me that the role of our working group might best be served by a consideration of those demographic or health responses which we would expect to find, given certain sociopolitical situations. (This alleviates the pressure for me to say something meaningful or profound here, and makes better use of *my* limited resources.) Therefore, I will make abbreviated comments on the issues of aggregation, stratification, cooperation-competition, and site depopulation or abandonment. Each of these implies to me a set of socio-cultural constructs involving decision making, and each suggests to me a set of demographic constructs and visually detectable disturbances that might appear both in the dirt and on a graph.

1. Aggregation: As stated, there are reasons to suggest both aggregation as response to stress, and also as a cause of stress. Which prevailed? Paleopathological analysis with some temporal controls could tell us a lot. Are the frequencies of stress-related pathologies higher in the early or the later phases? Will the answers suggest that resource limitations are present before, during, or after the process of aggregation?
2. Stratification: Social stratification by age and sex seems extremely likely. Although there are a lot of archaeologists who favor an egalitarian interpretation within sites, stratification has its share of proponents. Stratification with significant differential control of resources can lead to differing levels of stress and disease. High variance in stature estimations and pathology frequencies could be indicative of stratification, especially when age and sex are controlled.
3. Cooperation and Competition: Gumerman and Dean[12] argue that strategies favoring cooperation are more adaptive and review the literature on inferring alliances. Sites that appear defensive in nature are certainly prevalent in the later periods of several subregions, particularly the Colorado Plateaus. If competition results in overt hostilities, then traumatic lesions and other evidence of violence should be in high frequency. The presence of cooperation or competition suggests shortages, which may be reflected in nutritional indicators, mobility, and dispersion. If regional cooperation is working, the incidence of stress-related pathologies, nutritional diseases, and population abandonments should decline.
4. "Abandonment": Evacuation of a site or area is not a conservative response for agricultural groups with semipermanent architecture. Still, it beats mass starvation. It is measured relatively well by several archaeological indicators. Cycles of occupation, abandonment (I like "depopulation" better), and reoccupation are quite intriguing. Do these reflect cycles of resource depletion and

TABLE 1 Summary of skeletal indicators of stress (based on Goodman et al. 1988).

Indicators	Parts of the population and skeleton required	Groups at risk	Severity and Time of stress	General comments
Life tables and mortality schedules	Large and complete population or representative sample	All	Chronic, severe	Best indicator of overall adaptation. Accuracy of aging technique is critical.
Adult stature	Adult population, appendicular skeletons, especially lower extremities	Subadults	Summation of preadult factors	Short stature (small body size) may be a response to chronic undernutrition.
Growth curves Retardation Shape differences	Dental-aged subadults and subadult long bones	Subadults	Chronic Chronic (1 year)	Can aid in estimation of time of greatest stress in an individual's life.
Sexual dimorphism	Adult males and females; innominate used primarily, femur and other long bones secondarily	Subadults	Summation of preadult factors	Must consider genetic factors. Sexual dimorphism decreases with increased stress.
Harris lines	Adult or subadult radiographs of long bones (tibia, femur, and others)	Subadults Adults	Acute stress; reoccurring stress	Some evidence for inverse association with nutritional status; peak occurrence often near weaning.
Vertebral canal stenosis	Adult vertebrae	In utero to 3 yr.	Early chronic	Unclear association with early deprivations.
Skull base height	Adult cranium	In utero to 5 yr.	Early chronic	Unclear association with early deprivations.
Enamel hypoplasia and enamel microdefects	Any teeth; anteriors more sensitive	0.5 in utero to 7 yr.	Acute stress (w/chronic undernutrition)	Association with nutritional status and decreased longevity; peaks near weaning.

Dental asymmetry	Dentition	In utero	Early and severe	Measure of developmental noise. Required sample may be quite large.
Dental crowding	Maxilla and mandible with teeth in situ	Subadults	Chronic; severe	May be nutritional, but must not be confused with more common genetic causes.
Traumatic lesions	All bones	All	Acute	May differentiate between "violence" and oher causes of fractures.
Periosteal infection	All ages; long bones	All	Chronic	Some infections may not appear on bone.
Porotic hyperostosis and cribra orbitalia	Cranium, particularly orbital regions	Both sexes 0.5-8 yr. females 20-30 yr	Acute to severe	Related to iron-deficiency anemia; potential synergism with infection; highest prevalence before 5 yr.
Osteoporosis	Femur and rib cross sections commonly used	Juveniles; Reproductive females; Senile	Acute to severe Chronic to severe Chronic	Evidence for increase in females at reproduction; may be related to calcium or protein-energy malnutrition.

5. regeneration, or climatic cycles, or both? Examination of the skeletal evidence should suggest diminishing health leading up to abandonment if the nutrition had been compromised, but climatic fluctuation certainly does not have to be the cause. Soil depletion or exhaustion of wood fuel supplies can both result in serious population perturbations as we have seen in many parts of the world.

Consideration of these topics, with careful evaluation of the evidence, should lead to some very clear and testable retrodictive models (indeed, they already have in previous literature).

DATA

The types of demographic data available to us are well known to the participants in this conference and need not be elaborated upon here. Suffice to say the one behavior we probably would like most and have the most difficulty with is probably fertility. Likewise, with respect to health, Martin[24] has provided an excellent summary of the paleoepidemiological data that are available and there is no need to repeat it.

Nevertheless, it may be useful here to provide a summary of the skeletal indicators of stress that may be available to us in various Southwestern samples (see Table 1 from Goodman et al.[11]).

BLACK MESA

Although an acknowledged marginal area with a somewhat truncated chronology ending approximately A.D. 1150, Black Mesa provides a very good example of the range of demographic and health questions that have been posed in the Southwest. It is also an area with which I am most familiar and in which there has been a lot accomplished. It reflects well the tradition of ecological research and the refinements that have been possible. From Swedlund and Sessions'[38] phase-based estimates of population growth in the area, we gained a now all-too-familiar picture of positive population growth that underscored the "progressive" framework in which many of us were working. With his more sophisticated analysis, incorporating much additional data and more realistic models, Plog[32] elucidated the irregularities in population growth on Black Mesa (which he was able to generalize to other sites and subregions). He shows the deviation-amplifying tendencies apparent in some of these population histories, argues that they do not correlate with hydrological cycles—at least for Black Mesa—and states:

> While environmental changes such as droughts may have increased this disequilibrium and hastened local declines, they did not cause the abandonment. Rather, abandonments were a result of the long-term inability of populations in many parts of the northern Southwest to develop organizational frameworks, technological systems, and procurement systems that were sufficient to regulate both population and subsistence yields in a marginal environment for agriculture. (p. 254–255).

When these findings are coupled with the evidence for health changes on Black Mesa[24,25] (admittedly based on limited data), one sees strong evidence for a population that was consistently under marginal nutritional stress, in a marginal environment, and one in which stress markers and disease incidence tend to increase through time. Again, Martin[24] is able to generalize these findings to many other sites in the Southwest, several of which are in less marginal areas.

If the results inferred from the population and health data are valid, and there is lots of supporting evidence, then density-dependent factors were frequently, though intermittently, operating throughout the Pueblo Occupation Phases. This suggests that agriculture was probably always marginal in the Southwest—somewhere—and that hydrological cycles, although of established value, may not be uniformly good predictors of where. If Palkovich[29] is correct about Arroyo Hondo, and some other sites, undernutrition sometimes became serious malnutrition and even starvation, especially for infants and children.

CONCLUSIONS

Sometimes, in the attempts to discern signal from noise, objectives deteriorate into the study of noise. One thing that strikes me in a recent and somewhat random effort to quickly "retool" myself on current demographic research in the Southwest, is how several of the major themes from a decade or two ago seem to hold up to close scrutiny today. The refinements in method and increase in data points are astounding—overwhelming—but the major trends and explanatory variables used in the past continue to be invoked. The signal is still fuzzy but becoming increasingly clear: population is a variable with which to be reckoned.

Is population the driving variable which explains social organization and change? Yes! (I am a demographer, what do you expect me to say?) But, alas, also no. The inability of Southwestern groups to effectively regulate population, as suggested by Plog[32] and others, inevitably must have led to inappropriate social and political responses. However, many additional responses were probably quite effective, and led to seasonal and short-term migration, reductions in fertility, and no doubt increases in mortality that eventually allowed communities to persist in one form or another.

If density-dependent factors were stringent and of high frequency, then I think we would see a very different aggregate result for the Southwest, and Lee's[21] points must be considered here. If we were to take all of the population growth estimates for the various sites and subregions and superimpose them on the same time scale, then high-density dependence would result in a no-growth or very slow growth trend over time—just as the paleolithic model does in the aggregate as noted in the Introduction. Instead, we find that many subregions are simultaneously growing, probably with intermittent and apparently weak density-dependent effects, at least until relatively late in Pueblo times. So the Malthusian assumption at least partly captures the experience for those first several hundred years.

Health indicators qualify this picture somewhat. Martin[24] says that a reader of the paleopathological literature would come away with the following impression:

> There were major and persistent nutritional deficiencies resulting from a corn diet; crowded and unsanitary living conditions enhanced the chances of picking up communicable diseases such as gastroenteritis; dental problems including caries and periodontal disease were a major concern; most adults had arthritis and spinal degeneration from carrying heavy loads; parasites such as lice and helminths were common; and infant and childhood mortality was high. With respect to trends over time, a continuum of health problems suggests that there were changes in the patterns with an increase in diseases associated with large and aggregated populations (p. 22).

I can envision an Anasazi historian, writing (petroglyphing) in the twelfth century A.D. , "Our numbers continue to grow, but we do not feel very well." This observation gives one reminiscences of many parts of the Third World today. Population growth under nutritional stress is a common feature in many countries. Nutritional and health deficits can affect the total fertility rate while leaving the NRR well above zero.

I mentioned in the Introduction that, in spite of my own admonishments about simplistic, univariate thinking, I still tend to ask myself, "but what really went wrong in those times of decline?" I am still motivated, I think, by the old models of carrying capacity. But, when I do allow myself to ponder the critical resources, I am increasingly interested in wood. Kohler and Mathews'[20] article on fuel wood limitations in the northern Anasazi area presents to me a plausible model for a very serious resource limitation. And the inhabitants of Chaco were apparently willing to travel as far as 75 km for construction timbers.[1] I do not know the weaknesses in these arguments—they may well be there—but wood for fuel and construction is of major importance to preindustrialized populations and it can be exhausted very rapidly.[2] We all know about the almost global crisis in wood today. Slash-and-burn agriculture, development and construction, and fuel for cooking can quickly lead to

[2] F. Plog,[31] 286–287, would agree?

a devastated area. In parts of Nepal, women now spend, on the average, two out of seven days *just to collect enough fuel for cooking.*

> "In rural parts of the Himalayas and the African Sahel, women and children spend between 100 and 300 days a year gathering fuelwood. Boiling water becomes an unaffordable luxury, and quick-cooking cereals replace more nutritious but slower cooking foods, such as beans."[33]

What are the implications for health and nutrition? Cooking also eliminates many common pathogens. Even in historical New England, an area with which I am more familiar, many towns were virtually clear cut within the first 100 years of their existence and towns were going far afield for construction materials and limiting access to fuel wood for domestic use.[9] In a marginal environment like the Southwest, one can imagine this happening much more quickly. Do the site and areal abandonments and reoccupations correspond to regeneration cycles for timber? Is this a major intermittent disturbance? Might they be the result of other forces? One is reminded of stories about medieval communities having to abandon their castles because of over exploitation of gause and due to pollution by huma waste within the castle walls.

Whether or not wood, or something else, or several factors, are the critical resource is hardly the point of this platform paper, however; rather, my purpose has been to delineate the major issues with which the Demography and Health group must grapple. Some issues may be conspicuous by their absence, and some may be overstated excessively, and with some I may contradict myself. These "disturbances" give us a place to start.

ACKNOWLEDGMENTS

This overview of my thinking on the topic has had as much to do with who I have talked to over the years than to the references I have consulted. Of those individuals, I would give special mention to several former students—Stanton Green, Jim Moore, Sheryl Horowitz, Bob Paynter, Richard Meindl, Ann Magennis, and Debra Martin—and my colleagues and friends—George Armelagos, Brooke Thomas, Martin Wobst, Robert Euler, and George Gumerman. I would like to thank the organizers of this conference and the Santa Fe Institute for their encouragement and support, and Jeff Dean and Marcus Feldman for comments on the paper.

REFERENCES

1. Betancourt, Julio L., Jeffrey S. Dean, and Herbert M. Hull. "Prehistoric Long-Distance Transport of Construction Beams, Chaco Canyon, New Mexico." *Amer. Antiquity* **51** (1986): 370–375.
2. Buikstra, Jane E., and Lyle W. Koenigsberg. "Paleodemography: Critiques and Controversies." *Amer. Anthropologist* **87** (1985): 316–333.
3. Coale, A., P. Demeney, and B. Vaughan. *Regional Model Life Tables and Stable Populations.* New York: Academic Press, 1983.
4. Coale, A., and E. E. Kisker. "Mortality Crossovers: Reality or Bad Data." *Population Studies* **40** (1986): 380–401
5. Cohen, Mark N., and George J. Armelagos, eds. *Paleopathology at the Origins of Agriculture.* Orlando, FL: Academic Press, 1984.
6. Corruccini, Robert S., Elizabeth M. Brandon, and Jerome S. Hamdler. "Inferring Fertility from Relative Mortality in Historically Controlled Cemetery Remains from Barbados." *Amer. Antiquity* **54** (1989): 609–614.
7. Dean, Jeffrey S., Robert C. Euler, George J. Gumerman, Fred Plog, Richard H. Hevly, and Thor N.V. Karlstrom. "Human Behavior, Demography, and Paleoenvironment on the Colorado Plateaus." *Amer. Antiquity* **50** (1985): 537–554.
8. Euler, Robert C. "Demography and Cultural Dynamics on The Colorado Plateaus." In *The Anasazi in a Changing Environment*, edited by George J. Gumerman. Cambridge: Cambridge University Press, 1988.
9. Finison, Karl. *Wood Resource Use in the Connecticut Valley, 1650–1900.* Working Paper, Connecticut Valley Population Project. Manuscript, University of Massachusetts, Amherst, 1977.
10. Gage, Timothy B. "Bio-Mathematical Approaches to the Study of Human Variation in Mortality." *Yearbook of Phys. Anthropology* **32** (1989): 185–214
11. Goodman, Alan H., R. Brooke Thomas, Alan C. Swedlund, and George J. Armelagos. "Biocultural Perspectives on Stress in Prehistoric, Historical, and Contemporary Population Research." *Yearbook of Physical Anthropology* **31** (1988): 169–202.
12. Gumerman, George J., and Jeffrey S. Dean. "Prehistoric Cooperation and Competition in the Western Anasazi area." In *Dynamics of Southwest Prehistory*, edited by Linda Cordell and George Gumerman, 99–148. Washington, DC: Smithsonian, 1989.
13. Hassan, F. *Demographic Archaeology.* New York: Academic Press, 1981.
14. Hillerman, Tony. *A Thief of Time.* New York: Harper & Row, 1988.
15. Howell, Nancy. "Toward a Uniformitarian Theory of Human Paleodemography." *J. Human Evol.* **5** (1976): 25–40.
16. Jorde, Lynn B. "Precipitation Cycles and Cultural Buffering in the Prehistoric Southwest." In *For Theory Building in Archaeology*, edited by L. Binford, 385–396. New York: Academic Press, 1977

17. Keyfitz, Nathan, Wilhelm, and Fleiger. *Population: Facts and Methods of Demography.* San Francisco: W. H. Freeman, 1971.
18. Koenigsberg, Lyle W., Jane E. Buikstra, and Jill Bullington. "Paleodemographic Correlates of Fertility: A Reply to Corruccini, Brandon, and Handler and to Holland." *Amer. Antiquity* **54** (1989): 626–636.
19. Koenigsberg, Lyle W., and S. R. Frankenberg. "Estimation of Age Structure in Anthropological Demography." *Amer. J. Phys. Anthropology,* in press.
20. Kohler, Timothy A., and Meredith H. Mathews. "Long-Term Anasazi Land Use and Forest Reduction: A Case Study from Southwest Colorado." *Amer. Antiquity* **53** (1988): 537–564.
21. Lee, Ronald D. "Population Dynamics of Humans and Other Animals." *Demography* **24** (1987): 443–465.
22. Longacre, William A. "Changing Patterns of Social Integration: A Prehistoric Example from the American Southwest." *Amer. Anthropologist* **68** (1966): 94–102.
23. Longacre, William A. "Population Dynamics at the Grasshopper Pueblo, Arizona." In *Demographic Anthropology Quantitative Approaches,* edited by E. B. W. Zubrow, 169–184. Albuquerque: University of New Mexico Press, 1976.
24. Martin, Debra L. "Patterns of Health and Disease: Stress Profiles for the Prehistoric Southwest." SAR Conference Paper, Organization and Evolution of Prehistoric Southwestern Society.
25. Martin, Debra L., Alan H. Goodman, George J. Armelagos, and Ann L. Magennis. *Black Mesa Anasazi Health: Reconstructing Life from Patterns of Death and Disease.* Carbondale: Southern Illinois University Press, 1991.
26. Moore, James, Alan C. Swedlund, and George J. Armelagos. "The Use of Life Tables in Paleodemography." In *Population Studies in Archaeology and Biological Anthropology: A Symposium,* edited by Alan C. Swedlund, 57–70. American Antiquity Memoir 30, 1975.
27. Moore, James A. "The Trouble With Know-It-Alls: Information as a Social and Ecological Resource." In *Archaeogloical Hammers and Theories,* edited by James A. Moore and Arthur S. Keene, 173–191. New York: Academic Press, 1983.
28. Paine, Richard R. "Model Life Tables as a Measure of Bias in the Grasshopper Pueblo Skeletal Series." *Amer. Antiquity* **54** (1989): 820–824.
29. Palkovich, Ann M. "Agriculture, Marginal Environments, and Nutritional Stress in the Prehistoric Southwest." In *Paleopatholoy at the Origins of Agriculture,* edited by M. N. Cohen and G. J. Armelagos, 425–461. New York: Academic Press, 1984.
30. Plog, Fred. "Political and Economic Alliances on the Colorado Plateaus, A.D. 400 to 1450." *World Archaeoloy* **2** (1983): 289–330.
31. Plog, Fred. "The Sinagua and Their Relations." In *Dynamics of Southwest Prehistory,* edited by Linda Cordell and George Gumerman, 263–291. Washington, DC: Smithsonian, 1989.

32. Plog, Stephen. "Patterns of Demographic Growth and Decline." In *Spatial Organization and Exchange: Archaeological Survey on Northern Black Mesa*, edited by S. Plog, 224–255. Carbondale: Southern Illinois University Press, 1986.
33. Postel, S., and L. Heise. "Reforesting the Earth." In *State of the World: A Worldwatch Institute Report on Progress Towards a Sustainable Society.* New York: W. W. Norton, 83–100, 1988.
34. Slobodkin, L.B., and L. Rappaport. "An Optimal Strategy of Evolution." *Quart. Rev. Biol.* **49(3)** (1974): 181–200.
35. Stoddar, Ann L. W. "The Status of Bioarchaeological Research in the American Southwest." In *Human Adaptations and Cultural Change in the Greater Southwest*, edited by Simmons, Stodder, Dykeman, and Hicks, 167–190. Arkansas Archaeology Survey, Research Series 32, 1988.
36. Swedlund, Alan C. "Historical Demography as Population Ecology." *Annl. Rev. Anthropology* **7** (1978): 137–173.
37. Swedlund, Alan C., and George J. Armelagos. *Demographic Anthropology.* Dubuque, IA: William C. Brown, 1976.
38. Swedlund, Alan C., and Steven E. Sessions. "A Developmental Model of Prehistoric Population Growth on Black Mesa, Northeastern Arizona." In *Papers on the Archaeology of Black Mesa Arizona*, edited by George J. Gumerman and Robert C. Euler, 136–148. Carbondale: Southern Illinois University Press, 1976.
39. Thomas, R. Brooke, Bruce Winterhalder, and Steven D. McRae. "An Anthropological Approach to Human Ecology and Adaptive Dynamics." *Yearbook of Phys. Anthropology* **22** (1979): 1–46.
40. Weiss, Kenneth M. *Demographic Models for Anthropology.* Memoirs of the Society for American Archaeology 29, 1973.
41. Wood, J., H. C. Harpending, G. R. Milner, and K. M. Weiss. "The Osteological Paradox: Linking Mortality and Morbidity in Skeletal Populations." *Current Anthropology*, forthcoming.

Ben A. Nelson,* Debra L. Martin, Alan C. Swedlund,† Paul R. Fish,‡ and George J. Armelagos*****
*Department of Anthropology, State University of New York at Buffalo
**School of Natural Science, Hampshire College
†Department of Anthropology, University of Massachussetts at Amherst
‡Arizona State Museum, University of Arizona
***Department of Anthropology, Emory University, Atlanta, GA 30322

Studies in Disruption: Demography and Health in the Prehistoric American Southwest

INTRODUCTION

Issues of demography and health have long played a significant role in Southwestern archaeology. Many of the most significant events highlighted by archaeologists from the standpoint of culture history are also demarcated by significant demographic corollaries. Indeed, one of the major questions confronted by the earliest researchers—what had happened to the northern Anasazi—was framed in terms of population processes. Several cultural events in southwestern prehistory are characterized by their demographic manifestations, such as periods of "aggregation," "abandonment," and migration. Likewise, complexity itself has demographic resonances that imply measures of rate, density, and scale.

As with demographic variables, the health status of Southwestern prehistoric societies has long been recognized as a fundamental aspect of understanding and interpreting the past. From Mathews and colleagues,[55] Hrdlicka,[40] and Hooton,[39] we see early attempts to characterize health of the early Pueblos through their skeletal remains. Unfortunately, the descriptive nature of these and many later studies did not render the information useful for making inferences at the population level.

Understanding Complexity in the Prehistoric Southwest,
Eds. G. Gumerman and M. Gell-Mann, SFI Studies in the Sciences of
Complexity, Proc. Vol. XVI, Addison-Wesley, 1994

Early archaeologists, in the absence of adequate empirical data, visualized the prehistoric past relative to historical and contemporary counterparts. For example, Colton[13] was heavily influenced by his stay in a Hopi village where he witnessed communities with a high density of people living in close proximity to their middens. Using ethnographic analogy, he suggested that settled village life in prehistory must have been fraught with disease and sickness, since this is what he saw when visiting Hopi villages. Aberle[1] noted that child mortality at San Juan and Santa Clara Pueblos was twice the prevailing rate for "Whites" in 1930. The mortality rivaled the rates for the slums of U.S. cities 40 years earlier. While these observations probably say more about the deplorable neglect of the B.I.A. in attending to the needs of Native Americans in the early part of this century, they nevertheless do point to the potential crises that could occur prehistorically when infectious disease was present under crowded living conditions.

Many have argued[47,48] that disease and epidemics may have played a major role in the eventual depopulation of the Southwest. Few models of culture change currently rely on disease as the primary causal factor. More recent research[54] suggests that while infectious disease was probably a continual problem for Southwestern settled populations, it is not sufficient as a general explanation for all major cultural and demographic transitions. In what follows we aim at a broader interpretation of the role of demography and health in regional development in the Southwest. At this early stage in the research, it is not possible to offer a general model; what we are able to do is to refine the questions and test for plausible associations in a series of test cases.

In this paper we discuss several hypotheses regarding the conditions under which prehistoric Southwestern populations experienced biological disruption. Biological disruption is the collective, mutual impact of demographic disturbance and physiological disruption. It is broadly conceived as negative change in the demography and health of the local population indicative of problems ranging from marginal success in coping to complete failure. Evaluating biological disruption requires examination of a number of indicators such as patterns of morbidity, as inferred from skeletal traits, and rates of fertility and mortality, as estimated from the age and sex distribution of the burial population. Some of the operative questions in this reckoning of group viability are what made people sick, which groups in the population were most vulnerable to illness, and how long adults might have expected to live.

Our intent is to make the approach and interpretation accessible to archaeologists and biological anthropologists without reference to more sophisticated technical descriptions of cases and procedures. The cases were selected to meet a variety of criteria. First, they represent a range of organizational states, as defined below. Second, they are cases for which we have a fair amount of information about the overall archaeological context, so that we are not analyzing biological data in a cultural vacuum. Third, they are cases in which relatively large samples of skeletal material have been recovered, analyzed, and reported. Nevertheless, they represent first approximations based on preliminary data, and are offered here in the form of a heuristic. A goal will be to see wider and more rigorous testing of these models in the future.

The cases that we discuss are (in order of maximum settlement size): Black Mesa, Arroyo Hondo, Mesa Verde, Chaco Canyon, and Casas Grandes. In some instances, the samples can be divided into early and late. These are strategic cases; they cannot be taken as statistically representative, nor is the information about them as consistent as we might wish. Each case is described individually to evaluate the degree of biological disruption as measured by our projections of disease load, age at death, and fertility.

After describing each case individually, discussing the archaeological context and measures of biological disruption, we juxtapose the cases to test hypotheses about health and population structure. Each measure provides a quantitative means of ranking the cases, but the overall evaluation is necessarily qualitative and interpretive because of the unevenness of the data sets. The basic approach is to take into account all that is known of the biological data, including facts about how the data were analyzed, and develop a ranking of the relative health of people belonging to the five test cases and their subpopulations. That ranking allows us to judge which of the hypotheses best fits the data.

ORGANIZATIONAL STATES

Throughout this analysis we use the term "organizational state," which is defined almost as broadly as biological disruption. The organizational state of a population refers to its density as well as to its social composition. The four organizational states that we discuss are dispersal, aggregation, political centralization, and collapse. These states encompass much of the diversity of Southwestern prehistory, and we feel that they have important implications for demography and health because they entail different patterns of access to critical resources.

Dispersed populations were those in which no more than a few families resided in a single settlement, and the settlements were scattered widely across the landscape. Such populations subsisted on a mixture of wild and domesticated foods and tended to employ extensive rather than intensive systems of crop production. Their dispersal allowed them to use mobility as a mechanism for coping with resource shortages.

Aggregated populations were those in which several dozen or more families lived in a single settlement, resulting in single-site populations of as many as several hundred individuals. Population density was high enough to place limitations on mobility as an option for solving resource problems, and also led to the tilling of relatively marginal plots of land. At the same time, wild resources became less reliable sources of food because they could not tolerate intensive harvesting. As a result, large aggregates of people tended to intensify their agricultural systems, employing water-control techniques and other measures that required greater amounts of labor input in relation to the output of food crops. Periodic crop failures were probably more threatening to aggregated populations than to dispersed ones, because of the absence of backup strategies based on mobility or the use of local wild resources.

Politically centralized populations were those controlled by an internal decision-making elite. There has been much debate in Southwestern archeology about the extent and even the existence of political hierarchies.[1] Political centralization is thought to have occurred only in large populations with densities great enough to

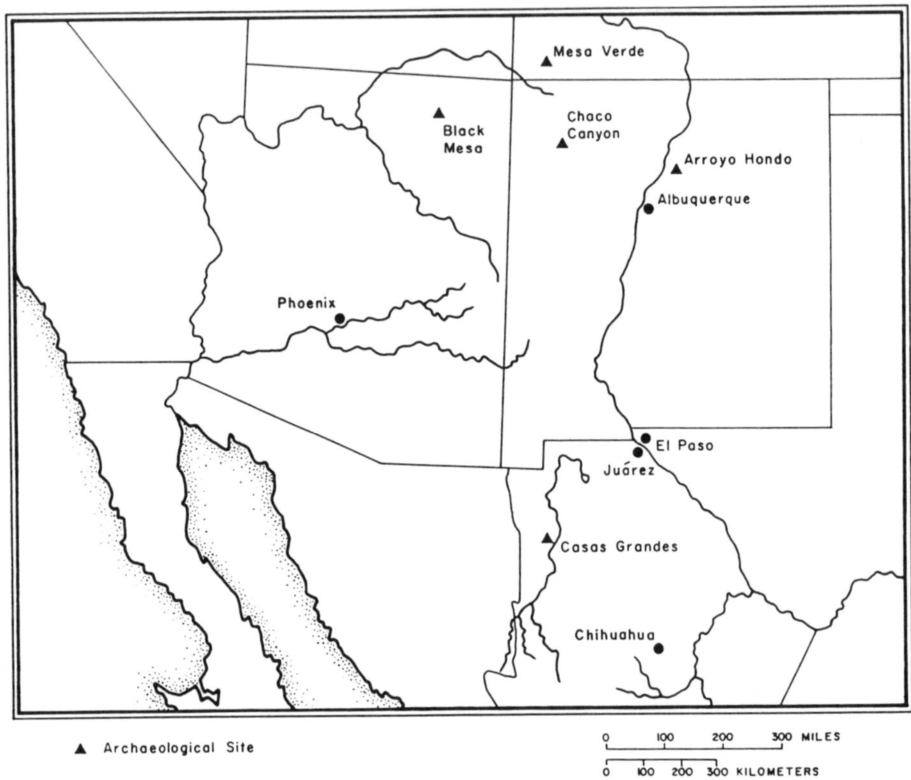

FIGURE 1 Locations of case studies.

[1]See Cordell,[15,124] Graves,[32] Grebinger,[34] Lightfoot,[52] Minnis,[64] Nicholas and Fienman,[68] Plog,[76,77] Reid,[85,125] Sebastian,[100] Upham,[109,110,111,112,108] Vivian,[114,115] and Whittlesey.[121,122]

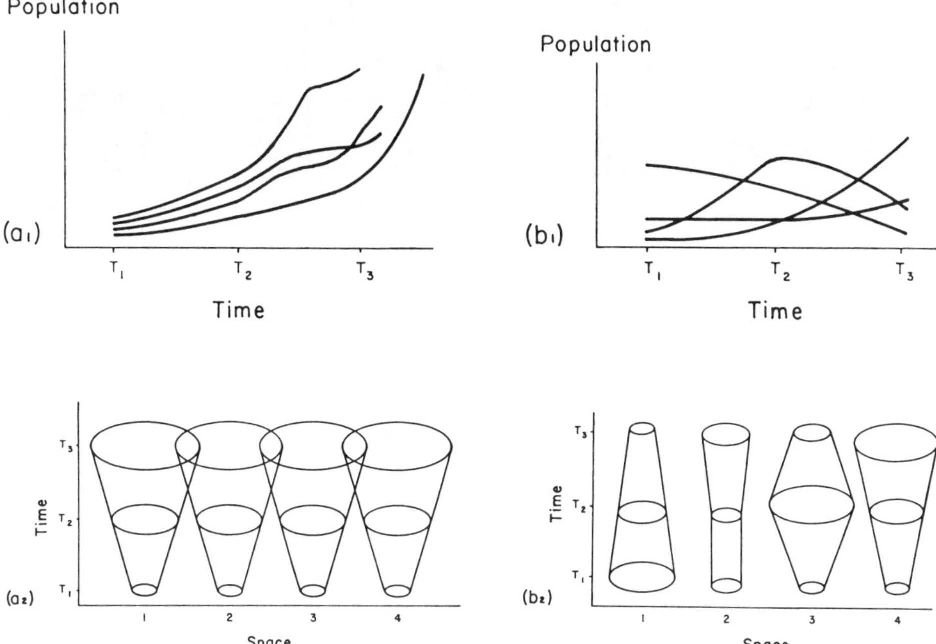

FIGURE 2 Population growth and resource impingement. Graphs (a1) and (a2) depict hypothetical local populations growing simultaneously to the point that they impinge upon each other's resource catchment areas. Graphs (b1) and (b2) depict local populations growing independently, largely avoiding impingement.

require specialization in the management of risks, conflicts, and resources.[52] Population growth is thus conceived as a selective force that favored political centralization. Figure 2 (parts (a1) and (a2)) shows abstractly how several groups living in different parts of the same region might have grown to the point that they impinged on one another's territories and created tensions relating to resources. Parts (b1) and (b2) of the figure indicate how populations might have grown independently, such that no impingement occurred most of the time. Incidents of synchronized growth in adjacent areas may be largely responsible for the appearance of more highly integrated political systems at various times and places in the prehistoric Southwest (Figure 3).

The moderately compelling case that has been made for the existence of political centralization is founded primarily on nonbiological evidence such as settlement pattern, site structure, and the distribution of prestige goods. Comparison of biological data from ostensibly centralized and noncentralized political systems is a powerful and underexploited source of inference about this issue.

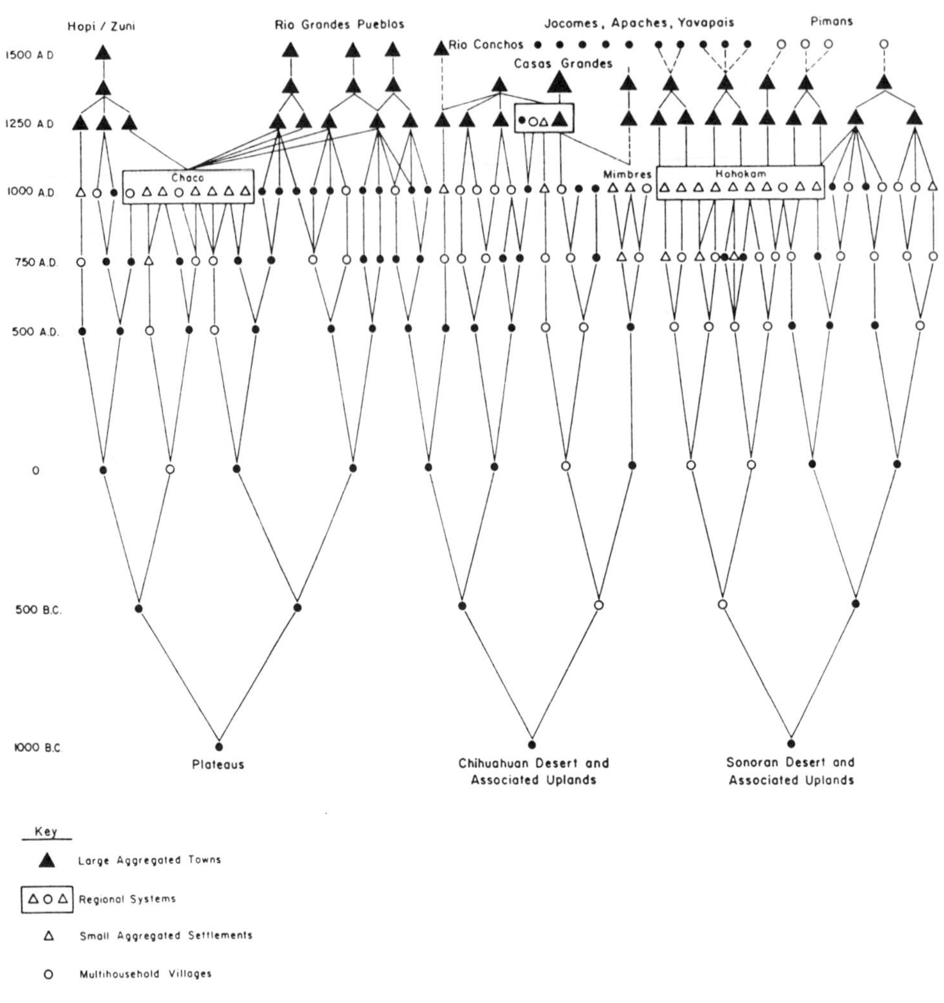

FIGURE 3 Cultural traditions of the Southwest.

Political centralization implies the existence of a group with privileged access to resources, including those that sustain health and prolong life. In contemporary traditional cultures, status differentiation is often marked by more and better food going to high-status individuals.[94] This pattern has been identified also in several prehistoric cases outside the Southwest. For example, anthropologists have found differences of stature (body height) between elite and nonelite burial populations at Tikal, Guatemala,[37] in the Ohio Hopewell region,[11] and at the Mississippian center of Moundville.[79] Powell further suggests, on the basis of strontium analysis, that elite males had a diet richer in meat than the rest of the population. Allison[5] shows that prehispanic Chilean and Peruvian shamans had a lower prevalence of arthritis and infection than commoners.

Such patterning often falls along lines of gender as well as status. M. L. Powell[79] shows that in the Moundville case all of the stature difference is attributable to males; there is no difference between elite and non-elite females. The difference in meat intake in this case also appears to be restricted to males. In the case of the Chilean and Peruvian shamans, commoner males had fewer nutritional and infectious pathologies than commoner females.[5] An occupational dimension also enters into the relationship between political centralization and health. Occupationally related stressors such as degenerative joint disease (osteoarthritis), trauma, and bone degeneration from wear-and-tear may be present if the elites did not participate in the daily labors of food production.[45] At Chucalissa, a chiefdom near Moundville, M. L. Powell[79] finds that elite males had more fractures than any other group. Allison's[5] shamans, however, had a lower frequency of fractures than high-status females or commoners of both sexes. Differences in these indicators of the degree of involvement with work and labor need also to be analyzed with respect to age and sex within a given population. Division of labor within a community may fall along gender or age lines, and in some cases occupational stress may not be related to status differences. Careful analyses of these indicators within and between groups is necessary to tease out the causal relationship with status.

The last organizational state that we mention is collapse. The term is actually an abbreviation for the interrelated processes of collapse, abandonment, and reorganization. Fish et al.[28] show that these processes were part of a yet more inclusive phenomenon that occurred in the late prehistory of the Southwest: the development of organizational modes that enabled larger territorial units, denser and more integrated populations, and intensified production systems. Populations that "collapsed," in other words, were probably reabsorbed into groups already occupying other localities, perhaps in different organizational states.

Archaeologists can seldom pinpoint the relocation of aggregates as independent entities in a more amenable region, so that health and demographic profiles can be compared before and after abandonment of the homeland. Biological data do, however, hold promise for determining whether biological disruption was a factor in prompting abandonment. If deficits of subsistence resources were a consistent problem, data from the terminal phases of occupation in a variety of settings should reveal deterioration of diet and health. Sources of evidence include subsistence remains, coprolites, and skeletal series.

HYPOTHESES

1. Biological disruption was independent of organizational state. Such lack of relationship would be expected if all Southwestern adaptations were about equally

successful. It might imply that the disparate forms of organization were fostered by the need to maintain a constant relationship between population and resources, and that such balance was in fact continuously achieved. Such an outcome seems unlikely, considering all of the population change, environmental disturbance, and organizational variability that we know to have existed. For this hypothesis to be accepted, the indicators of health must be uniformly distributed across the test cases. Uniform distribution is not meant in the simplistic sense that all populations manifest the same kinds of health problems, but in the stricter sense of highly patterned associations among conditions such as iron deficiency and infection.

2. Biological disruption was associated with and dependent on organizational state. More realistically, we might expect a complex relationship. The phenomena that archaeologists label "aggregation" or "political centralization" were attempts to maintain biological homeostasis while coping with changing natural and social conditions. Southwestern populations presumably found ways to remain relatively well most of the time, provided that key parameters such as crop yields and access to wild resources remained within tolerable limits. When physical needs were not met, health would decline along with fertility. We expect that cultural changes followed: social structures were remodeled, production increased, territories expanded, or alliances extended. Such organizational transformations have archaeologically visible consequences, such as the establishment of large pueblos where small, scattered villages existed before.
   One major possibility, therefore, is that each of the various organizational states had distinctive patterns of demography and health. Although it is difficult to predict the impacts of each state precisely, it is not unreasonable to suspect that each state may have presented people with a different set of selective pressures, resulting in a unique profile of disease, mean age at death, and fertility. For example, two communities of the same size, one dispersed and one aggregated, might have had quite different disease profiles because contagious disease is known to be much higher among children living in densely settled populations. This hypothesis implies that it might not be possible to predict the degree of biological disruption by any underlying variable such as maximum community size, yet each organizational state might manifest its own pattern of disruption.
3. Biological disruption was density dependent. This view proceeds from what might be called a "Walden Pond premise" that health is found in isolation. Even though it is a commonsense assumption that crowding begets disease, systematic reasoning reaches the same conclusion, as is noted above and has been noted since Colton's time. We also suggest that a concomitant of increased mortality may have been increased fertility, a point discussed further below. If the relationship between biological disruption and density was direct, the ranking of cases by biological disruption should be the same as their ranking by degree of aggregation.
4. Biological disruption was density dependent with class or gender differentiation. This variant of the last hypothesis takes into account the privileges of access to

resources enjoyed by special categories of individuals. For example, if political hierarchies developed in some of the larger systems such as Chaco Canyon and Casas Grandes, commoners may have been subject to the effects of density while elites were buffered against them. With or without political hierarchy, females may have been deprived of access to nutrition as in some of the cases discussed in a previous section. Low-status people who suffered from protein deficiencies, for example, may have attained shorter stature and suffered more nutrition-related infections than their higher status counterparts. This hypothesis implies a more or less direct relationship between aggregation and biological disruption, but with a special divergence for elites at the upper end of the curve. If elites cannot be distinguished in analysis for reasons of poor archaeological visibility or inadequate sample size, then the effect would be to depress the curve at the upper end, creating the appearance of a threshold in the relationship.

5. Biological disruption was time dependent. Another scenario is that biological disruption occurred late in the duration of any organizational state, regardless of what the organizational state might have been. The premise in this hypothesis is that systems organized themselves to solve problems that were ultimately biological; decisions to aggregate or centralize, for example, were responses to threats or opportunities. Although such responses presumably increased the well-being of the population in the short term, they may have introduced or failed to solve other problems that manifested themselves over longer periods. This hypothesis has some empirical support in that all Southwestern adaptations appear to have "failed" after some period of time. If this hypothesis is to be accepted, the data from test cases where there is good chronology should show evidence of biological disruption in the later phase of each case, independently of organizational state.
6. Biological disruption was dependent on environmental marginality. This hypothesis is based on the premise that agriculturalists in marginal environments would be more subject to crop failure than would farmers where growing conditions were favorable. Environmental marginality is difficult to quantify,[123] but hinges on the limited availability of soil, water, frost-free days, or combinations thereof. The paleoclimatic record of the Southwest testifies to periodic droughts of varying severity.[26] It is conceivable that at times those droughts afflicted the people living in marginal environments, while elsewhere people were buffered from the worst effects by the quality of their arable land. The data would be in conformity with this hypothesis if the ranking of biological disruption in the test cases corresponded with their ranking by environmental marginality. A ranking by environmental marginality on a case-by-case basis is more than we can actually accomplish, but we can specify which cases probably represent the extremes.

THE ENTANGLEMENTS OF LIFE TABLE ANALYSIS

The appropriateness of life tables in paleodemographic settings is an issue that periodically waxes and wanes in the anthropological literature. During the mid- to late 1980s a number of articles appeared which are basically cautionary, if not outright negative, regarding the interpretation of life table values based upon skeletal populations. It is not necessary here to fully explore this issue (see Johansson and Horowitz[41] and Swedlund and Meindl[105] for more complete discussions), but some preliminaries are in order.

Paleodemographic life tables are the product of multiple sampling processes, and are therefore subject to many questions of validity and reliability. For example, we must assume that the data are a real reflection of the actual population which deposited them insofar as age and sex are concerned; we must also assume that the age and sex assessment of those skeletons is accurate within a reasonable margin of error, and, normally, we assume that the population in question was a closed population with respect to migration and undergoing no growth (i.e., stationary) over the period represented by the material. No small order.

Moreover, even assuming that the skeletal series is a reasonable facsimile of the population and the death rates to which they were subjected, we must contend with another important reality: skeletal age distributions and the mortality profiles that they imply are the product of two demographic forces, mortality and fertility. We seldom know the fertility regimes under which a prehistoric population lived, and a very wide range of combinations of birthrates and deathrates can create the same apparent age at death distribution. No small problem.

Although the importance of population growth rate (i.e., primarily fertility) has been noted for many years in the paleodemographic literature,[7,9,66] a major contribution from the articles of the past few years has been to demonstrate more explicitly the precise impact of fertility on age at death distributions. Notably Sattenspiel and Harpending,[95] Buikstra et al.,[12] and Johannson and Horowitz[41] show in a variety of related, analytical ways that fertility, rather than the underlying death rate, is often the primary determinant of mean age at death in mortality distributions. In skeletal populations the estimation of life expectancy at birth by conventional life table procedures can thus be very misleading and is better thought of as mean age at death for the skeletal series (the two are equivalent under stationary conditions, but potentially quite different under stable or quasi-stable conditions). Moreover, in many situations the paleodemographer should interpret mean age at death as an indicator of fertility and growth rather than mortality.

These caveats have caused some to retreat from the use of paleodemographic life tables altogether while others have regarded it as a challenge that can be met in a variety of ways (for an example of the latter, see van Gerven and Armelagos[113]). What are the implications of these debates for the present study and for ongoing research on the demography and health of prehistoric populations in the Southwest? We argue that the Southwest is a prime example of an archaeological setting in which much can be gained by the use of skeletal series for interpretation of demography and health. The availability of well-preserved remains, coupled with good dating and independent inferences on population growth and decline, aid considerably in addressing questions that go unasked or unanswered in other regions.

Recent criticisms are based on valid concerns and a deeper appreciation of demographic parameters and their estimation than was characteristic of earlier studies. However, they sometimes fail to take into account the fact that a paleodemographer's sources include the archaeological context as well as the paleopathological profile of the series. If, for example, a high proportion of subadult and maternal-aged female deaths are associated with indicators of marked nutritional stress and infection, then it is not unreasonable to assume that risks of morbidity and mortality were high. If, on the other hand, mortality of the young "appears" high, but no indicators of stress are present, then one might reasonably assume that high fertility is driving the age at death distribution, or that virulent and epidemic infectious disease caused children to die before an indicator appeared on the bone tissue. If an independently derived growth trend can be discerned from analysis of architecture, settlement pattern, and artifact accumlation, the resolution improves another degree.

Finally, just as we must remember that an age at death series is a product of two forces—births and deaths, we must also remember that high-fertility, rapid growth populations are also frequently accompanied by high mortality in the observed world populations today. There is a subjective element to how we interpret these situations that provides a lesson for paleodemographic research. For example, areas like modern Bangladesh have been confronted with very high fertility and growth accompanied by excessively high mortality for over 100 years. Development agencies historically assessed the problem as either fertility that was too high or inadequate care of the young, and far less often acknowledged the systemic nature of the problem. Students of paleodemography should not make the same mistake. Often the low mean age at death in paleodemographic populations is probably a product of both high fertility and high mortality, in spite of the fact that the value for mean age at death may be more highly correlated with the birth rate under the conditions specified by most demographic models.

The goal of paleodemographic research should be to provide a plausible set of scenarios that reflect the likely experience of prehistoric populations. A careful consideration of multiple sources of inference should provide reasonable explanations under favorable research conditions. These conditions appear to be relatively well met in the prehistoric Southwest.

Given all of these concerns, our basic analytical procedure is to bring multiple sources of inference to bear on the age at death distributions. Rather than attempting to fit the distributions to some reference model to estimate parameters (such as life expectancy), we simply point to what the respective distributions suggest in terms of health and demography for each case study. In so doing we are restricted to estimates based on the actual observed deaths at each age and, given abridged life tables and paleodemographic methods, the mean age at death values should be viewed in relative terms as opposed to interpreting them as absolute values in years. Each of the tables was taken from the raw data published in the literature; since computational errors were discovered in some of the original tables, each was recalculated by the same procedure using a spreadsheet program. In all cases the sexes are combined for the tables to maximize sample size and, as noted above, these represent first approximations with many data limitations that will hopefully be resolved in the near future.

Because a good deal is known about each of the regions from which the skeletal material has come, we have some *a priori* expectations about what will be observed. In general, material from sites or areas which were undergoing population growth should show relatively low mean age at death; those in which the material represents a period of decline could reflect low mean age at death as a function of low life expectancy, which should be coupled with indicators of deteriorating health. Stable or growing populations in comparatively abundant resource catchment areas, or in areas of low population density, should have correspondingly higher mean age at death and fewer stress indicators. From numerous sources, as summarized by Martin,[54] we know that the overall picture for prehistoric agriculturalists in the Southwest should be one of endemic nutritional stress that is chronic but not necessarily severe, resulting nevertheless in relatively high levels of infectious disease and low life expectancies.

BLACK MESA

ARCHAEOLOGICAL CONTEXT

Black Mesa is a physiographically defined upland segment of northeastern Arizona with a diameter of approximately 120 km. The Mesa is composed of several smaller plateaus rising 600 m above Monument Valley to the north and sloping toward the Hopi Mesas to the the south. The surface is generally covered by dense piñon and junipers at lower elevations and small islands of ponderosa and firs at higher ones; however, locations with deep alluvium are characterized by expanses of sagebrush. Surface water is scarce, occurring as canyon-head springs. Agriculture is dependent on combined supplies of winter and summer rainfall, both of which are in short supply.

On the basis of ceramics and other aspects of material culture, Black Mesa is one of several subareas for the Kayenta Branch of the Western Anasazi. Extensive dendrochronological research on Black Mesa in particular and for the Kayenta region in general permits refined dating of the archaeological sequence. Dendrochronology, in conjunction with large-scale studies of pollen and geomorphology, has resulted in perhaps the most detailed environmental sequence that can be correlated with prehistoric culture change anywhere in the world.[18,35] The Black Mesa environment is highly risky, with few good locations for agriculture, and wide fluctuations in climatic and hydrologic variables essential for successful harvests.

A number of rockshelters and 35 open sites have been dated to the Basketmaker II (560 B.C.–A.D. 550) time period on Black Mesa; all but one of these dated sites has yielded maize.[2] The sites range from isolated hearths to small, possibly permanent hamlets of five to six dwellings with facilities for significant storage.[101] On some portions of the Mesa, the initiation at this time of an enduring pattern of upland-lowland biseasonal mobility appears consistent with available data.[80]

From these earlier Archaic or Basketmaker beginnings until abandonment around A.D. 1150, the Black Mesa sequence follows a trajectory paralleling that of the Kayenta area as a whole. Small, dispersed settlements characterize the entire period from Basketmaker III through the Pueblo II abandonment, with even the largest villages never exceeding 20 or so rooms. Settlement occupations are typically brief, seldom persisting beyond a generation, and even multicomponent sites have a significant hiatus between short-term occupations.[36]

Population reconstructions generally depict curves that peak sharply during the early decades of the twelfth century.[25] Extended habitation ceases on Black Mesa sometime after A.D. 1150 and this is correlated with a decrease in water tables and precipitation and an increase in soil erosion. Populations are believed to have moved to neighboring locations with more predictable agricultural and domestic water such as the Klethla-Longhouse Valley to the northwest and the Hopi Mesas to the south. These movements are also correlated with larger and more aggregated villages and towns in these better watered locations. By the late fifteenth century, only the large Hopi towns, located in the most optimal agricultural area of northeastern Arizona, remain.

Social ranking and complex, stratified organization does not characterize Black Mesa society nor probably later ones in the larger region. Although a case can be made for "hereditary oligarchies" controlling ritual knowledge[109] and pan-Western Anasazi political alliances[76,77] after A.D. 1300, persuasive counterarguments have been made that the dramatic social and ideological changes reflected by material culture are modifications of the continuing theme of competition-retribution-cooperation.[3]

The transition to agriculture over 1500 years ago on Black Mesa appears to have been rapid and to have initiated dramatic changes in human-land relationships. Resources for hunters and gatherers on Black Mesa and across the Plateau are patchy, widely separated, and unpredictable with the potential for supporting very few people; agriculture may well have been a necessity in this region to produce significant and archaeologically visible populations.[30] Richard Ford[30,126] notes that the complex of cultigens and disturbance plants first documented during Basketmaker times changes little in the ethnobotanical record of a relatively complete Western Anasazi sequence including historic Hopi.

[2] See Gumerman and Dean,[36] pp. 109–111.

[3] See Adams,[2] and Gumerman and Dean,[36] pp. 134–137.

The burials from Black Mesa ($n = 172$) were excavated between 1967 and 1983 by the Black Mesa Archaeological Project of Prescott College and Southern Illinois University. Burials were divided into early Pueblo (A.D. 800–1050) and late Pueblo (A.D. 1050–1150) phases. The sample sizes are 51 and 121 respectively. No statistically significant differences in the frequency of pathologies was discerned between the two groups, although the break between early and late Pueblo represents a meaningful culture change based on the kind and amount of greyware ceramics found and the cessation of building activity between A.D. 1030–1070. Between A.D. 980 and 1070, there appears to have been a hiatus in the occupation of Black Mesa.

The skeletal population is the result of both complete site excavations and limited testing of sites during the 1981–1983 field seasons. While some cultural factors may have played a role in biasing the sample towards increased numbers of late Pueblo burials and midden burials, there is sufficient representation of the early Pueblo and non-midden burials to ensure balance if not statistical validity. Distribution of individuals by age and sex further supports the notion that the skeletal population is fairly representative of nonindustrialized, agrarian populations.[118]

PATHOLOGIES

No statistically significant trends exist to suggest that health declined in the later period on Black Mesa. Iron deficiency anemia is present on 87.7% of the Black Mesa individuals that could be analyzed, and this includes both active and healed lesions. Although this may seem extraordinarily high, 64.3% of the cases are slight or mild in involvement, 25.2% are moderate, and only 10.3% show severe manifestations. The great majority of the cases of porotic hyperostosis for Black Mesa fall into the category of healed or remodeled, while only 12.9% of the cases were active lesions at the time of death. This suggests that iron deficiency on Black Mesa was possibly episodic or seasonal and widespread, but not particularly severe.

The group most affected by iron deficiency on Black Mesa was infants under the age of two years. In all cases, these individuals show not healing and have active lesions at the time of death, suggesting that for this youngest age category, iron deficiency may have been a contributing factor to mortality.

Nutritional anemia and infection co-occur in 61.9% of the subadult population. Infection is high in infants aged newborn to 6 months; its prevalence suggests that there may be an interaction between mother and fetus regarding acquired infections. Alternatively, prolonged or obstructed labor also increases the likelihood that mother or infant may become infected. It should be emphasized that we are considering the individuals who died as youth, and that such a sample is not necessarily representative of a living Anasazi cohort. There is evidence to suggest that infectious, transmissible disease found a more vulnerable host in the anemic individual.

For the whole population, 24% exhibit some type of infectious disease and in young females it reaches 17%.

The Black Mesa children aged 6–10 also show signs of infectious disease. This is unusual because other reports commonly cite the age group of 1–4 as being the most afflicted with infections.[61] One interpretation of this finding is that there were repeated but nonlethal bouts of infection affecting the children. Chronic, but relatively mild, transmissible diseases may have been endemic on Black Mesa. The infections were persistent enough to *reinfect* older children, but mild enough for them to show recovery.

In terms of growth disruption, 85% of the individuals from Black Mesa have dental defects (hypoplasias). This is among the highest reported frequencies for prehistoric populations. The peak occurrence at ages 2–4 corresponds to probable times of weaning.

A summary of the health status based on the paleopathological analyses suggests that nutritional stress was ubiquitous on Black Mesa throughout the occupation. However, the health data do not imply severe malnutrition. Instead, a picture emerges of endemic, mild to moderate nutritional stress that had an impact on almost all age and sex groups. The generally mild nature of the iron deficiency, the pervasiveness of childhood growth disruption, and the clustering of pathologies around infancy and weaning, all point to a difficult existence, but also to an ability to respond and recover from bouts of poor health.

TABLE 1 Black Mesa Composite Life Table, Early and Late Pueblo Groups.

X	Dx	dx	lx	qx	Lx	Tx	e′x
0–0.9	17.22	104.4	1000.0	0.1044	853.0	25170.16	25.17
1–4.9	29.35	177.9	895.6	0.1986	3226.7	24317.13	27.15
5–9.9	10.08	61.1	717.7	0.0851	3435.9	21090.44	29.39
10–14.9	15.22	92.3	656.6	0.1405	3052.5	17654.56	26.89
15–19.9	8.13	49.3	564.4	0.0873	2698.7	14602.07	25.87
20–24.9	8.76	53.1	515.1	0.1031	2442.7	11903.41	23.11
25–29.9	8.88	53.8	462.0	0.1165	2175.4	9460.69	20.48
30–34.9	9.98	60.5	408.2	0.1482	1889.6	7285.28	17.85
35–39.9	7.91	47.9	347.7	0.1379	1618.5	5395.65	15.52
40–44.9	4.52	27.4	299.7	0.0914	1430.2	3777.12	12.60
45–49.9	8.28	50.2	272.3	0.1843	1236.2	2346.95	8.62
50–59.9	36.65	222.1	222.1	1.0000	1110.7	1110.74	5.00
	165						

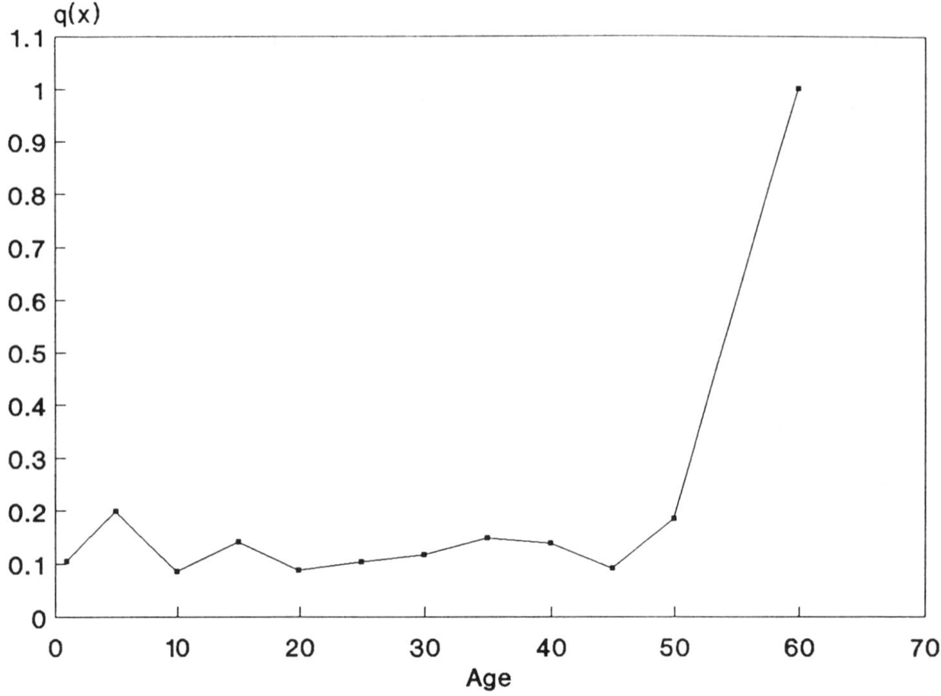

FIGURE 4 Mortality curve for Black Mesa. Definition of life table parameters: (a) x is the year or age interval being used in the table. (b) d'_x is, in paleodemographic tables, the actual number of deaths in each interval. (c) d_x is the number dying based on a proportion of the total sample, converted to a base of 10. (d) 1_x is the number of living in each interval out of 1000 alive at age 0. In paleodemographic tables

$$1_x = 1_{x-1} - d_{x-1}\,.$$

(e) q_x is the probability of dying at age x: $q_x \;=\; d_x/1_x$. (f) L_x is the number of individuals alive between age x and age $(x + 1)$. If we can assume that mortality changes between x and $(x + 1)$ are simply linear functions of s_x, then L_x may be given, for grouped data, by:

$$*L_x = (1_x + 1_{x+1})/2\,.$$

Where the age intervals are greater than single years, L_x is given by:

$$L_x = n[(1_x + 1_{x+1})/2]$$

where n is the number of years in age interval x. (g) T_x is the TOTAL number of years lived by all the members of the population at L_x.

FIGURE 4 (cont'd.)

$$T_x = L_x + L_{x+1} + L_{x+2}, \ldots, L_{x+n}\,.$$

(h) e_x is the expectation of life, or mean life-time remaining to those individuals attaining age x. $e_x = T_x/1_x$.

LIFE TABLES

Of the 172 skeletons, 165 are sufficiently known to permit inclusion in the life table. There is both an early and a late series for Black Mesa, covering the periods A.D. 800–1050 and 1050–1150, but the early series only consists of 49 burials and therefore cannot be treated independently. The combined periods could still be characterized as a time of moderate growth and comparatively low population density in contrast to some of the other cases.

The mean age at death for the combined sample is 25.17 years (Table 1). The pattern of age-specific mortality (Table 1, Figure 4) shows a slight elevation of mortality after the first year of life (1–4.9); given that infant mortality was at least 10 percent, this is consistent with the argument of a fairly high level of weanling deaths. Beyond this one distinct pattern, the rest of the mortality curve is relatively smooth and predictable.

Given that the relatively high mean age at death was occurring in the presence of population growth, it is tempting to conclude that the value has some true relationship to life expectancy. Black Mesa appears to have been a population that was battling daily with environmental limitations that predisposed them to mild-to-moderate ill-health, but was still faring reasonably well in comparison to its counterparts in other regions.

ARROYO HONDO

ARCHAEOLOGICAL CONTEXT

Arroyo Hondo pueblo is located approximately 8 km south of Santa Fe in the upper Arroyo Hondo drainage. Lacking broad, arable floodplains and significant perennial streams, the area has been described as having second-order carrying capacity.[127] Only 129 ha of potentially arable land are located within 2 km of Arroyo Hondo and less than half of this would be classified as good agricultural land according to ethnographic standards.[4]

[4]See Wetterstrom,[120] p. 38.

Prior to A.D. 1300, this area supported a small population with only short-term habitation sites. Population in the district grew rapidly after A.D. 1300, first at two sites and finally at one large pueblo.[21] The founding of Arroyo Hondo pueblo corresponds with a dramatic population increase witnessed throughout the northern Rio Grande at the end of the thirteenth century, probably resulting from amalgamations of migrants from the northern San Juan–Mesa Verde area.[5] A clear understanding of the social processes leading to an incorporation of migrants into the region is lacking and is critical for illuminating Arroyo Hondo demography. Perhaps immigrant groups were forced to occupy more marginal land such as that surrounding Arroyo Hondo because the most productive territory was already taken by local populations.

At its peak around A.D. 1330, Arroyo Hondo was an impressive settlement with 24 roomblocks clustered around 10 plazas. For an interval of about 25 years after A.D. 1345, the pueblo was nearly abandoned. A more limited second phase of settlement on the ruins of the previous village began after A.D. 1370 and reached its peak shortly after A.D. 1400. This second phase involved approximately 200 rooms in only 9 roomblocks around three plazas. A catastrophic fire destroyed the village shortly after A.D. 1420.[120] Cycles of growth and decline at Arroyo Hondo have been interpreted as closely correlated with increases and decreases in effective moisture.[93]

A maximal population estimate for the early phase (A.D. 1300 to 1345) of Arroyo Hondo occupation is difficult. Over 1,000 rooms date to this period, but a clear sequence of occupation and abandonment cannot be determined. Wetterstrom,[6] using various assumptions regarding numbers of persons per room and numbers of contemporaneous rooms, estimates that population could have ranged between 400 and 1,300 individuals. By evaluating Arroyo Hondo agricultural potential, she[7] concludes that 50% of the calories for a population of 600 could have been supported by fields within 2 km of the pueblo. Even in the best of times, then, little surplus would have been available at early population levels.

Detailed study of food remains from Arroyo Hondo clearly documents that agricultural products in general and corn in particular dominated the diet.[120] While the range of wild resources exploited is impressive, their total yield must have been modest.[8] Furthermore, little difference can be discerned in the subsistence of the large early population of the pueblo and the much smaller late one (Figure 5). In a location where agricultural potential was low by regional standards and opportunities for intensification nonexistent, the risk of subsistence stress must have been particularly great for the numerous earlier residents at Arroyo Hondo.

[5] See Cordell,[124] p. 323.

[6] See Wetterstrom,[120] p. 46.

[7] See Wetterstrom,[120] p. 50.

[8] See Wetterstrom,[120] p. 84.

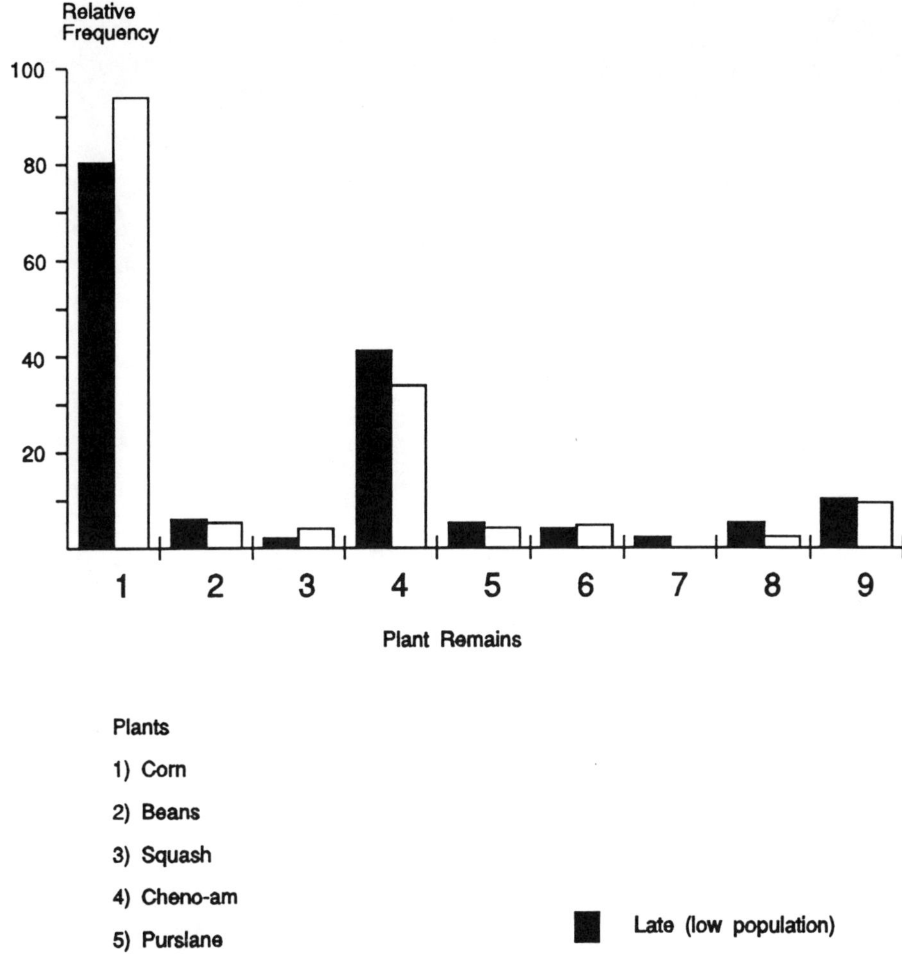

FIGURE 5 Temporal comparison of diets at Arroyo Hondo. Data from Wetterstrom.[120]

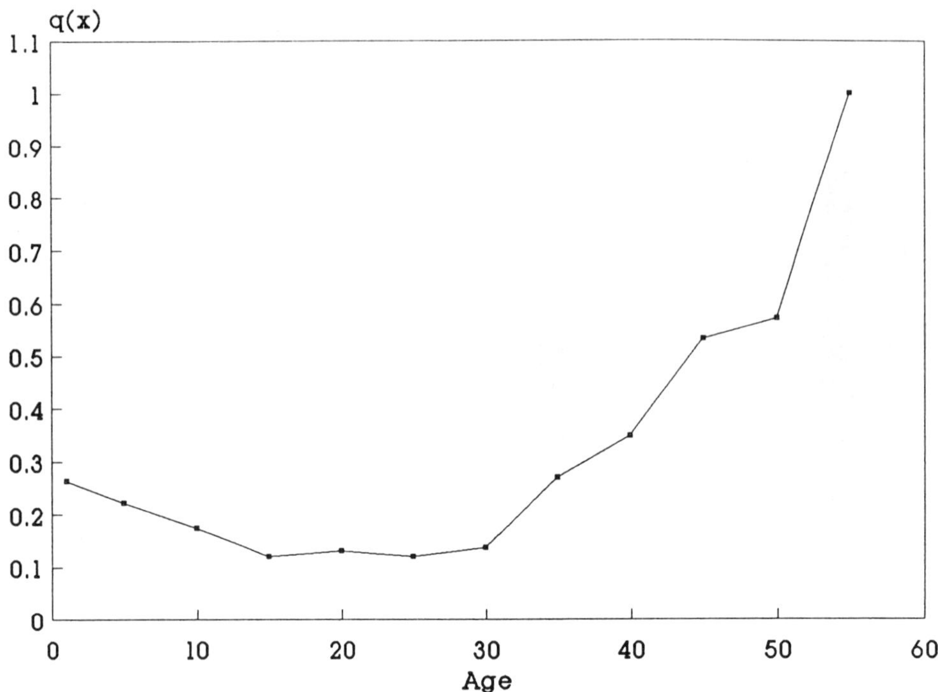

FIGURE 6 Mortality curve for Arroyo Hondo. See Table 4 for explanation of parameters.

Based on archaeological reconstruction, ethnobotanical analyses, and human skeletal remains, there is a strong case for a population undergoing extreme nutritional stress. With a dependence on maize agriculture, malnutrition and related diseases are documented in frequencies that are very high for the youngest children in the population. Both Wetterstrom and Palkovich suggest that periodic food shortages caused by local droughts most likely tipped the delicate balance that desert farming is based on. Using World Health Organization calorie and protein standards for children, the Arroyo Hondo diet has been estimated to have been "barely adequate during years of average precipitation."[9]

One hundred twenty burials were excavated between 1970 and 1974 by the School of American Research. The excavation covered a wide range of functional and spatial proveniences. Burials were recovered from the floors of rooms and pits, in plazas, and from middens. In view of the systematic excavation of the entire site, it is likely that the human remains, although small in number, are relatively

[9] See Palkovich,[72] p. 434.

representative of the population living at Arroyo Hondo. Component I yielded 108 burials, and Component II only 12. All ages and both sexes are well represented for Component I individuals, but Component II is represented only by infants and four older adults. Because there were no statistical differences between the two groups in pathologies or metrics, most studies conducted on these burials collapse the two groups into one.[72]

PATHOLOGIES

Of the 120 individuals represented by the skeletal remains, almost all individuals were afflicted with some form of pathology relating to poor nutritional status or problems with growth and development. Approximately 40% of the individuals show some involvement of cranial lesions relating to iron deficiency anemia. The incidence of active infection for the population is 8.5%, but most of these are located on the youngest children and co-occur with active cases of anemia. Trauma (dislocations and fractures) is found on 12% of the adult individuals.

For children aged between birth and 10, 21% show signs of active and advanced cases of porotic hyperostosis indicative of iron deficiency anemia. These individuals also had a high co-occurrence of active periosteal reactions; for infants this is an unusual pattern suggesting that the iron deficiency and infection acted synergistically. Normally, infants receive automatic protection from nutritional and infectious diseases from the mother if she herself is healthy. Palkovich[74] speculates that Arroyo Hondo infants had immediately acquired infections from their mothers, implying that maternal health was greatly compromised during pregnancy. Another feature of this population shows a high frequency of bowed long bones, about 15%, again suggesting a population undergoing periodic or endemic undernutrition.

A summary of the health status based on the paleopathological analyses of Palkovich[71,74] indicates that this population was suffering from endemic malnutrition that was greatly affecting subadult mortality.

LIFE TABLES

Of the 120 skeletons in the the Arroyo Hondo sample, 110 could be used for life table construction. This borders on a sample too small to be useful but, in cases of extensive excavation and careful diagnosis, we believe these samples should be included in studies. Palkovich's treatment of this series is one of the most thorough in the Southwest; when combined with the nutritional-ethnobotanical report of Wetterstrom, it provides important data.

Arroyo Hondo is one of the latest sites studied and all indications are that it was a population in severe decline. Mean age at death is by far the lowest encountered in any Southwestern series at 16.16 years (Table 2). Again, fertility may have some important contribution to this value, but the fact that the population was in decline

TABLE 2 Arroyo Hondo Composite Life Table, Smoothed.

X	dlx	dx	lx	qx	Lx	Tx	e′x
0–1.0	29.00	263.5	1000.0	0.2635	868.3	16162.49	16.16
1–4.9	18.00	163.5	736.5	0.2221	2553.5	15294.24	20.77
5–9.9	11.00	99.9	573.0	0.1744	2614.9	12740.78	22.24
10–14.9	6.25	56.8	473.0	0.1201	2223.1	10125.84	21.41
15–19.9	6.02	54.7	416.2	0.1314	1944.4	7902.73	18.99
20–24.9	4.79	43.5	361.5	0.1204	1698.8	5958.34	16.48
25–29.9	4.79	43.5	318.0	0.1369	1481.2	4259.49	13.39
30–34.9	8.15	74.1	274.5	0.2698	1187.3	2778.26	10.12
35–39.9	7.67	69.7	200.4	0.3477	828.0	1590.95	7.94
40–44.9	7.67	69.7	130.7	0.5330	479.5	762.99	5.84
45–49.9	3.84	34.9	61.1	0.5714	218.1	283.48	4.64
50–54.9	2.88	26.2	26.2	1.0000	65.4	65.42	2.50
	110.1						

leads us to believe that there is some relationship here to life expectancy as well. Infant and childhood mortality is extremely high (Table 2, Figure 6), and there are no unusual blips on the age-specific mortality curve with the possible exception of the 30–35 age group. However, given the small numbers this could easily be due to sampling. In sum, Arroyo Hondo should be regarded as the extreme for low survivorship in prehistoric Southwestern societies.

MESA VERDE

ARCHAEOLOGICAL CONTEXT

The Mesa Verde region covers a broad portion of the upper San Juan Basin in southwestern Colorado and southeastern Utah. Much of the relatively rugged upland region is well watered by numerous small streams and has lush, more heavily forested vegetation than much of the Colorado Plateau. Although a number of earlier Archaic sites have been identified, agricultural populations are first identified during Basketmaker II times and are continuously present through the Pueblo III period, when the region is abandoned around A.D. 1300.

Numerous tree-ring dates, associated with a well-studied sequence of change in material culture styles, result in one of the most refined regional chronologies in the Southwest. Cultural continuity in ceramics and styles of other items of material culture is well documented for the Mesa Verde region. Although a few possible great kivas may serve functions of community integration during Basketmaker III, small surface pueblos with kivas and great kivas appear during Pueblo I times at about A.D. 720.[70] Experiments in new and rapidly changing forms of public architecture are symbolized by the appearance of towers and very large kivas (late Pueblo II) followed by the addition of specialized ceremonial sites such as Fire Temple and Sun Temple.

Clusters of sites called "communities" are present throughout the Puebloan occupation.[91,92] These communities appear to equate with what might be called a local group whose affiliation is determined not by kinship but by spatial proximity that brings members into face-to-face contact.[10] Variation in the size and functional diversity of such interrelated sites becomes greater through Pueblo III times, when smaller settlements continue in conjunction with towns of up to 2500 inhabitants.[92] Great kivas at several larger Pueblo I sites may mark initial differentiation in communities of sites. More complex patterns of settlement appear to begin in Pueblo II, culminating with a probable hinterland arrangement of smaller settlements and more specialized site types in outlying territory about large towns.[11]

Population aggregation and increased socipolitical complexity appears periodic in at least portions of the Mesa Verde region during Pueblo I times. This instability has been correlated with escalating subsistence costs connected with population increase, environmental degradation reducing both agricultural and nonagricultural resources, and climatic fluctuations.[12] For instance, investigators at Cedar Mesa have concluded that the cumulative effects of slash-and-burn agriculture and the loss of soil fertility from continuous cultivation may account for the instability of individual settlements as well as the episodic nature of occupation on the Mesa.[56]

Survey evidence suggests a smooth and gradual curve for population growth for the region as a whole.[38,91] Maximum size of sites and their complexity also increases steadily through time.[13] Population densities appear high when compared with surrounding areas such as Chaco Canyon.[14][14] Rohn[15] estimates that at least 30,000 people were present in the upper San Juan by Pueblo III times. Regional abandonment was complete by the end of the thirteenth century and marks one of the major instances of prehistoric population relocation in the Southwest.

Studies of Mesa Verde diet from the perspectives of plant macrofossils recovered from a wide range of archaeological contexts,[58] analyses of human coprolites,[64,98,102] and isotopic data[20,57] indicate no significant change in plant resource emphases from

[10]See Rohn,[91] p. 216.

[11]See Rohn,[92] pp. 156–163.

[12]See Orcutt et al.,[70] p. 211.

[13]See Rohn,[92] p. 163.

[14]See Hayes,[38] p. 20.

[15]See Rohn,[92] p. 166.

Basketmaker III through Pueblo III (see Figure 7). Furthermore, these investigations show that plants from agricultural contexts, both cultigens and weedy disturbance plants, form a consistent "dietary core" throughout this sequence.[16] A shift to intensive use of agricultural products must have occurred in either Basketmaker II or earlier in the Archaic. Increasing use of soil and water control devices from Pueblo II through Pueblo III occupations is correlated with aggregation, strongly suggesting agricultural intensification, i.e., increases in labor input relative to productive output.[90,92]

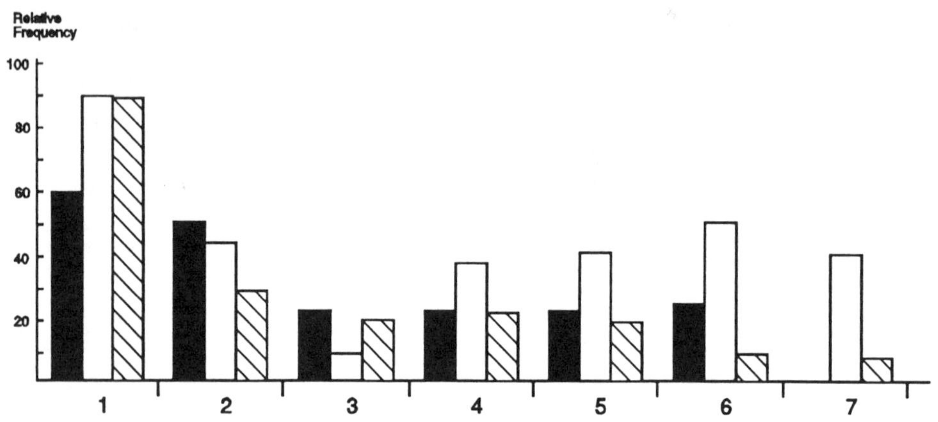

FIGURE 7 Temporal comparison of Four Corners diets. Adapted from Minnis.[64]

[16]See Minnis,[64] p. 559.

Stodder[104] provides a thorough summary of the ecological and cultural variables pertaining to diet and nutrition. The subsistence strategy in the region was relatively homogeneous with dry-farming horticulture producing maize, beans, and squash. During the late ninth century, there was an increase in storage facilities and pioneer (semidomesticated plants such as amaranth, goosewood, and beeweed) plant remains relative to wild plant remains. The botanical evidence documents increasing agricultural intensification through to the Pueblo III period with some sites demonstrating water management systems.

There are indications that the protein content of the diet may have been deficient, especially towards the latter time period. Consumption of corn and semidomesticated weedy plants as well as rodents, deer, and rabbits (who are attracted to fallow agricultural fields) provides an adequate amount of calories, but is low in usable protein, especially for a growing population which may have more children to feed. Large fauna such as American elk and bighorn, piñon nuts, and beans appear to have decreased significantly over time.

In general, the changes in diet and nutrition over time when viewed within the context of a growing population suggest a negative correlation. As population size and density increased, dietary variability and nutritional quality decreased.

There are a total of 466 individual skeletons used in the single best study of Mesa Verde remains to date, that by Stodder.[104] The human remains were recovered from a number of sites including the Dolores River Valley, Mesa Verde National Park, Mancos, Johnson and Lion Canyons, and two sites near the town of Yellow Jacket. The sample represents two time periods: an early period with 168 burials (Basketmaker III–Pueblo II, A.D. 600–975) and a late period with 276 burials (Pueblo II–III, A.D. 975–1300). Although these burials represent all age classes and both sexes, it is difficult to assess how representative they are of the regional population since they were excavated by numerous different crews and under different burial recovery strategies.

PATHOLOGIES

Stodder[104] finds no significant differences between the earlier Basketmaker III–Pueblo II and later Pueblo II/III groups in terms of frequencies of pathologies. She argues that the burials are representative enough to track diachronic changes in the region, and that the lack of significant differences in rates of pathologies reflect behavioral and cultural adjustments made during the later, more stressful, periods.

Of the 466 individuals represented by the skeletal remains, approximately 69% show some involvement with the cranial lesions that result from iron deficiency anemia. Cases of infectious disease are much lower, with 12% of the population showing active cases. The rate of trauma for the Mesa Verde population is 13%.

For children aged between birth and 16 years, 84% show signs of iron deficiency anemia. No cases of bowed long bones were found, suggesting that the nutritional problems were not as severe as those at Arroyo Hondo. However, Stodder also provides an extensive analysis of dental enamel defects (hypoplasias) which are formed between the ages of birth and 6 and are related to general physiological disruption. For Mesa Verde, 60% of the individuals have at least one defect, and the peak occurrence of defects for most individuals occurs at around age 3. This is related most likely to weaning stress. Although there were no differences in frequency between early and late individuals, there was a slight difference in the patterning of peak occurrence of defects. The early group showed a bimodal distribution of defect occurrences at ages 2.5 and 4. The late group showed a peak to occur at age 3 and to last to age 5. This suggests that during the later time period, children were breast-fed longer, and during the early period, children were weaned earlier. This makes sense when viewed in light of population growth and increased fertility during the early period, and population growth and decline in the late period.

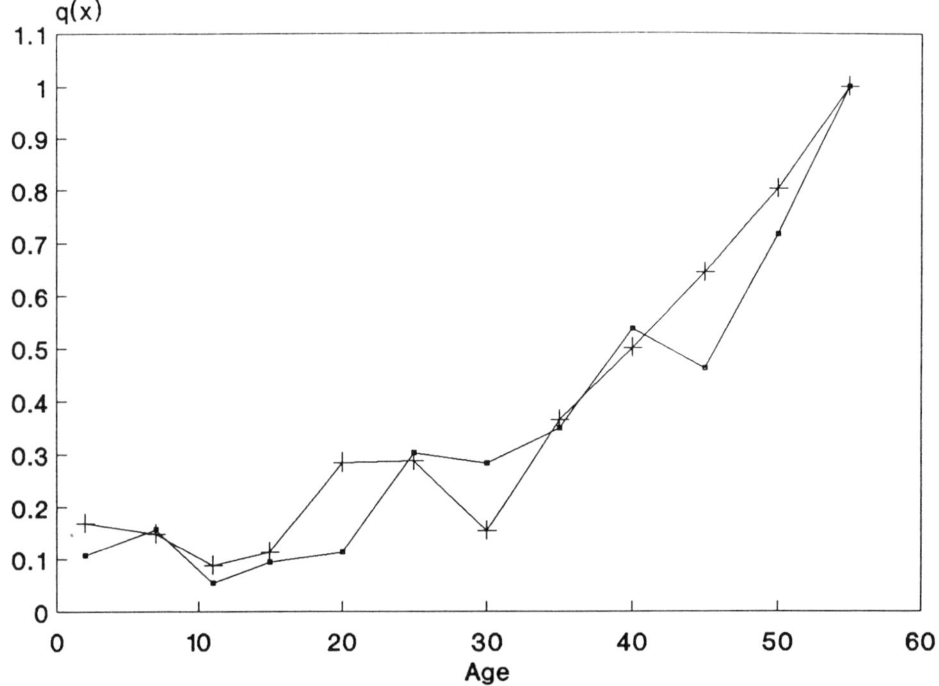

FIGURE 8 Mortality curves for Mesa Verde. See Table 4 for explanation of parameters.

TABLE 3 Mesa Verde Composite Life Table, Early Sample.

X	Dx	dx	lx	qx	Lx	Tx	e′x
0–2	16	106.7	1000.0	0.1067	1893.3	21246.67	21.25
3–7	21	140.0	893.3	0.1567	4116.7	19353.33	21.66
8–11	6	40.0	753.3	0.0531	2933.3	15236.67	20.23
12–15	10	66.7	713.3	0.0935	2720.0	12303.33	17.25
16–20	11	73.3	646.7	0.1134	3050.0	9583.33	14.82
21–25	26	173.3	573.3	0.3023	2433.3	6533.33	11.40
26–30	17	113.3	400.0	0.2833	1716.7	4100.00	10.25
31–35	15	100.0	286.7	0.3488	1183.3	2383.33	8.31
36–40	15	100.0	186.7	0.5357	683.3	1200.00	6.43
41–45	6	40.0	86.7	0.4615	333.3	516.67	5.96
46–50	5	33.3	46.7	0.7143	150.0	183.33	3.93
51–55	2	13.3	13.3	1.0000	33.3	33.33	2.50
	150						

TABLE 4 Mesa Verde Composite Life Table, Late Sample.

X	Dx	dx	lx	qx	Lx	Tx	e′x
0–2	30	168.5	1000.0	0.1685	1831.5	18353.93	18.35
3–7	22	123.6	831.5	0.1486	3848.3	16522.47	19.87
8–11	11	61.8	707.9	0.0873	2707.9	12674.16	17.90
12–15	13	73.0	646.1	0.1130	2438.2	9966.29	15.43
16–20	29	162.9	573.0	0.2843	2457.9	7528.09	13.14
21–25	21	118.0	410.1	0.2877	1755.6	5070.22	12.36
26–30	8	44.9	292.1	0.1538	1348.3	3314.61	11.35
31–35	16	89.9	247.2	0.3636	1011.2	1966.29	7.95
36–40	14	78.7	157.3	0.5000	589.9	955.06	6.07
41–45	9	50.6	78.7	0.6429	266.9	365.17	4.64
46–50	4	22.5	28.1	0.8000	84.3	98.31	3.50
51–55	1	5.6	5.6	1.0000	14.0	14.04	2.50
	178						

A summary of the health status based on Stodder's[104] analysis indicates that this population never enjoyed freedom from mild to moderate, and sometimes severe, nutritional stress. Although diet may have been somewhat better in the early periods, children suffered from nutritional anemia and growth defects all the same.

LIFE TABLES

Mesa Verde provides sufficient temporal variability and numbers of skeletons to permit Stodder to divide the population into an early and late group. Of the 466 skeletons, 150 could be adequately aged for the early group and 178 for the late. The early period is one of growth in the region and the late period is characterized as one of decline.

Interestingly, the early sample provides a mean age at death of 21.25 years (e′0 value) while the late sample drops to 18.35 (Tables 3 and 4). Since the late period is one of decline, we would think this consistent with the interpretation that the decline paralleled a condition of population-resource imbalance in the late period. Since the late period is also characterized by a declining population size for the region, the attribution of the lowered mean age to real declines in life expectancy is reasonable. Since fertility would be expected to decline during a population downturn, the expected outcome would be to see an increase in the mean age at death because the population would be expected to consist of fewer young and more old people. The estimated age-specific mortality rates (qx) presented in Figure 8 are relatively smooth given the small sample sizes for each group. The drop in the 26–30 age group for the late sample and the 41–45 age for the early sample are indicative of likely underenumeration of deaths at those ages, but their impact is in the opposite direction from the one expected if the mean age at death differences were solely due to bias.

CHACO CANYON

ARCHAEOLOGICAL CONTEXT

Chaco Canyon is located in the San Juan Basin of northwestern New Mexico on the arid steppe lands of the Colorado Plateau. Its main dates of occupation are from A.D. 900–1140, with a peak of construction activity in the largest sites during the late eleventh and early twelfth centuries.[50] This expansion in the canyon closely parallels the changes in site and room counts throughout the San Juan Basin.[43]

Population estimates for the canyon itself range from 2000 to 10,000.[43] Among three equally well-reasoned analyses, Schelberg[96] concludes that Chaco Canyon never held more than about 3700 people, Lekson[50] arrives at a range of 2100–2700, and Vivian[115] concludes that 5000–5500 is a valid estimate. The settlements in the canyon include large, preplanned towns with elaborate masonry architecture, great kivas, and plazas, as well as smaller villages, which appear less formal.[114] Some of the difficulty in estimating population for the canyon is linked to uncertainty about the functions of the large sites. While some researchers believe that they were elite dwellings,[3,99] others see them as essentially vacant ceremonial or storage facilities.[43,107]

A prominent element of Chacoan material culture is the road system, which radiates in several directions from Chaco Canyon toward points as far as 80 km away.[46,69] Many of the settlements connected by the roads are similar to the large, formal towns within the canyon. Seventy to 80 of these "outliers," which also occur without apparent road connections, are scattered over an area of about 50,000 sq km surrounding Chaco Canyon.[53,80] Some of the well-documented outliers are surrounded by clusters of smaller sites, apparently comprising politically integrated local communities. Small villages also are found in high density in and around the Chaco Canyon.

Although the canyon is in a hydrologically advantageous situation compared to its surroundings, there is debate about whether the local environment was productive enough to support the numerous villages and towns. Winter[123] and Schelberg[96] propose that the canyon's inhabitants could have survived only by importing food from outlying areas; Reher[84] and Judge et al.[42] maintain that the climate was favorable enough during the occupation that local food production would have been ample to support the population. Vivian[115] argues that despite the natural poverty of the area, the water control system (which consisted of dams, canals, headgates, and bordered gardens) would have sustained crop yields sufficient to support even the relatively large population that he envisions.

A common theme in the subsistence ecology of the Chacoan system is variability in rainfall, both in specific places and among localities. Even during years of relatively good average rainfall, some areas would have received little precipitation. Subsistence security may have been chronically low, and punctuated by periods of significant stress. Virtually all researchers agree that the late end of the occupation, from A.D. 1050–1130, was a time of greater than average summer precipitation, except for the early 1080s and most of the 1090s.[17] That generally good period was followed by one of decreased available moisture from 1130–1180. Most feel that this downturn, coupled with apparently increasing population, contributed to the collapse of Chaco Canyon and its satellites.

The subsistence risk implied by variability in rainfall, along with the social connectivity suggested by the road system, have allowed some archaeologists to infer that the people of Chaco Canyon coordinated a large-scale food redistribution system.[42] This inference is further supported by the observation that the large rooms in the towns lacked habitation features, and might have been used to store surplus crops.[51,53] Others doubt that the rooms were empty, or that importation of food was a necessary or even feasible solution to variability in crop yields.[38,100,115]

[17] See Judge,[43] 213.

Toll[107] and Judge[43] argue that Chaco Canyon was predominantly a ritual center, and secondarily served to distribute goods such as ceramics and turquoise.

Whatever the specific interpretation of Chaco Canyon's role, there is general agreement that it was the core of a politically centralized social system. This conception is founded mainly on the centrality of the place as expressed in the road system, the monumentality of architecture in the towns, and the rank-size distribution of settlements.[97] It is also frequently surmised that the inhabitants of the great towns, such as Pueblo Bonito, were political elites.[100,97,80] Some amount of status differentiation appears to be indicated by the distribution of grave goods, e.g., the thousands of beads and other ornamental items found with burials in Room 33 of Pueblo Bonito,[75] compared with the relative absence of grave goods in the village sites.[3]

For purposes of this study of biological disruption, we acknowledge the possibility that some form of elite-commoner distinction existed. That possibility seems to be rather well supported by architectural patterns and artifactual distributions. An important question then is how much economic inequality accompanied this instance of political centralization. If the existence of a political elite implies differential access to resources, then we might expect several correlates in demography and health, as already indicated. In particular, a contrast should be apparent between the residents of the great towns, such as Pueblo Bonito, and the lesser villages.

Hundreds of burials have been excavated at Chaco Canyon during the past century, yet few systematic studies have been done on these remains. Akins's[3] monograph on Chacoan burials in general and Palkovich's[73] summary of Pueblo Bonito burials are the most comprehensive examinations to date. Akins's work focuses on a collection of 135 individuals from 31 Chaco sites; no single site has a large enough skeletal collection to support analysis when considered alone. Given that the sample comes from a diversity of site types and depositional contexts, it is likely that the collection is not representative of a living cohort. The burials were made during a span of about 250 years, although most are probably associated with the apogee of Chaco Canyon from about A.D. 1050–1115. They include primary and secondary burials from room fill, subfloor, and midden contexts in both large "towns" and small "villages," as well as some burials from isolated areas that are not part of sites.

PATHOLOGIES

Approximately 84% of the subadults in Akins's subsample of 135 exhibit lesions indicative of iron deficiency anemia. Because subadults are greatly underrepresented in this population, it is difficult to know if this figure is accurate. However, it is similar to the rates reported for Mesa Verde and Black Mesa. Cases of infectious disease are found in 17% of the population and the frequency of trauma is also 17%. Dental enamel defects as indicators of general physiological disruption are high in frequency. Approximately 88% of the total population shows at least one defect.

Palkovich[72,73] reports frequencies of some pathologies for the subsample of burials from Pueblo Bonito. This analysis is important for present concerns because it allows us to compare a sample of burials from potentially elite residences with burials from the rest of the population. The reported frequency for iron deficiency anemia for subadults aged newborn to 10 is 25%. Palkovich[73] also notes that one subadult shows lesions that appear to be skeletal tuberculosis. From these data Palkovich concludes that "subadult dietary inadequacy clearly had a significant morbidity/mortality impact even among these high-ranking Chacoan lineages."[18]

We question this interpretation, however, because when compared to other Southwestern skeletal series, the frequency of 25% for iron deficiency must be considered low. As Chaco Canyon was densely populated, it is possible that tuberculosis was present and that does indeed suggest something about sanitation and transmissibility of infectious disease; however, we argue that the low incidence of iron deficiency anemia in the children at Pueblo Bonito does not support Palkovich's interpretation of severe biological stress at Pueblo Bonito.

Akins's data further suggest that diet at Pueblo Bonito may have been better than that of the contemporary small sites. In examining attained stature for adults from Pueblo Bonito vs. the smaller sites, she finds a statistically significant trend supporting increased height of Pueblo Bonito individuals. Pueblo Bonito males and females were on average 4.6 cm taller than sex-matched individuals from small sites. Akins also provides an unpublished observation from Hrdlicka's original notes that suggests that he saw few or no pathological conditions in the Pueblo Bonito skeletons; he did note that approximately 21% of the subadults showed lesions of the cranium—most likely porotic hyperostosis.[19] Akins's conclusion is very different from that of Palkovich. While acknowledging the indications of some iron deficiency anemia among the subadults, she argues that the low frequency of this pathology combined with the greater attained adult stature suggest a group that was enjoying better diet and health than its small-site counterparts.

LIFE TABLES

The only aged and sexed skeletal series available to us for life table construction is the 94 skeletons from Pueblo Bonito studied by Palkovich.[72] The Pueblo Bonito series is again very small and of concern for this reason, but is also very interesting because it may represent the mortality experience of an elite subpopulation of Anasazi society. Because population was growing until about A.D. 1115, about 15 years before the apparent end of the occupation,[43] the Chacoan skeletal series should be characterized primarily by population growth.

[18]See Palkovich,[73] p. 11.

[19]See Akins,[3] p. 137.

TABLE 5 Pueblo Bonito Composite Life Table, Smoothed.

X	Dx	dx	lx	qx	Lx	Tx	e′x
0–1	1.00	10.7	1000.0	0.0107	994.6	26365.38	26.37
1–4.9	6.67	71.5	989.3	0.0722	3718.9	25370.74	25.65
5–9.9	8.67	92.9	917.8	0.1012	4357.0	21651.85	23.59
10–14.9	8.00	85.7	825.0	0.1039	3910.6	17294.86	20.96
15–19.9	8.33	89.2	739.3	0.1207	3473.2	13384.31	18.10
20–24.9	10.67	114.3	650.0	0.1758	2964.4	9911.09	15.25
25–29.9	11.67	125.0	535.7	0.2334	2366.1	6946.71	12.97
30–34.9	11.00	117.8	410.7	0.2869	1759.0	4580.61	11.15
35–39.9	8.00	85.7	292.9	0.2926	1250.1	2821.64	9.63
40–44.9	6.67	71.5	207.2	0.3449	857.3	1571.51	7.59
45–49.9	5.67	60.7	135.7	0.4475	526.8	714.25	5.26
50–54.9	7.00	75.0	75.0	1.0000	187.5	187.47	2.50
	93.35						

The most apparent problem in the life table (Table 5) is the highly probable underenumeration of infant deaths. Some have argued that Pueblo Bonito was a central place and elite site and, therefore, may not have had a normal age structure, but some additional infant deaths would be expected. If the value for infant deaths is arbitrarily set at 10% (q0 of .10), the mean age at death changes from 26.37 to 24.09. This difference suggests that Pueblo Bonito would still fall in the upper range for mean age at death for these populations. Beyond the infant mortality deficit, the age-specific mortality curve appears reasonably smooth for such a small sample, leading us to conclude that, again, it is probably meaningful as regards the actual experience of this population (Figure 9).

The relatively high mean age at death, even after controlling for infant underenumeration and in the presence of population growth, may well indicate that Pueblo Bonito enjoyed a relatively favorable status concerning access to resources. Whether this came from tribute or local production, it also seems to have occurred in one of the most densely occupied localities in the American Southwest.

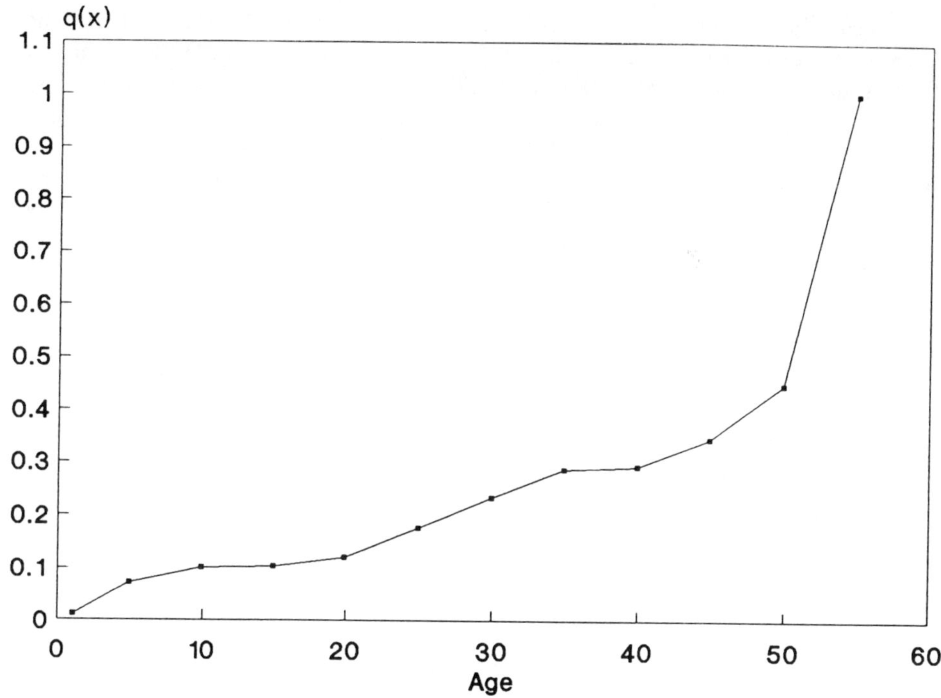

FIGURE 9 Mortality curve for Pueblo Bonito. See Table 4 for explanation of parameters.

CASAS GRANDES

ARCHAEOLOGICAL CONTEXT

The ruins of Casas Grandes, also known as Paquime, are located in northwestern Chihuahua, Mexico, in a desert grassland flanking the Sierra Madre Occidental. The site is about 250 km southwest of El Paso, Texas. Archaeologists consider this area part of the Greater American Southwest, even though it is in Mexico, because it appears to have shared more tradition with Southwestern societies than with Mesoamerican ones. The northern boundary of prehispanic Mesoamerica lay 600 km further south, much nearer to the geographic center of modern-day Mexico than to the international border.

Casas Grandes archaeology has been debated less intensely than that of Chaco Canyon because far fewer archaeologists have worked in the area. DiPeso, who conducted massive excavations at the site in 1958–1961, argues that Casas Grandes was a trading outpost of the Toltecs.[23] Based on the numerous occurrences of imported pottery, DiPeso suggests that the inhabitants maintained extensive commercial relationships throughout the Southwest and managed the acquisition of valuable raw materials such as turquoise for the Toltecs. Consistent with this Toltec-outpost argument, DiPeso holds that the main period of occupation at the site was from A.D. 1040–1340.

Further discussion of the chronology[10,49,50,82] has led to the conclusion that the main occupation, known as the Medio Period, dated to A.D. 1200–1400.[82] This dating places the occupation after the collapse of Chaco Canyon and contemporary with the Classic Hohokam. It also rules out a Toltec connection, as the Toltec ascended between about A.D. 900–1100. Although there are named phases within the Medio Period, their dating is unclear; as noted below most of the burials appear to belong to the Paquime phase, which represents the apogee of the site.

The multistoried structures of Casas Grandes were made of thick adobe and clustered around open plazas. The site had a sophisticated reservoir and drainage system, a series of huge stone-lined ovens, a subterranean walk-in well, and a variety of special rooms of unusual sizes and shapes. Macaw pens were located in one of the plazas. The periphery of the site was dominated by ceremonial facilities including several platform mounds and a ball court. Although no road system has been documented, there was apparently an extensive system of signalling stations where fires could be used to communicate messages among Casas Grandes and numerous outlying sites. Many of the larger outlying sites within a radius of 130 km also had ceremonial facilities and some had macaw pens.[62,64,65]

Casas Grandes probably held a larger aggregate of population than any other single site in the prehistoric Southwest; DiPeso[20] estimates that over 2,200 people lived in its 1,600 rooms during the Paquime phase. Including the large sites on nearby drainages would at least double this estimate. The site and its hinterlands apparently experienced a period of relatively rapid aggregation prior to the Paquime phase and a period of collapse after it; unfortunately those periods are not represented in the skeletal sample.[6]

Minnis[64] argues that Casas Grandes was also the most politically centralized community in the Southwest. Although he disputes DiPeso's[23] inference that the site was mainly occupied by elite foreign merchants, he cites considerable evidence for social ranking, including architectural symbolism, the distribution of prestige goods, and indications of economic specialization. Equally indicative of a status hierarchy is Ravesloot's[83] analysis of mortuary variability. He finds that

> "a system of social ranking based on inheritance probably integrated Casas Grandes society during the Medio period. Vertical social distinctions were

[20] See DiPeso,[23] p. 377.

marked by qualitative differences in mortuary treatment such as special burial locations, expensive grave facility constructions, variability in post-mortem processing of the body, and the relative cost of grave accompaniments as measured in terms of procurement and production."[21]

We suggest that Casas Grandes and Chaco Canyon may have been politically centralized to roughly the same degree, although Casas Grandes seems to have the strongest evidence for ranking. Comparisons are difficult to make because of obvious qualitative differences in organization. For example, at Casas Grandes numerous people were aggregated into a single massive settlement, while at Chaco Canyon they were spread among several smaller settlements, which were nonetheless in close proximity to one another and appear to have comprised a single polity. Also, Chaco Canyon appears to have had segregated housing for elites and commoners, while dwellings of the two classes were intermingled at Casas Grandes.

Expectations about the degree of biological disruption at Casas Grandes are therefore similar to those for Chaco Canyon. We might expect that density-dependent disruption might be somewhat intensified at Casas Grandes, because of the larger size and more nucleated nature of the central site. We might also expect slightly more pronounced differences in health and demography between elites and commoners because the artifactual evidence of status differentiation is somewhat clearer at Casas Grandes than at Chaco. However, even if our measures of biological disruption are sensitive enough to detect such differences, we will not be able to discern them in the sample as presently documented, because we do not have separate data on age, sex, and pathologies for elites vs. commoners. This means that the analysis must be based on a composite population.

Excavations carried out in the 1950s by the Amerind Foundation unearthed 641 skeletal remains. The preservation is very poor and has limited the ability of researchers to accurately age, sex, and analyze the collection. These burials "represent one quarter or less of the probable total number of remains at the site."[22] However, all ages and both sexes are well represented in the collection, and may be somewhat representative of the living population at Casas Grandes.

PATHOLOGIES

Unfortunately, information on nutritional status is not available in the skeletal reports.

There are currently two published studies on the Casas Grandes collection. Benfer[6] conducted a metric study of the adults only to analyze for biological differences in size among the time periods. Although some unusual pathologies were noted, no systematic study was done. Benfer states that "except for normal arthritic changes associated with aging, frequent dental caries, and occasional trauma, the

[21]See Ravesloot,[83] p. 68.

[22]See DiPeso,[22] and Benfer,[6] p. 3.

Casas Grandes population does not exhibit a great deal of pathology."[23] Weaver[116] analyzed only the subadult portion of the collection for two pathologies, iron deficiency anemia and infections. He noted that 48% of the subadults showed signs of cranial lesions resulting from iron deficiency anemia. Furthermore, 68% of the individuals with iron deficiency also had cases of infection suggesting a strong co-occurrence of the two pathologies.

LIFE TABLES

The Casas Grandes data include descriptions of 612 burials that have sufficient information to be aged for the life table (Table 6). Although this series is the largest among our case studies and can be divided into at least three temporal periods, according to Benfer[6] the bulk of the skeletal material comes from the Medio II period (Paquimé phase) and there are insufficient data to do legitimate life tables on the other periods. The Medio II period probably covers a period of population growth characterized by some decline in the later decades.

TABLE 6 Casas Grandes Composite Life Table, Smoothed, Al 1 Periods.

X	Dx	dx	1x	qx	Lx	Tx	e′x
0–0.9	62	101.3	1000.0	0.1013	854.4	20223.69	20.22
1–4.9	113	184.6	898.7	0.2055	3225.5	19369.28	21.55
5–9.9	23	37.6	714.1	0.0526	3476.3	16143.79	22.61
10–14.9	43	70.3	676.5	0.1039	3206.7	12667.48	18.73
15–19.9	41	67.0	606.2	0.1105	2863.6	9460.78	15.61
20–24.9	83	135.6	539.2	0.2515	2357.0	6597.22	12.23
25–29.9	102	166.7	403.6	0.4130	1601.3	4240.20	10.51
30–34.9	64	104.6	236.9	0.4414	923.2	2638.89	11.14
35–39.9	20	32.7	132.4	0.2469	580.1	1715.69	12.96
40–44.9	11	18.0	99.7	0.1803	453.4	1135.62	11.39
45–49.9	11	18.0	81.7	0.2200	363.6	682.19	8.35
50–59.9	39	63.7	63.7	1.0000	318.6	318.63	5.00
	612						

[23]See Benfer,[6] p. 31.

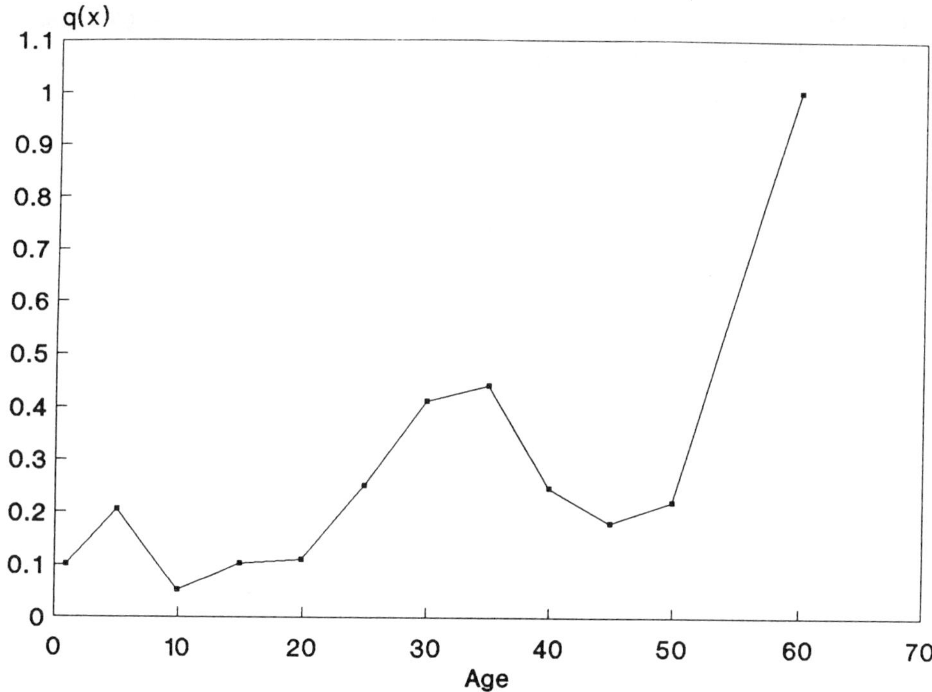

FIGURE 10 Mortality curve for Casas Grandes. See Table 4 for explanation of parameters.

Casas Grandes values indicate a mean age at death for the population of 20.22 years, which is probably a reasonable estimate given the large series. There is probably some underenumeration of infants (characteristic probably of all of the samples), but perhaps the most striking feature of this series is the large number of deaths occurring in the 29–34 age group (Figure 10). This bump is clearly not simply the result of sampling bias. The first hypothesis that came to mind was the possibility that this could be due to a large number of male deaths at the time of conquest of Casas Grandes.[60] However, on inspection of the time periods and the sex of the individuals involved, it became apparent that this is due to very high female mortality, over twice the male rate. Assuming that the skeletons are sexed correctly, this would be suggestive of very high maternal mortality during the Medio II period. It would be interesting to know if these deaths clustered at any particular time during the phase.

As to the question of whether or not the mean age at death bears some reflection on life expectancy, it would be our sense that it does, although it is probably a much better reflection of growth processes, given the fact that Benfer does not suggest a high level of pathology in this population. He also notes that a reasonably high proportion of individuals survived past 50.

RANKING BY PATHOLOGIES

It is extremely difficult to estimate the "wellness" or health of prehistoric populations because any measure that does so must take into account a number of diagnostic features as well as their complex interactions. The individual features that paleopathologists use are at best proxy measures of the day-to-day quality of people's lives and are not totally reliable indicators. The attempt to rank groups across space and time is equally vexing. Yet a careful reading of the frequencies of pathologies by category, age, sex, severity, duration, and state of healing does permit an approximation of biological disruption, especially if the demographic parameters of the group are also available for consideration. Measures of physiological disruption are virtually the only data that provide some, albeit limited, indication of the quality of people's lives on a daily basis.

The analysis of pathologies suggests the following qualitative ranking of degree of biological disruption, from least to greatest:

- Black Mesa
- Chaco Canyon (Pueblo Bonito)
- Mesa Verde (early)
- Chaco Canyon (small sites)
- Mesa Verde (late)
- Arroyo Hondo

Casas Grandes is not included in the ranking because of insufficient data.

Physiological disruption on Black Mesa can be characterized as omnipresent, but it does not seem to have played out in large-scale death and demise. Although Black Mesa has relatively high frequencies of infectious disease and iron deficiency anemia, the manifestation of the lesions is generally mild to moderate and often shows signs of healing. Dental defects are likewise high, but these combined pathologies most likely had only an indirect effect on mortality. They signal an environment that presented challenges in the form of periodic or seasonal food shortages, and also suggest the presence of contagious pathogens that were passed around during certain times of the year when people were in closed quarters. There is no evidence at Black Mesa for disease and death rates suggestive of a population on the verge of extinction. Nutritional and infectious disease was ubiquitous on Black Mesa, but the manifestation of the lesions does not imply severe malnutrition. Instead, a picture emerges of endemic, mild to moderate nutritional and infectious disease that certainly had an effect on morbidity. One further note is that people apparently bore little hostility to one another; the prevalence for trauma is the lowest in the case studies (4%).

At Chaco Canyon, it is possible to ask questions about the effects of social ranking on health and longevity. Although researchers have not systematically assessed indicators of health and demographic parameters for the inhabitants of Pueblo Bonito (the postulated elites) and the small-site occupants (the commoners), a cursory examination suggests that the elites were doing quite well. With only 25% of the children showing signs of iron deficiency anemia, Pueblo Bonito has among the lowest frequencies reported. Furthermore, growth is one of the more sensitive indicators of nutritional status, although it is a highly nonspecific and cumulative indicator. Tanner[106] has gone so far as to suggest that growth in height can be used to judge the equality of conditions within a society. The striking dissimilarity between attained adult stature at Pueblo Bonito compared with all the other case studies deserves close scrutiny.

Gray and Wolfe[33] computed the mean stature of males and females based on a broad survey of 216 societies. Interestingly, their mean heights for males and females (163.5 cm and 151.9 cm respectively) are nearly identical to the mean heights for Black Mesa, Mesa Verde, Arroyo Hondo, and Casas Grandes. Both achieved heights and the relative difference in heights between males and females are similar to many other prehistoric and historic groups living in traditional and marginal societies. These data suggest at least a mild to moderate degree of nutritional stress, as is likely to have been experienced by most of the groups sampled by Gray and Wolfe.

For children from the smaller sites of Chaco Canyon, iron deficiency anemia was rampant (84%) compared with the prevalence of 25% for Pueblo Bonito children. Furthermore, dental defects are the highest reported for the case studies, and suggest frequent physiological and metabolic disruptions for growing children. Although incomplete, the preliminary data collected from the small sites of Chaco Canyon places these groups towards the lower end of the health spectrum. Although the Chaco Canyon "commoners" do not seem to have experienced the malnutrition that is in evidence at Arroyo Hondo, neither do the data indicate sustained good health.

People living at Mesa Verde during the early occupation most likely experienced nutritional anemias and infectious diseases. Stodder[103] characterizes dietary stress at Mesa Verde as acute in nature since the highest frequency of severe lesions is in very young infants. During the earlier years when the population was growing, fertility was apparently high, and it is possible that this had a negative impact on reproductive aged females. With a medium life expectancy when compared across case studies, Mesa Verde offered a relatively good environment that could, for a time, support population increases, but with some food shortages and dietary challenges.

However, over time the situation at Mesa Verde may have changed rather dramatically with population size exceeding the food potential of the area. Reported frequencies of nutritional anemias for subadults and adults respectively in the early period are 86% and 31%. These frequencies rise to 93% and 74% in the later period, suggesting a dramatic increase in nutritional problems, especially for children. As discussed below, mean age at death drops by two years in the later period as well. Because there is no indication that fertility decreased during the later period, if it remained high in the face of nutritional inadequacies, it served to heighten the nutritional problems, and most likely worsen health.

The occupants of Arroyo Hondo were presented with a challenging environment and limited resources, and had few options for maintaining an adequate food base. Approximately 15% of the individuals who died at Arroyo Hondo show evidence of bowed long bones indicative of moderate to severe malnutrition. The very low achieved height of females (148 cm) may also reflect a group undergoing severe dietary constraints. Although fertility appears relatively low compared with other case studies, mean age at death is likewise very low and suggests a population in demographic retrenchment.

Palkovich[71] demonstrates the high co-occurrence of nutritional anemia and infection in the very youngest infants. This suggests that infants were acquiring infection from their mothers. Such a compromising condition only served to further undermine the viability of the population. Ill mothers giving birth to sick and dying infants suggests a population under great duress.

RANKING BY FERTILITY AND MORTALITY

We now return to the life tables and mortality curves to see what can be said about them as a group. One observation that is not directly drawn from the present analysis but derives from some work in progress is the fact that all of these tables share similarities and may be characteristic of a common family of life tables for the Southwest.[31] Indeed, they show a pattern that is distinct from paleodemographic life tables drawn from the prehistoric Midwestern United States and from the prehistoric Sudan. So there is reason to believe that real population processes are being measured. Moreover, the values for mean age at death and age-specific mortality, when compared for their relative differences, seem to make sense in terms of the known archaeological record and paleoepidemiological contexts.

Sattenspiel and Harpending[95] and Buikstra et al.[12] have suggested and improvised methods for inferring fertility levels directly from the mean age-at-death values derived from paleodemographic life tables. In addition, they have shown that the correlation between these estimated values and the mean age-at-death values is in excess of .9 for both methods. The correlation between mean age at death and life expectancy tends to be lower, probably closer to .7,[12] although the relationship between mean age at death and life expectancy tends to be consistent and interpretable.[41] These results and their interpretation rest on model life table assumptions about standard relationships between fertility and mortality which are theoretically valid and empirically correct, but which may not always obtain in human populations since they require closed, stable situations. In fact, many prehistoric populations probably deviated from these expectations considerably in their short-term demographic behavior.

These methods provide us with an additional means for comparison of our life table data. We select here the method proposed by Buikstra et al.[12] who suggest that the ratio of the individuals in the D30+/D5+ age at death groups is inversely proportional to the fertility rate. That is, the lower the value, the higher the presumed fertility under model life table assumptions. We have calculated these values for each of the life tables discussed above and compared them to the mortality values for each population as reflected in the mean age at death (Table 7). We then rank order each population according to their fertility and mortality estimates.

Two notes regarding the data are in order here. First, the mean age-at-death values have not been adjusted for the assumed underlying growth trajectories. If this were done, then the life expectancies would be slightly higher for those sites where growth is positive, and slightly lower for those where population was in decline. Efforts are underway to estimate valid growth rates for each of these locales, but the general trend is probably accurate as portrayed here.[18] Secondly, we would have preferred to also include the life expectancies at other critical ages and also the proportion dying (qx) at early ages for comparisons. However, the fact that Stodder used different age categories for Mesa Verde makes these comparisons meaningless and extrapolation very speculative. This problem is also being addressed.

The comparison of mortality and fertility suggests some interesting interpretations. In general, the mean age-at-death values are much more consistent in ranking with the expected relationships given what is known independently about the growth rates and the archaeological context. Arroyo Hondo is clearly the worst, Mesa Verde Late should also be in poor health, Pueblo Bonito is ranked highest which would be true if its residents comprised a privileged class, and the others fall in between at levels that are predictable given other sources of inference.

In contrast, the fertility estimates do not track the expected pattern as well. Mesa Verde Late has the highest fertility, however, we expect it to be in severe decline and not growing (remembering that a low value indicates high fertility). Black Mesa has the lowest fertility, but we know that it comes from a period of moderate and consistent growth and it should show relatively high fertility. Mesa Verde and Arroyo Hondo might be expected to cluster together but are separated by three other populations. On the other hand, Casas Grandes and Mesa Verde Early would be expected to have comparatively high fertility probably, and that is where they rank (although Casas Grandes is somewhat enigmatic since Medio II is somewhat late).

In sum, while it is generally true mathematically that fertility has a more important effect on mean age-at-death than mortality, there is sufficient evidence here to suggest to us that the unmodified and untransformed mean age-at-death rates are informative of real morbidity and mortality processes occurring in these respective areas. If that is true, then it reflects the fact that the archaeological record of deaths in settings like the Southwest affords us the opportunity to infer something real and substantial about the experience of the populations who provided them.

EVALUATION OF THE HYPOTHESES

As one might anticipate, the measures of biological disruption do not automatically and unanimously generate a "clean" ranking of the cases. These seemingly equivocal results are to be expected given the quality of the data and the complexity of biocultural systems. Yet if we reject the ranking suggested by fertility, as proposed above, we find that the pathology and mortality profiles produce a remarkably similar ordering of the cases. Combining the latter two lines of evidence, the following approximate ranking emerges (from least to greatest degree of biological disruption):

- Black Mesa
- Chaco Canyon (Pueblo Bonito)
- Mesa Verde (Early)
- Casas Grandes
- Chaco Canyon (small sites)
- Mesa Verde (Late)
- Arroyo Hondo

We can now weigh each of the hypotheses presented above, reviewing the prominent points in support and opposition. To begin, we can immediately reject the notion that biological disruption was constant; the groups examined in this study show significant variation in attributes of health and demography. Although there is redundancy in the kinds of diseases suffered by the various populations, the manifestations differ conspicuously in their prevalence, severity, duration, amount of healing, and associations with one another.

Secondly, the hypothesis that biological disruption was directly dependent on environmental marginality is also unsupported. As already noted, our data on environmental variability are insufficient for a strict ranking, but it is not too difficult to designate the best and the poorest environments. Considering length of growing season, moisture availability, and soil quality, Casas Grandes is clearly the most favorable environment among those discussed here; at the other extreme are Black Mesa and Arroyo Hondo. Casas Grandes falls in the middle of the ranking, and Black Mesa and Arroyo Hondo fall at opposite ends. The idea of a direct relationship between biological disruption and environmental marginality is obviously simplistic. Yet environmental limitations clearly constrained the subsistence systems and must have played some role in the relationship, as suggested below in reference to Arroyo Hondo.

The prediction that biological disruption is time dependent also appears to lack empirical foundation, at least in this data set. Health appears to have declined toward the end of the occupation at Mesa Verde, but not at Black Mesa or Arroyo Hondo, the other two cases for which early vs. late data are at least partially available. This is a highly significant finding, because it seems to suggest that the "boom and bust" cycles so commonly seen in the Southwest did not necessarily culminate in biological disaster. This is not to suggest that food provisioning never became an issue, but rather that Southwestern populations were generally able to recognize the signs of impending crisis and reorganize and relocate themselves before their means of food acquisition and storage failed them completely.

The hypothesis that biological disruption was dependent on organizational state receives some support in the data, but also leaves some discrepancies unresolved. The more dispersed cases (Black Mesa and Early Mesa Verde) appear to have experienced less disruption than the more aggregated ones (Late Mesa Verde and Arroyo Hondo), but the politically centralized cases do not seem to belong to a distinct pattern as the hypothesis would predict.

At first glance, the hypothesized relationship between biological disruption and density also appears to be only weakly supported by the data, in that the relationship leaves a number of cases unexplained. If, however, the politically centralized cases are removed (Chaco Canyon and Casas Grandes), the remaining cases are in almost the order that would be predicted by their population densities, as indicated by the sizes of their largest communities (Black Mesa, Mesa Verde Early, Arroyo Hondo, and Mesa Verde Late). The difference is that according to the biological measures, Arroyo Hondo occupies the position of ultimate disruption instead of the penultimate one predicted by community size. It is important to note here that we are not referring to density of a local population relative to earlier periods in the same area, but rather to an absolute scale of density that applies to all areas simultaneously. In general, this finding does not agree with a model that holds local "carrying capacity" to be critical, but rather suggests some sort of absolute threshold of community size beyond which health maintenance becomes increasingly problematic. However, the transposition of Arroyo Hondo and Mesa Verde Late may show that Arroyo Hondo was in double jeopardy, subject to the impacts of both large community size and environmental marginality.

The final hypothesis is that biological disruption was density dependent with class or gender differentiation. As already noted, the cases in which local communities may have been hierarchically organized do not conform to the density relationship just discussed, but seem to be governed by some other principle. At Chaco Canyon, where it is possible to distinguish the potentially elite segment of the population from the rest and measures of both demography and health are available, the elite group is strikingly better off than the remainder of the population. The Pueblo Bonito group appears to have experienced only slightly more biological disruption than the isolated, autonomous groups at Black Mesa. At Casas Grandes, even though there is strong evidence in mortuary treatment for the existence of an elite group, we are unable to link physiological measures to individuals, and thus are forced to deal with parameters that represent the commoner and elite populations together. This mixed measurement of high- and low-status individuals may account for the intermediate position of Casas Grandes in the ranking, perhaps not coincidentally between the elites and commoners of Chaco Canyon.

CONCLUSION

Even though we suspect that our case studies have underenumerated infant deaths, and the diagnoses of pathologies are limited to the signatures on bone and are affected by the variable interpretations of investigators, a pattern has emerged and relative comparisons are provocative. Interpretations have shifted in recent years from using demography and health as univariate, major independent variables, to viewing them in interaction and as only part of a suite of variables that account for the complexity of prehistoric populations in the Southwest. Similarly, analyses of biological variables in the past were often limited and health variables were kept on the independent side of the equation. In this study, hypotheses were framed to see if the archaeological context could be predictive of observed biological processes.

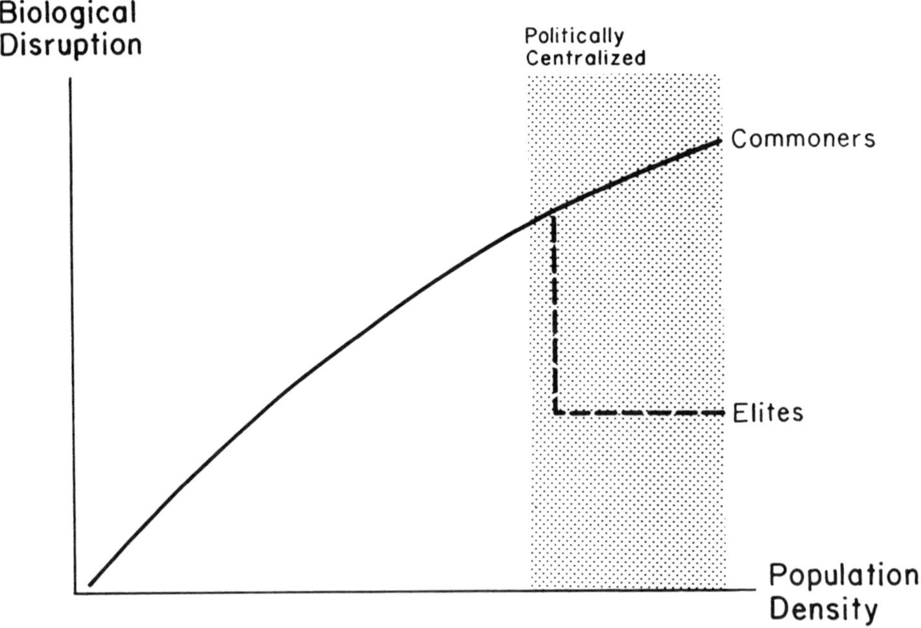

FIGURE 11 Biological disruption determined by density and status differences.

For the cases in the present sample, an interaction of population density, political economy, and environmental marginality seems to account best for the degree of biological disruption (Figure 11). The groups most free from biological disruption were those who were politically autonomous, whether their autonomy was a product of belonging to a small, isolated community, or of belonging to a politically privileged group within a larger and less isolated community.

The data in these case studies do not support a scenario of rapid population growth repeatedly leading to widespread suffering. With the exception of Mesa Verde, we find no evidence of significant worsening of health toward the ends of occupational sequences. We suggest that for the most part, ancient Southwestern people were able to foresee the limitations of their subsistence systems. Ethnographic observations indicate that they commonly laid by two to seven years of surplus food.[67] With a surplus available, periods of prolonged crop failure were not necessarily characterized by unmitigated misery, but perhaps more by planning for relocation and sometimes reorganization into different economies. This view is supported by analyses of abandonment[28] and population change,[19] which show that abandonment of one region was usually accompanied by new aggregation in another.

The above is not to suggest that human suffering was unknown or played no part in Southwestern prehistory. The cold and static archaeological facts cited here were undoubtedly the objects of much emotion for the once-living humans whose remains we study. The loss of an infant to a family, or an epidemic episode to the larger group, presented concrete experiences requiring ideological and adaptive adjustments. No doubt the people's ritual solutions and appeals to the supernatural, in response to the biological disruption they suffered, often became incorporated as enduring elements of Puebloan culture. A particularly poignant case is the kachina cult, which apparently originated in conjunction with the abandonment of the Mesa Verde region and the great increase in population along the Upper Rio Grande.[2,59] It is also possible that the reputed aversion to status differentiation among the Western Pueblo[24] stems from social inequities shouldered by their ancestors, e.g., the residents of Chaco Canyon in the eleventh and twelfth centuries.

ACKNOWLEDGMENTS

We thank George Gumerman, the Santa Fe Institute (especially Murray Gell-Mann), and the School of American Research (especially Douglas Schwartz) for providing us with the opportunity to work together on this paper. Thanks are also due to Richard Ford for permission to cite unpublished observations on Black Mesa ethnobotany. Jim Paxon and Bob Cordts, of the U.S. Forest Service, Black Range District, generously provided work space for the senior author during preparation of the manuscript. We would also like to thank Jeff Dean, Richard Meindl, and Timothy Gage for comments on various interpretations presented, although any errors or ommissions are our own. The hand-drawn illustrations are by Dana Slawson.

REFERENCES

1. Aberle, S. B. D. "Child Mortality Among Pueblo Indians." *Amer. J. Phys. Anthropology* **16** (1932): 339–348.
2. Adams, E. Charles. *The Origin and Development of the Pueblo Katsina Cult.* Tucson: University of Arizona Press, 1991.
3. Akins, N. J. *A Biocultural Approach to Human Burials from Chaco Canyon, New Mexico.* Reports of the Chaco Center No. 9. Sante Fe, NM: National Park Service, 1986.
4. Akins, N. J., and John D. Schelberg. "Evidence for Organizational Complexity as Seen from the Mortuary Practices at Chaco Canyon." In *Recent Research on Chaco Prehistory*, edited by W. James Judge and John D. Schelberg, 89–102. Reports of the Chaco Center No. 8. Albuquerque: National Park Service, 1984.
5. Allison, M. J. "Paleopathology in Peruvian and Chilean Populations." In *Paleopathology at the Origins of Agriculture*, edited by Mark N. Cohen and George J. Armelagos, 515–530. New York: Academic Press, 1984
6. Benfer, Robert A. "An Analysis of a Prehistoric Skeletal Population, Casas Grandes, Chihuahua, Mexico." Ph.D. Dissertation, University of Texas at Austin, 1968.
7. Bennett, K. A. *The Indians of Point of Pines, Arizona: A Comparative Study of Their Physical Characteristics.* Anthropology Papers No. 23. Tucson: University of Arizona Press, 1973.
8. Bennett, K. A. "On the Estimation of Some Demographic Characteristics on a Prehistoric Population from the American Southwest." *Amer. J. Phys. Anthropology* **39** (1973): 723–731.
9. Bocquet-Appel, J., and C. Masset. "Farewell to Paleodemography." *J. Human Evol.* **11** (1982): 231–333.
10. Braniff, Beatriz. "Ojo de Agua, Sonora, and Casas Grandes, Chihuahua: A Suggested Chronology." In *Ripples in the Chichimec Sea: New Considerations of Mesoamerican-Southwestern Interactions*, edited by Frances Joan Mathien and Randall H. McGuire, 70–80. Carbondale: Southern Illinois University Press, 1986.
11. Buikstra, J. E. "Contributions of Physical Anthropologists to the Concept of Hopewell: A Historical Perspective." In *Hopewell Archaeology: The Chillicothe Conference*, edited by D. S. Brose and N. Greber, 220–233. Kent, OH: Kent State University Press, 1979.

12. Buikstra, Jane E., Lyle W. Konigsberg, and Jill Bullington. "Fertility and the Development of Agriculture in the Prehistoric Midwest." *Amer. Antiquity* **51** (1986): 528–546.
13. Colton, H. S. "The Rise and Fall of the Prehistoric Population of Northern Arizona." *Science* **84** (1936): 337–343.
14. Cordell, Linda S. *Prehistory of the Southwest.* Academic Press, New York, 1984.
15. Cordell, Linda S. "Northern and Central Rio Grande." In *Dynamics of Southwestern Prehistory*, edited by Linda S. Cordell and George J. Gumerman, 293–336. Washington, DC: Smithsonian Institution, 1979.
16. Cordell, Linda S., and Fred Plog. "Escaping the Confines of Normative Thought: A Reevaluation of Puebloan Prehistory." *Amer. Antiquity* **44** (1979): 405–429.
17. Cordell, Linda S., Steadman Upham, and S. L. Bock. "Obscuring Patterns in the Archaeological Record: A Discussion from Southwestern Archaeology." *Amer. Antiquity* **52** (1989): 565–577.
18. Dean, Jeffrey. Personal communication, 1985.
19. Dean, Jeffrey S., William H. Doelle, and Janet D. Orcutt. "Adaptive Stress: Environment and Demography." Paper presented at the Advanced Seminar entitled "The Organization and Evolution of Prehistoric Southwestern Societies," School of American Research, Santa Fe, New Mexico, 1990.
20. Decker, Kenneth W., and Larry L. Tiezen. "Isotopic Reconstruction of Mesa Verde Diet from Basketmaker III to Pueblo III." *The Kiva* **55** (1989): 33–46.
21. Dickson, D. Bruce. *The Arroyo Hondo Survey.* Santa Fe: School of American Research Press, 1979.
22. DiPeso, Charles C. Personal communication.
23. DiPeso, Charles C. *Casas Grandes: A Fallen Trading Center of the Gran Chichimeca.* Flagstaff: Northland Press, 1974.
24. Eggan, Fred. *Social Organization of the Western Pueblos.* Chicago: Chicago University Press, 1970.
25. Euler, Robert C. "Demography and Cultural Dynamics on the Colorado Plateaus." In *The Anasazi in a Changing Environment*, edited by George J. Gumerman, 192–229. Cambridge: Cambridge University Press, 1988.
26. Euler, Robert C., George J. Gumerman, Thor N. Karlstrom, Jeffrey S. Dean, and Richard Hevly. "The Colorado Plateau: Cultural Dynamics and Paleoenvironment." *Science* **205** (1979): 1089–1101.
27. Fish, Paul R., Suzanne K. Fish, George Gumerman, and J. Jefferson Reid. "Towards an Explanation of Southwestern 'Abandonments.'" Paper presented at the Advanced Seminar entitled "The Organization and Evolution of Prehistoric Southwestern Society," School of American Research, Santa Fe, 1990.
28. Fish, Suzanne K., Paul R. Fish, and John Madsen. "Sedentism and Settlement Mobility in the Tucson Basin Prior to A.D. 1000." In *Perspectives on*

Southwestern Prehistory, edited by P. Minnis and C. Redman, 76–91. Boulder, CO: Westview Press, 1990.

29. Ford, Richard. Personal communication.
30. Ford, Richard I. "Ecological Consequences of Early Agriculture in the Southwest." In*Papers on the Archaeology of Black Mesa, Arizona*, Vol. 2, edited by Stephen Plog and Shirley Powell, 127–138. Carbondale: Center for Archaeological Investigations, Southern Illinois University, 1984.
31. Gage, T. B., and A. S. Swedlund. Personal communication.
32. Graves, Michael W., and J. Jefferson Reid. "Social Complexity in the American Southwest: A View from East Central Arizona." In *Recent Research in Mogollon Archaeology*, edited by S. Upham, F. Plog, D. G. Batcho, and B. E. Kauffman, 266–275. Museum Occasional Papers No. 10. Las Cruces: New Mexico State University, University, 1984.
33. Gray, P. J., and L. D. Wolf. "Height and Sexual Dimorphism of Stature Among Human Societies." *Amer. J. Phys. Anthropology* **53** (1980): 441–456.
34. Grebinger, Paul S. "Prehistoric Social Organization in Chaco Canyon, New Mexico: An Alternative Reconstruction." *The Kiva* **39** (1973): 3–23.
35. Gumerman, George J. *The Anasazi in a Changing Environment.* Cambridge: Cambridge University Press, 1988.
36. Gumerman, George J., and Jeffrey S. Dean. "Prehistoric Cooperation and Competition in the Western Anasazi Area." In *Dynamics of Southwestern Prehistory*, edited by L. S. Cordell and G. J. Gumerman, 99–148. Washington, DC: Smithsonian Institution Press, 1989.
37. Haviland, William A. "Stature at Tikal, Guatemala: Implications for Ancient Maya Demography and Social Organization." *Amer. Antiquity* **32** (1967): 316–325.
38. Hayes, Alden C. "A Survey of Chaco Canyon Archaeology." In *Archaeological Surveys of Chaco Canyon, New Mexico*, edited by A. C. Hayes, D. M. Brugge, and W. J. Judge, 1–68. Chaco Canyon Studies Publications in Archaeology 18A. Washington, DC: National Park Service, 1981.
39. Hooton, Ernest A. *The Indians of Pecos Pueblo: A Study of Their Skeletal Remains.* Papers of the Southwestern Expedition, Vol. 4 . New Haven: Yale University Press, 1930.
40. Hrdlicka, Ales. "Physiological and Medical Observations Among the Indians of the Southwestern United States and Northern Mexico." *Bureau of American Ethnology Bulletin* **37** (1908): 103–112.
41. Johansson, S. Ryan, and S. Horowitz. "Estimating Mortality in Skeletal Populations: Influence of Growth Rate on the Interpretation of Levels and Trends During the Transition to Agriculture." *Amer. J. Phys. Anthropology* **71** (1986): 233–250.
42. Judge, W. James., W. B. Gillespie, S. H. Lekson, and H. W. Toll. "Tenth Century Developments in Chaco Canyon." In *Papers in Honor of Erik Reed*, edited by A. Schroeder, 65–98. Archaeological Society of New Mexico Papers No. 6. Albuquerque: Albuquerque Archaeological Society, 1981.

43. Judge, W. James. "Chaco Canyon-San Juan Basin." In *Dynamics of Southwestern Prehistory*, edited by L. Cordell and G. Gumerman, 209–262. Washington: Smithsonian Institution Press, 1989.
44. Kelley, N. Edmund. *The Contemporary Ecology of Arroyo Hondo, New Mexico.* Santa Fe, NM: School of American Research, 1980.
45. Kennedy, K. A. R. "Skeletal Markers of Occupational Stress." In *Reconstruction of Life from the Skeleton*, edited by M. Y. Iscan and K. A. R. Kennedy, 129–160. New York: Alan R. Liss, 1989.
46. Kincaid, Chris ed. *Chaco Roads Project: Phase I.* Albuquerque: Bureau of Land Management, Albuquerque District Office, 1983.
47. Kunitz, Stephen J. "Disease and Death Among the Anasazi." *El Palacio* **76** (1970): 17–24.
48. Kunitz, Stephen J., and Robert C. Euler. *Aspects of Southwestern Paleoepidemiology.* Anthropological Reports No. 2. Prescot, AZ: Prescott College, 1972.
49. LeBlanc, Steven A. "The Dating of Casas Grandes." *Amer. Antiquity* **45** (1980): 799–806.
50. Lekson, Stephen H. "Dating Casas Grandes." *The Kiva* **50** (1984): 55–60.
51. Lekson, Stephen H., Thomas C. Windes, John R. Stein, and W. James Judge. "The Chaco Canyon Community." *Sci. Amer.* **256(7)** (1988): 100–109.
52. Lightfoot, Kent G. *Prehistoric Political Dynamics: A Case Study from the American Southwest.* DeKalb: Northern Illinois University Press, 1984.
53. Marshall, Michael P., John R. Stein, Richard W. Loose, and Judith E. Novotny. *Anasazi Communities of the San Juan Basin.* Santa Fe: Public Service Company of New Mexico and Historic Preservation Bureau, Planning Division, Department of Finance and Administration of the State of New Mexico, 1979.
54. Martin, D. "Patterns of Health and Disease: Stress Profiles for the Prehistoric Southwest." Paper presented at the Advanced Seminar entitled The Organization and Evolution of Prehistoric Southwestern Societies, School of American Research, Santa Fe, New Mexico, 1990.
55. Mathews, W., J. L. Wortman, and J. S. Billings "Human Bones of the Hemenway Collection in the United States Medical Museum." *Memoirs of the National Academy of Sciences* **7** (1893): 141–286.
56. Matson, R. G., William Lipe, and William Hasse. "Adaptational Continuities and Occupational Discontinuities: The Cedar Mesa Anasazi." *J. Field Archaeology* **15** (1988): 245–264.
57. Matson, R. G., and Brian Chisholm. "Basketmaker II Subsistence: Carbon Isotope and Other Dietary Indicators from Cedar Mesa, Utah." *Amer. Antiquity*, in press.
58. Matthews, Meredith. "Botanical Studies: Nature and Status of the Data Base." In *Dolores Archaeological Program: Studies in Environmental Archaeology*, compiled by Kenneth Petersen, Vickie Clay, Meredith Matthews, and

Sara Neusius, 41–62. Denver: Bureau of Reclamation, United States Department of Interior, 1985.

59. McGuire, Randall H. "Breaking Down Cultural Complexity: Inequality and Heterogeneity." In *Advances in Archaeological Method and Theory*, edited by Michael B. Schiffer, 91–142. New York: Academic Press, 1983.
60. McGuire, Randy. Personal communication, 1990.
61. Mensforth, R. P., C. O. Lovejoy, J. W. Lallo, and G. J. Armelagos. "The Role of Constitutional Factors, Diet and Infectious Disease on the Etiology of Porotic Hyperostosis and Periosteal Reactions in Prehistoric Infants and Children." *Med. Anthropology* **2** (1978): 1–59.
62. Minnis, Paul E. "Peeking Under the Tortilla Curtain: Regional Interaction and Integration on the Northeastern Periphery of Casas Grandes." *Amer. Archaeology* **4** (1984): 181–193.
63. Minnis, Paul E. *Social Adaptation to Food Stress: A Prehistoric Southwestern Example.* Chicago: University of Chicago Press, 1985.
64. Minnis, Paul E. "Prehistoric Diet in the Northern Southwest: Macroplant Remains from Four Corner Feces." *Amer. Antiquity* **54** (1989): 543–563.
65. Minnis, Paul E. "The Casas Grandes Polity in the International Four Corners." In *The Sociopolitical Structure of Prehistoric Southwestern Societies*, edited by Steadman Upham, Kent G. Lightfoot, and Roberta A. Jewett, 269–306. Boulder, CO: Westview Press, 1989.
66. Moore, J. A., A. C. Swedlund, and G. J. Armelagos. "The Use of Life Tables in Paleodemography." In *Population Studies in Archaeology and Biological Anthropology*, edited by A. C. Swedlund, Society for American Archaeology, Memoir 30 (1975): 57–70. *Amer. Antiquity* **40(2)**.
67. Nelson, Ben A. "Cultural Responses to Population Change: A Comparison of Two Puebloan Occupations of the Mimbres Valley, New Mexico." Ph.D. Dissertation, Department of Anthropology, Southern Illinois University at Carbondale, 1980.
68. Nicholas, Linda M., and Gary M. Feinman. "A Regional Perspective on Hohokam Irrigation in the Lower Salt River." In *The Sociopolitical Structure of Prehistoric Southwestern Societies*, edited by S. Upham, K. G. Lightfoot, and R. A. Jewett, 199–236. Boulder, CO: Westview Press, 1989.
69. Obenauf, Gretchen. "The Chacoan Roadway System." M.A. Thesis, Department of Anthropology, University of New Mexico, 1980.
70. Orcutt, Janet, Eric Blınman, and Timothy Kohler. "Explanations of Population Aggregation in the Mesa Verde Region Prior to AD 900. In *Perspectives on Southwestern Prehistory*, edited by Paul Minnis and Charles Redman, 196–212. Boulder, CO: Westview Press, 1990.
71. Palkovich, Ann M. *The Arroyo Hondo Skeletal and Mortuary Remains.* Santa Fe, New Mexico: School of American Research Press, 1980.
72. Palkovich, Ann M. "Agriculture, Marginal Environments, and Nutritional Stress in the Prehistoric Southwest." In *Paleopathology at the Origins of*

Agriculture, edited by M. N. Cohen and G. J. Armelagos, 425–461. Orlando: Academic Press, 1984.

73. Palkovich, Ann M. "Disease and Mortality Patterns in the Burial Rooms of Pueblo Bonito: Preliminary Considerations." In *Recent Research on Chaco Prehistory*, edited by W. James Judge and John D. Schelberg, 103–114. Reports of the Chaco Center No. 8. Albuquerque: National Park Service, 1984.
74. Palkovich, Ann M. "Endemic Disease Patterns in Paleopathology: Porotic Hyperostosis." *Amer. J. Phys. Anthropology* **74** (1987): 527–537.
75. Pepper, George R. "The Exploration of a Burial Room in Pueblo Bonito, New Mexico." In *Putnam Anniversary Volume*, edited by His Friends and Associates, 196–252. New York: G. E. Stechert, 1909.
76. Plog, Fred. "Political and Economic Alliances on the Colorado Plateau, A.D. 400-1450." *Advances in World Archaeology* **2** (1983): 289–292.
77. Plog, Fred. "Exchange, Tribes, and Alliances: The Northern Southwest." *Amer. Archaeology* **4** (1984): 217–223.
78. Powell, Mary Lucas. "Health, Disease, Social Organization in the Mississippian Community at Moundville." Ph.D. Dissertation, Department of Anthropology, Northwestern University, 1985.
79. Powell, Mary Lucas. *Status and Health in Prehistory.* Washington, DC: Smithsonian Institution, 1988.
80. Powell, Shirley. *Mobility and Adaptation: The Anasazi of Black Mesa, Arizona.* Carbondale: Center for Archaeological Investigations, Southern Illinois University, 1983.
81. Powers, Robert P., William B. Gillespie, and Stephen H. Lekson. *The Outlier Survey: A Regional View of Settlement in the San Juan Basin.* Reports of the Chaco Center No. 3. Albuquerque: National Park Service, Division of Cultural Research, 1983.
82. Ravesloot, John C., Jeffrey S. Dean, and Michael S. Foster. "A New Perspective on the Casas Grandes Tree-Ring Dates." Paper presented at the Fourth Mogollon Conference, University of Arizona, October 16–17, 1986.
83. Ravesloot, John C. *Mortuary Practices and Social Differentiation at Casas Grandes, Chihuahua, Mexico.* Tucson: University of Arizona Press, 1989.
84. Reher, Charles A. "Settlement and Subsistence Along the Lower Chaco River." In *Settlement and Subsistence Along the Lower Chaco River: The CGP Project*, edited by Charles A. Reher, 7–109. Albuquerque: University of New Mexico Press, 1977.
85. Reid, J. Jefferson. "Measuring Social Complexity in the American Southwest." In *Status, Structure and Stratification: Current Archaeological Reconstructions*, edited by M. Thompson, M. T. Garcia, and F. J. Kense, 167–174. Calgary: The Archaeological Association of the University of Calgary, 1985.
86. Reid, J. Jefferson. "A Grasshopper Perspective on the Mogollon of the Arizona Mountains." In *Dynamics of Southwest Prehistory*, edited by L. Cordell

and G. Gumerman, 65–98. Washington, DC: Smithsonian Institution Press, 1989.

87. Reid, J. Jefferson., Michael B. Schiffer, Stephanie M. Whittlesey, Madeleine J. Hinkes, Alan P. Sullivan III, Christian Downum, William A. Longacre, and H. David Tuggle. "Perception and Interpretation in Contemporary Southwestern Archaeology: Comments on Cordell, Upham, and Brock." *Amer. Antiquity* **54** (1989): 802–814.
88. Reinhard, K. J. "Cultural Ecology of Prehistoric Parasitism on the Colorado Plateau as Evidenced by Coprology." *Amer. J. Phys. Anthropology* **77** (1988): 355–366.
89. Reinhard, K. J. "Parasitism at Antelope House, a Puebloan Village in Canyon de Chelly." In *Health and Disease in the Prehistoric Southwest*, edited by C. F. Merbs and R. J. Miller, 220–233. Anthropological Research Papers No. 34. Tempe: Arizona State University, 1985.
90. Rohn, Arthur H. "Prehistoric Water and Soil Conservation on Chapin Mesa." *Amer. Antiquity* **28** (1963): 441–455.
91. Rohn, Arthur H. *Culture Change on Chapin Mesa.* Lawrence: The Regents Press of Kansas, 1977.
92. Rohn, Arthur H. "Northern San Juan Prehistory." In *Dynamics of Southwest Prehistory*, edited by L. S. Cordell and G. Gumerman, 149–178. Washington, DC: Smithsonian Institution Press, 1989.
93. Rose, Martin, Jeffrey Dean, and William Robinson. *The Past Climate of Arroyo Hondo, New Mexico, Reconstructed from Tree Rings.* Santa Fe: School of American Research Press, 1981.
94. Rosenberg, E. M. "Demographic Effects of Sex-Differential Nutrition." In *Nutritional Anthropology*, edited by N. Jerome, R. Kandel and G. Pelto, 183–203. New York: Redgrave, 1980.
95. Sattenspiel, L., and H. Harpending. "Stable Populations and Skeletal Age." *Amer. Antiquity* **48** (1983): 489–449.
96. Schelberg, John D. "Economic and Social Development as an Adaptation to a Marginal Environment in Chaco Canyon, New Mexico." Ph.D. Dissertation, Department of Anthropology, Northwestern University, 1982.
97. Schelberg, John D.. "Analogy, Complexity, and Regionally-Based Perspectives." In *Recent Research on Chaco Prehistory*, edited by W. James Judge and John D. Schelberg, 5–22. Reports of the Chaco Center No. 8. Albuquerque: National Park Service, Division of Cultural Research, 1984.
98. Scott, Linda. "Dietary Inferences from Hoy House Coprolites: A Palynological Interpretation." *The Kiva* **44** (1979): 257–281.
99. Sebastian, Lynne. "Leadership, Power, and Productive Potential: A Political Model of the Chaco System." Ph.D. Dissertation, Department of Anthropology, University of New Mexico, 1988.
100. Sebastian, Lynne. "Sociopolitical Complexity and the Chaco System." In *Chaco and Hohokam: Prehistoric Regional Systems in the American Southwest*, 109–134. Santa Fe: School of American Research Press, 1991.

101. Smiley, Francis E. "The Chronometrics of Early Agrictltural Sites in Northeastern Arizona." Ph.D. Dissertation, Department of Anthropology, University of Michigan, 1985.
102. Stiger, M. A. "Ansazai Diet: The Coprolite Analysis." M.A. Thesis, University of Colorado, Boulder, Colorado, 1977.
103. Stodder, Ann Wiener. "Complexity in Early Anasazi Mortuary Behavior: Evidence from the Dolores Archaeological Program." Paper presented at the 51st Annual Meeting of the Society for American Archaeology, New Orleans, 1986.
104. Stodder, Ann Wiener. "The Physical Anthropology and Mortuary Practices of the Dolores Anazazi: An Early Pueblo Population." In *Local and Regional Context, Dolores Archaeological Program: Settlement and Environment*, compiled by K. L. Peterson and J. D. Orcutt, 339–504. Denver, CO: United States Department of Interior, Bureau of Reclamation Engineering and Research Center, 1987.
105. Swedlund, A. C., and R. Meindl. "Can You Get There from Here?: Recent Controversies over the Method and Theories of Paleodemography." *Ann. Rev. Anthropology*, in preparation.
106. Tanner, J. M. "Growth as a Mirror of the Condtion of Society: Secular Trends and Class Distinctions." In *Human Growth: A Multidisciplinarv Review*, edited by A. Demirjian, 3–34. London: Taylor and Francis, 1986.
107. Toll, H. Wolcott. "Pottery Production Public Architecture and the Chaco Anasazi System." Ph.D. Dissertation, Department of Anthropology, University of Colorado, Boulder, 1985.
108. Upham, Steadman, Kent G. Lightfoot, and Gary Feinman. "Explaining Social Determined Ceramic Distributions in the Prehistoric Plateau Southwest." *Amer. Antiquity* **46** (1981): 822–833.
109. Upham, Steadman. *Polities and Power: An Economic and Political History of the Western Pueblo*. New York: Academic Press, 1982.
110. Upham, Steadman. "Archaeological Visibility and the Underclass of Southwestern Prehistory." *Amer. Antiquity* **53** (1988): 245–261.
111. Upham, Steadman. "East Meets West: Hierarchy and Elites in Pueblo Society." In *The Sociopolitical Structure of Prehistoric Southwestern Societies*, edited by S. Upham, K. G. Lightfoot, and R. A. Jewett, 77–102 Boulder, CO: Westview Press, 1989.
112. Upham, Steadman, and Gail M. Bockley. "The Chronologies of Nuvakwewtaqa: Implications for Social Processes." In *The Sociopolitical Structure of Prehistoric Southwestern Societies*, edited by S. Upham, K. G. Lightfoot, and R. A. Jewett, 447–490. Boulder, CO: Westview Press, 1989.
113. van Gerven, D. P., and G. J. Armelagos. "Farewell to Paleodemography? Rumors of Its Death Have Been Greatly Exaggerated." *J. Human Evol.* **12** (1983): 353–360.

114. Vivian, R. Gwynn. "An Inquiry into Prehistoric Social Organization in Chaco Canyon, New Mexico." In *Reconstructing Prehistoric Pueblo Societies*, edited by W. A. Longacre, 59–83. Albuquerque: University of New Mexico Press, 1970.
115. Vivian, R. Gwynn. "Chacoan Subsistence." In *Chaco and Hohokam: Prehistoric Regional Systems in the American Southwest*, 57–76. Santa Fe: School of American Research Press, 1991.
116. Weaver, D. S. "An Osteological Test of Changes in Subsistence and Settlement at Casas Grandes, Chihuahua, Mexico." *Amer. Antiquity* **46** (1981): 361–364.
117. Weaver, D. S. "Subsistence and Settlement Patterns at Casas Grandes, Chihuahua, Mexico." In *Health and Disease in the Prehistoric Southwest*, edited by C. F. Merbs and R. J. Miller, 119–127. Anthropological Research Papers No. 34. Tempe: Department of Anthropology, Arizona State University, 1985.
118. Weiss, K. M. "Demographic Models for Anthropology." *Memoir of the Society for American Archaeology,* No. 27. American Antiquity, Vol. 38, No. 2, Part II, 1973.
119. Weiss, K. M. "Demographic Theory and Anthropological Inference." *Ann. Rev. Anthropology* **5** (1975): 351–381.
120. Wetterstrom, Wilma. *Food, Diet and Population at Prehistoric Arroyo Hondo Pueblo. New Mexico.* Santa Fe: School of American Research Press, 1986.
121. Whittlesey, Stephanie. "Uses and Abuses of Mogollon Mortuary Data." In *Recent Research in Mogollon Archaeology*, edited by S. Upham, F. Plog, D. G. Batcho, and B. E. Kauffman, 285–293. The University Museum Occasional Papers No. 10. Las Cruces: New Mexico State University, 1984.
122. Whittlesey, Stephanie. "Review of Prehistoric Political Dynamics: A Case Study from the American Southwest, by Kent G. Lightfoot." *The Kiva* **51** (1986): 211–214
123. Winter, Joseph C. "Human Adaptation in a Marginal Environment." In *Human Adaptation in a Marginal Environment: The UII Mitigation Project*, edited by James A. Moore and Joseph C. Winter, 483–520. Albuquerque: Office of Contract Archeology, University of New Mexico, 1980.

Ben A. Nelson,* Timothy A. Kohler,† and Keith W. Kintigh‡
*Anthropology, State University of New York at Buffalo
†Department of Anthropology, Washington State University
‡Department of Anthropology, Arizona State University

Demographic Alternatives: Consequences for Current Models of Southwestern Prehistory

INTRODUCTION

We all see problems in estimating prehistoric Southwestern populations, problems of such a magnitude that they seriously affect the plausibility of many models. One response to such a difficulty is to proceed with model building as if we could, in fact, estimate population accurately. Another response is to "construct around" the problem, i.e., to work with models that do not rest on demographic premises or require us to establish absolute demographic parameters. Yet another response is to continue pressing ahead with the basic data collection that will perhaps ultimately allow more accurate population estimates.

All of those responses have been and will be employed by Southwestern archaeologists. Ultimately the first and second responses, ignoring and constructing around, are unsatisfactory because of the inherent interest and probable systemic importance of demographic variables. The third response, collecting more data, may bear fruit eventually, but only if we have a clear grasp of why the data are being collected and what are the pitfalls, so that data collection proceeds with maximum efficiency.

One response that has not been commonly adopted is to ask how much difference our measurement error makes. In this paper we are concerned with three

Understanding Complexity in the Prehistoric Southwest,
Eds. G. Gumerman and M. Gell-Mann, SFI Studies in the Sciences of
Complexity, Proc. Vol. XVI, Addison-Wesley, 1994

particular aspects of the demographic problem: (1) What is the nature (depth, extent, severity) of the problem? (2) How, exactly, are current models of Southwestern prehistory affected by our inability to estimate population reliably? (3) What solutions are available? To define the nature of the population-estimation problem, we discuss some specific areas (Dolores, Sand Canyon, Mimbres, and Zuni) where survey and excavation data are relatively good and yet difficulties of estimation persist. We then consider the impact of those difficulties upon substantive understanding, not just in these particular cases, but in the Southwest more generally.

The latter part of the paper is an exercise where, given a range of demographic possibilities (which our current methods are incapable of narrowing), we try to specify the corresponding range of possible implications for various models. The models include those currently proposed for the adoption of agriculture and the emergence and elaboration of sociopolitical complexity. A question that lingers beyond or behind all of these considerations, and is addressed only briefly, concerns the extent to which our view of population as a critical variable is affected by the modern human condition.

PROBLEMS

In this section we discuss efforts to estimate population in four areas (Figure 1) in order to illustrate some limitations in our current methods. Absolute population sizes are notoriously difficult to reconstruct from the archaeological record, even in the Southwest where structures are relatively well preserved and visible from the surface and limits of sites are unusually well defined. To estimate population from these structural remains, there must be a way to determine the number of people represented by structures of various type and size; the duration of use of such structures; and the time of their use (both in terms of season and date). As the following cases illustrate, state-of-the-art methodology has large enough margins of error that quite different substantive conclusions can be drawn from the same sets of data.

DOLORES AREA

In the Dolores area as in most of the Southwest, population is estimated by counting rooms. One challenge, typical to the region, is that the kind of room that must be counted to estimate population changes through the main occupation. From the initial occupation at A.D. 600 until around 780, households resided in semisubterranean "pit structures" of various size and shape. Surface structures found north

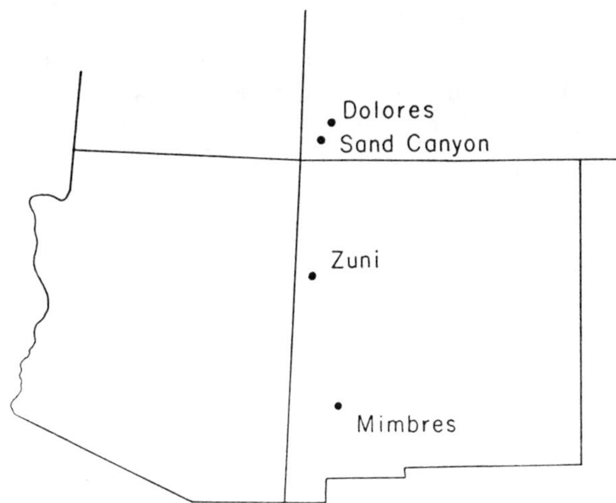

FIGURE 1 Areas mentioned in text.

of the pit structures before about A.D. 780 are small and lack hearths, suggesting they were used primarily for storage. After about A.D. 780, however, most households began to live in more substantial suites of surface rooms built north of the pit structures, and the pit structures appear to have been shared by two or three such households, probably related to each other as extended families. Drawing on ethnographic studies of household size in the Puebloan Southwest, Schlanger[82,83] estimates household size at about five people per living room throughout the occupation of the Dolores area.

Schlanger[82,83] suggests that habitations lacking evidence of remodeling were used for 15 years or less. This estimate is based on cross-cultural regularities in the use-lives of various types of structures, median intervals between remodeling episodes of excavated structures that are datable by tree rings, and superposition of structures whose construction dates can be estimated from tree rings. Population estimates for the Dolores Archaeological Project (DAP) assumed that each unremodeled structure had a 15-year use-life; a lower estimate of use-life would result in lower population estimates. Frequency of structure remodeling in each period is estimated for all structures using data from excavated structures in each period.

For unexcavated sites dating to periods when residence is believed to have been solely or primarily in surface structures, Schlanger estimated numbers of households based on regression of the rubble area of excavated sites against the number of households inferred for those sites. This yielded an approximate estimate of 50 m^2 of rubble/household. For periods when primary residence is believed to have been in pit structures, survey counts of pit-structure depressions were used to estimate household numbers. The key assumptions in this approach should be pointed out: (1) function of structures and suites of rooms belonging to a household can be correctly identified; (2) household size (persons per domicile) can be accurately estimated; (3) households can be accurately assigned a date; (4) survey can accurately estimate rubble area and can recognize pit-structure depressions; (5) duration and continuity of occupation can be accurately assessed.

Duration of structure use and site occupation are probably the most contentious points in such reconstructions, not just for Southwestern archaeology, but for Neolithic archaeology in general (see, for example, Whittle's[1] discussion of settlement permanence and continuity for Neolithic Europe). Our best hope to make progress on these issues seems to be through study of the accumulation of materials. Materials enter the archaeological record more rapidly than structures accumulate at a site, with the result that the quantities of chipped stone, broken ceramics, and food refuse can tell us something about how long a site was in use. We return to this point below.

Schlanger[84] estimates that the maximum momentary population for the 65.5 sq km of the Dolores area was about 800 in the period from A.D. 840–880. (In an earlier study, Schlanger[2] arrived at a higher figure of 1250 for the same period.) A potential natural growth rate (of about 0.02) can be calculated based on Mesa Verde skeletal series, taking into account the length of female reproductive period, child spacing, proportion of female births, and proportion of women who survive to mean age of childbearing[3]; the actual rate of growth certainly exceeded the potential natural growth in the mid-800s, and possibly exceeded it in the mid-700s.

Using another reasonable set of assumptions about how to measure population from the archaeological record, Kane[4] estimates at least 3,000 people in the Dolores area by the peak of occupation. Kane measures room block length (Schlanger uses rubble area), which is converted to number of households, taking into account the lengths of front (habitation) rooms, the ratio of two- to three-room households, and the changes in each through time. Two-room households are assigned 3.5 people; three-room households, 5 people. Kane computes estimates at both 75% and 100% occupancy rates; we report the more conservative 75% occupancy estimates. Where they overlap, Kane's estimates reveal the same trends in population size through time as do Schlanger's, but are absolutely higher. Kane's technique is not, however, applicable to the earliest portions of the sequence, when households were located principally in pit structures.

This factor-of-three difference between Kane's and Schlanger's estimates probably reflects the maximum degree of accuracy that we may expect for prehistoric population estimates under the best of circumstances. In fact, the most recent reworking of these data, by Wilshusen and Blinman[5] results in population estimates

[1]See Whittle,[98] pp. 59, 79.

[2]See Schlanger,[83] p. 505.

[3]See Schlanger,[82] p. 149 and Hassan,[33] p. 140.

[4]See Kane,[41] p. 370.

[5]see Wilshusen and Blinman,[103] pp. 256–258.

intermediate between those of Schlanger and Kane. Taking Schlanger and Kane's figures as minimum and maximum estimates, there were less than 1.2 to 4.3 ha of acceptable, fairly reliable agricultural land per person at the time of maximum population. The true figure is likely to be lower than the suggested upper bound; the computation for land incorporates a 1-km buffer around the project area. That buffer may have been used by people who are not counted in the population estimates. Taken at face value, the higher (but not the lower) population estimate suggests the Dolores area was approaching the minimum per capita hectarage of 0.8 ha/person with a preference for 1.2 to 1.6 ha per person, that can be calculated from Hopi data for per capita field requirements.[6] Moreover, simulation studies[49] demonstrate considerable competition for productive land adjacent to the large villages that sprang up in the mid-800s by the peak of occupation.

SAND CANYON AREA

A briefer example from an adjacent portion of the Four Corners area reveals a similar predicament. In what probably is the most detailed and rigorous examination of the balance between population and agricultural productivity in the Southwest to date, Van West[93] constructs a series of 400 maps of agricultural productivity, one for each year from A.D. 901 to 1300, for a large portion of the Mesa Verde Region. In these maps, developed using geographic information systems technology with a spatial resolution of 4 ha, Van West employs tree-ring data to estimate Palmer Drought Severity Indices (PDSIs) that are particular to various elevations and soil categories in her study area. Maize productivity estimates can be derived from these because of the overlap of the last portions of the tree-ring-derived PDSIs with the earliest modern records of agricultural productivity.

Although arduous to construct, these temporally and spatially fine-grained estimates of productivity can be applied to an almost endless series of interesting problems in areas with appropriate archaeological surveys. The vicinity of the large, late PIII Sand Canyon Pueblo is one such area. Various local surveys have been compiled and translated into demographic estimates by Adler.[1] Adler, in fact, constructs two estimates, one assuming an average site longevity of 20 years, the other assuming an average site longevity of 50 years. Van West's productivity figures suggest the probability of local inadequacies in maize production at the time of abandonment of this site (coincident with the abandonment of the Four Corners) if the 50-year assumption for site longevity is correct.

In both the Dolores and Sand Canyon examples, entire intellectual edifices invoking different causes for aggregation, differing pressures for competition over and control of agricultural lands, and divergent scenarios for abandonment appear to depend on which of two plausible estimates of prehistoric population levels is the more accurate. The Sand Canyon example also serves to illustrate that one of

[6] See Schlanger,[82] pp. 173–178.

the weakest links in our ability to derive these estimates is our ignorance of site longevity.

MIMBRES VALLEY

The Mimbres Valley is another area in which a concerted effort has been made to reconstruct the prehistoric demographic sequence.[7,13,54] The Mimbres area is of interest because it is one of the first places in the Southwest where population aggregation clearly occurred. We are interested in understanding the demographic factors associated with that aggregation. Steven LeBlanc and his colleagues[7] surveyed the Mimbres Valley in southwestern New Mexico with demographic reconstruction as a major goal. A key variable for this research was the estimated floor area occupied during each period. Based on a carefully designed probability sample survey of the valley, they estimate: 12,904 m^2 for the 350-year-long Early Pithouse period; 23,151 m^2 for the 450-year-long Late Pithouse period; and 80,405 m^2 for the 150-year-long Classic Mimbres period.

In translating these floor area estimates into population figures, they take into account the lengths of the chronological periods, structure use-life, and floor area per person. Structure use-life is estimated at 75 years for both pithouses and pueblos, while room area per person is assumed to be 4 m^2 per person for the pithouse periods and 6 m^2 for the pueblo architecture in the Classic Mimbres period. Because of the length of the periods, they fit a mathematical model to estimate contemporaneously occupied floor areas and populations.

Based mainly on the Early Pithouse and Classic period data, these investigators conclude that in the 100 m^2 there is steady annual population growth averaging about 0.3% during the Early and Late Pithouse periods, and that this growth continues or accelerates slightly during the Classic Mimbres. The population is estimated to grow from 290 at the beginning of the Early Pithouse period to 830 at the beginning of the Late Pithouse Period to 3,200 at the beginning of the Classic Mimbres to 5,133 at the end of the Classic Mimbres.

Whereas Blake and his co-authors realize that the observed Late Pithouse period floor area is only about 42% of that predicted by their continuous growth model, they feel that this value may be seriously underestimated for that period. Although Classic Mimbres sites with Late Pithouse period ceramics are assumed to be underlain by Late Pithouse period pithouses, in some cases, the Classic Mimbres period occupations may have completely obscured the Late Pithouse component.

More recent work has suggested that the structure use-lives used by Blake et al. are much too long; use-lives on the order to 15 years are now commonly accepted.[2,13,82,83] We will use this occasion to reevalute the earlier conclusions through an analysis intended to delineate the *range* of reconstructions that are consistent with what we know. This reanalysis utilizes the same cumulative floor areas by period, the same period lengths, and same floor area per person used by Blake, but differs in assuming much shorter use-lives of 15 years of pithouses

(Ahlstrom[7] suggests 15–20 years) and pueblo rooms. It *assumes* a constant annual growth rate within, but not between periods.

For a single period, the cumulative total floor area (observed) is a function of the structure use-life (fixed), the floor area per person (fixed), the period length (fixed), the rate of population growth (unknown), and the initial population (unknown). It turns out that any number of combinations of initial population and rate of growth can result in a given total floor area. Thus, for the Early Pithouse period, an initial population of 134 with no growth would produce the observed floor area, as would an initial population of 2 with 1.7% growth rate, an initial population of 75 with a 0.3% growth rate, or an initial population of 112 and a 0.1% growth rate.

However, with two independent observations on a constant-growth trajectory, it is possible to fix both the initial population and the growth rate. If we take both Early and Late Pithouse floor areas at face value and assume constant growth across the two periods, then an initial population of 115 and a growth rate of 0.087% fits both totals well and yields a terminal population of 231. However, if the Late Pithouse floor area is in fact underestimated by half, as Blake et al. suspect, then the best fit is provided by an initial population of 85 and a growth rate nearly three times larger, 0.250%, and a terminal population of 553.

These alternative reconstructions can be taken as starting points for understanding growth in the Classic Mimbres period. Using the curve fit to the observed floor area data, the Classic Mimbres floor area is consistent with a growth rate of 0.807%, yielding a terminal population of 3,390. Assuming the Late Pithouse floor area is underestimated by a factor of 2 provides a growth rate of 0.974% and a terminal population of 2,367. In either case, growth during the Classic Mimbres must have been explosive; the required rate of growth is so high as to indicate that in-place population growth is probably not responsible for the Classic Mimbres floor area.

However, if one disregards the Late Pithouse data altogether (it is not possible to fit a constant growth model to the three observed floor area values), the Early Pithouse and Classic Mimbres data are fit by a constant growth model with an initial population of 71, a growth rate of 0.333% and a terminal population of 1671. However, in this case, the Late Pithouse period floor area would be predicted to be 64,744 m^2, 2.8 times the observed.

Clearly, there are major problems in reconstructing the demographic picture in this important prehistoric case, and we are not the first to recognize them. Cameron[13] and Lekson[54] apply different assumptions to the Mimbres data and obtain results that are significantly different from those described here as well as those of Blake et al. [7] Growth rates considered seriously vary from less than 0.1% to nearly 2.0%. Assuming only the 15-year structure use-life, absolute population estimates vary by a factor of 2; inclusion of a wider, but reasonable, range of estimates for floor area per person and structure use-life would widen this range dramatically.

[7]See Ahlstrom,[2] pp. 633–639.

However, we should not lose sight of the fact that Blake and his co-authors have created one of the very best data sets available for regional reconstruction of demographic data. The critical limitations are those imposed by the archaeological record and associated time and money constraints. Despite the variance among the differing interpretations that reasonable archaeologists might consider consistent with these data, we know a *great* deal more than we would without the data. What we do know is best assessed through a quantitative consideration of differing models as is done above (and to a significant extent, in the original article). Such a consideration allows us to reject many interpretations that are not plausible given reasonable assumptions and the actual data. It also allows us to direct out attention to the assumptions and data values that are critical for additional refinement of our understanding.

ZUNI AREA

The Zuni area is perhaps more typical of other parts of the Anasazi and Mogollon worlds in the sense that for most of prehistory, we do not have data available that permit any sort of quantified demographic estimates for substantial areas and time periods. We hope, however, that for the Zuni area, we know enough to have a sense of what we do know.

Most of the known settlements (and most systematic archaeological survey) in the Zuni River valley has been on the floors and slopes of the many canyons tributary to the main valley. What is most striking about the results of the intensive systematic surveys that have been done is the tremendous variability in the density of sites of different periods, even in topographically similar nearby locales. There are substantial contrasts in density even for the later periods when most structures are pueblos that have relatively high archaeological visibility. A great deal of true variation in site density for earlier periods is probably compounded by the greatly reduced archaeological visibility of pit structures and over a variety of topographic settings, some of which are characterized by sand dunes.

Whereas a significant body of systematic survey data has accumulated over the last 20 years, there is probably no sensible way to aggregate these data to produce area-wide demographic estimates. This is because the observed variability is great and because the structure of that variability is not sufficiently well understood that it can be controlled for. Thus, there is no real evidential basis for *any* demographic reconstruction (including those based only on room counts) for the Zuni area for any time prior to the mid-1200s.

The problems discussed in this chapter (see also Powell[77] and Christension) doubtless apply to many other areas. Where we have good systematic data (e.g., Mimbres and Dolores), the problems in constructing credible demographic estimates are evident. Nonetheless, archaeologists are typically willing to accept demographic scenarios based on (to put it generously) sparse data. In these cases we must harbor doubts as to whether the reconstructions are more a reflection of what is *expected* to happen than what really happened, and whether the similar intra- and interregional patterns[25] result more from shared expectations than from similar archaeological records.

CONSEQUENCES

Most models of Southwestern prehistory are rooted in ecological functionalism,[62] which views cultures as elaborate devices for maintaining homeostasis between human populations and their physical environments. Cultural institutions, in this paradigm, are systematic solutions to the problems of capturing energy reliably and distributing it more or less equitably within a society. Social organization and religion arise epiphenomenally to assure smooth flows of matter and information, promoting risk minimization and sound resource management. Cultural institutions evolve in response to changes in the population-resource equilibrium, either as a result of environmental perturbation, within- or between-group competition, or changes in human population levels.

To test hypotheses that arise from such a paradigm, it is obviously crucial for archaeologists to have analytical control over population size and other demographic parameters. Some aspects of the ecological-functional paradigm are currently under vigorous attact, and justifiably so.[34,35,63] Yet we are convinced that demographic issues will still be a central part of what is left in the aftermath. Population levels, as argued in the above cases, do make a difference. Our intention in this section is to see what difference they make to a series of models that are at the core of our current understanding of Southwestern prehistory.

ADOPTION OF AGRICULTURE

Perhaps the classic expression of the assumption that cultural changes are demographically driven is found in explanations of the adoption of agriculture. Archaeologists generally see this adoption as a coping response by small groups whose subsistence options became increasingly limited owing to gradual population increase.[15] Populations growth led to overexploitation of fragile wild resources and the filling of alternative territories by new groups. The necessity of getting a living from a more limited territory made people turn to food production by manipulation of local resources in order to enhance their survival probability.

This population-pressure model follows logically from our understanding of global demographic history in the post-Pleistocene period, when human population first spilled into the New World. In very general terms, the inhabitable portions of the globe were gradually filled with human populations, who then had to search for ways to increase food production in order to cope with continued population growth and shrinking foraging territory.[6,14,80]

In sorting out the role of population growth at the subglobal level, however, it is probably important to distinguish adoption from origins or invention. Not all populations adopted agriculture and some did so much later than others. People in Mesoamerica had knowledge of plant cultivation techniques for four millennia before Southwestern populations began to follow their example at ca. 1500–1000 B.C.[27,64,101] Southwestern archaeologists, however, have tended to assume that the processes that led to the adoption of agriculture were the same as in Mesoamerica. This choice stems partly from the perspective of the American Southwest as an extension or periphery of Mesoamerica, which is not unreasonable in view of numerous cultural continuities between the two regions. But in reality we have no documentation of parallel demographic processes, e.g., a bulge of population beginning in the central Mexican highlands at ca. 5000 B.C.and gradually welling outward until it affected the Southwest four thousand years later.

Given the disparity in timing, it would seem advisable to learn more about population trends and their timing with respect to agricultural beginnings in the two regions. It may be that the pressures that led to agricultural dependence were different in Mesoamerica, where it originated, from those in the Southwest, where it was essentially borrowed. We return to this point below.

Flannery[27] questions the role of population pressure in the origins of agriculture in Mesoamerica itself. He argues that at the time that plant cultivation began, Archaic population in the Valley of Oaxaca could not have been high enough to engender chronic shortages of wild resources. Flannery's conclusion is based upon exhaustive comparison of archaeological site distribution with extensive botanical census data, supported by Reynolds' simulation of variability in yields and the effectiveness of various potential human responses to that variation. This conclusion might actually strengthen the argument that agricultural adoption in the American Southwest stemmed from a chain reaction of population pressure that began with the more advanced neighbor to the south. By suggesting that population growth was a consequence of agriculture rather than a cause, Flannery's model allows for a much later explosion of population in Mesoamerica than the old population-pressure model.

Turning back to the Southwest, Hunter-Anderson[37] and Wills[101] suggest that, paradoxically, early cultivation may have been a strategy for improving the efficiency of foraging for wild resources. These fresh views of the process have agriculture co-evolving not only with population growth, but also with hunting and gathering behavior. Both scholars picture Archaic populations as initially seasonally transhumant, living mainly in the deserts in the winter and spring, and using the uplands as resource areas in the summer and fall. Both view low-intensity plant cultivation as a possible mechanism for improved upland exploitation because it would generate food supplies to span periods of low resource availability in the mountains. With stored maize available, humans could either overwinter in the uplands[37] or monitor future resource availability there in the spring in preparation for the productive summer and fall seasons.[101]

Matson[61] also reviews the adoption of agriculture in the Southwest, to discern whether the earliest agricultural populations were immigrants or *in situ* descendants of earlier Archaic peoples. He is unable to resolve this issue for the Southwest as a whole because of the sparsity of data and conflicting indications from different parts of the region, and turns instead to modeling changes in planting strategies and plant characteristics that would have enabled the northward spread of agriculture.

This brief discussion indicates that the ambiguity of early demographic history in the Southwest allows us to construct a number of contradictory scenarios. We need to eliminate some demographic alternatives to reach conclusions about whether agricultural adoption was caused by, a consequence of, or a co-evolutionary process with, population growth.

EMERGENCE AND ELABORATION OF SOCIOPOLITICAL COMPLEXITY

During the past ten years, archaeologists have been debating the existence of sociopolitical inequality in prehistoric Southwestern societies. The major question is whether those societies had decision-making hierarchies controlled by elites who enjoyed privileged access to information, labor, and other resources. If one grants that such decision-making hierarchies existed, additional questions follow about how pronounced the hierarchies were and what processes governed their emergence, elaboration, maintenance, and collapse.

It is not our purpose to evaluate the merits of arguments about the existence or nonexistence of hierarchies. Rather we wish to show that substantially different interpretations of social structure may be supported by some different, but equally plausible, demographic assumptions about the same sets of data.

Most of the arguments for social hierarchy begin from premises about organizational needs of growing populations.[16,39,57,76,90,92] Many archaeologists suspect that as community sizes increased from small bands of close kin to villages of hundreds or over a thousand people, internal tensions emerged that required systematizing control over resources and information. By analogy with many chiefly societies, an elite component would have served to adjudicate disputes, to allocate land, and above all to cope with subsistence risks by assuring food provisions during lean times. Managing the external affairs of the community also required the designation of people with the authority to act on the group's behalf.

The model of community organization that Upham[90,92] constructs for Nuvaqeotaka illustrates the crucial role of demographic assumptions in these arguments. Upham considers the community of Nuvaqeotaka to consist not only of the large site itself, but also a large number of much smaller outlying settlements. He views the small outliers as subordinate to the central place, occupied by people of lower status and wealth. In a provocative extension of the Nuvaqeotaka model to the wider Southwest, Upham[91] suggests that there may have existed an "underclass" of nomadic people who, precisely because of their social status, are nearly invisible archaeologically. This underclass would consist of the occupants of many of the small, ephemeral sites that we often classify as "limited activity settlements."

Part of the criticism that others have directed at the Nuvaqeotaka model centers on the role of the small outliers.[78] The critics suggest that the outliers may not have been contemporary with the large site, or that if they were, they may not have been residences but field houses instead. Upham[91] has begun to address the contemporaneity question at one site, but his results cannot necessarily be extended to the whole population of outliers and, in any case, do not bear upon the residency question. A similar criticism may apply to the generic underclass model. The archaeological expectations that one may derive from this model are seemingly identical to those that could be derived from competing models. For example, the limited-visibility deposits of the postulated underclass could equally well result from the hunting-and-gathering or farmstead components of small egalitarian societies, as has been assumed traditionally.

The unfortunate fact about Nuvaqeotaka and all other apparent clusters of sites in the Southwest is that we have not developed reliable methods for establishing the contemporaneity, occupation lengths, and functions of small sites, even when we have relatively good samples of them. This deficiency allows us to construct models that may well be true, but cannot be falsified. With questions as basic as those about contemporaneity and residency unanswered, we can construe the same set of evidence to mean subordinate communities or farmsteads, underclass or hunting-gathering stations.

Such problems of how Southwestern communities were organized, and how many people they embraced, have implications that relate to the history of integration at the local, regional, and macroregional levels. Although we are demonstrably ignorant about demography, and our methods and data do not always afford us the privilege of empirical tests, we can learn about the consequences of our ignorance by toying with some alternate assumptions. This "science fiction" may show us the potential payoff of solving some of the problems in modeling demographic change.

We shall make these assumptions in two sets of extremes, which essentially imply the highest and lowest conceivable orders of integration. First, let us assume the following:

1. all small sites were residences;
2. all small sites in the vicinity of a large village were contemporary; and
3. all large villages in a region were contemporary with one another.

Let us call each of these nuclear arrangements a *large population aggregate*, and the process of their joint growth, *macronucleation*.

The "alliances" envisioned by Upham[90] and Plog[76] are one specific form of imaginable relationship among such large population aggregates. The highest conceivable order of integration is that all such phenomena throughout the Southwest occurred simultaneously, or in an order that implies mutual causal interdependence. We already know that macronucleation, if it occurred, did not occur simultaneously throughout the Southwest; we see different timing of appearance of large population aggregates in regions such as the Hohokam, Chaco, Casas Grandes, the Western Pueblo, and the Rio Grande. The models of colonization and dependency proposed by Kelley and Kelley,[43] Weigand et al.,[97] and LeBlanc,[55] are attempts to see sequential Southwestern macronucleation as a product of external economic forces.

Let us now take a much different set of assumptions, the other extreme, which implies the lowest conceivable order of integration:

1. most small sites either were not full-time residences or were not contemporary with a nearby large village site; and
2. the large villages in a region were not contemporary with one another.

Under these assumptions we might see each region as inhabited by a much smaller population, which either moved periodically from one large village to another, using the small sites as logistical stations, or oscillated between aggregated and dispersed settlement patterns, shifting its territorial range from time to time. Nelson and LeBlanc[55] use the label "short-term sedentism" to describe one variant of the latter pattern.

Between these two sets of extreme assumptions is a range of others too long and complex to enumerate. The point here is not to consider all possible models of organization. Rather we want to suggest that current organizational models are really research questions, not conclusions to be drawn from understanding that exists about most Southwestern settlement systems. What is at stake in understanding the demography of the Southwest is the ability to recognize and reliably separate models such as regional alliance formation (or some other macro-nucleative process) from less complexly linked ones such as short-term sedentism.

We could go on to discuss the impacts of our inability to estimate population on other questions, such as the processes of aggregation and abandonment. The above issues amply demonstrate, though, that demographic assumptions make a difference. Let us turn instead to the question of what can be done to narrow this methodological impasse.

SOLUTIONS

One solution to the problems enumerated above is to conduct more 100% surveys of cultural universes. The Valley of Oaxaca[9] and the Kincaid settlement system[65] are examples of areas where 100% surveys have resulted in quantum leaps of understanding. It is clear that probability samples of land surfaces do not provide the kinds of data required for reliable population estimates.[26,40] On the other hand, the Dolores case illustrates that 100% surveys do not in and of themselves provide such data. We suggest that complete surveys must be coupled with further efforts to understand assemblage formation processes, broadly defined to include architectural variability (e.g., Diehl[24]). However, there appears to be a pressing need to expand the study of accumulation processes, currently centered on architecture, to include artifacts with shorter use-lives such as ceramics.

TABLE 1 Sites in Dolores Probability Sampling Program.[1]

Site	Name	Date Range (Years A.D.)	Sample Proportion	Sample Size (2 × 2 m units)
Villages:				
*5MT23	Grass Mesa	750-910	.010	56
*5MT2182	Rio Vista	750-910	.030	57
5MT4477	Masa Negra	850-975	.021	13
Hamlets:				
*SMT2161	Prince Hamlet	750-850	.050	10
5MT2194	Casa Bodega	750-850	.030	20
*5MT2215	Sundance Pueblo	1050-1200	.050	12
*5MT2226	Dovetail Hamlet	825-875	.034	12
*5MT2336	Kin Tl'iish	775-1200	.017	13
5MT2854	Aldea Sierritas	600-850	.018	20
5MT4545	Tres Bobos	600-700	.042	20
5MT4614	Prairie Dog	700-760	.080	35
*5MT4650	Hanging Rock	860-880	.070	13
Seasonal or Special Use:				
*5MT2729	Paintbrush House	1025-1125	.056	14
*5MT4751	Pinyon House	1050-1125	.058	12
Limited Activity:				
5MT2242	Ridgeline Camp	Archaic/ Anasazi	.030	30
5MT4681	Hawk House	Archaic/ Anasazi	.075	6
5MT4789	Quasimodo Cave	(Sundial)	.050	4
5MT2199	Horse Bone Camp	Archaic/ Anasazi	.171	20

1 *Indicates stratified sample. No asterisk indicates simple random sample of squares.

TABLE 2 Population Total Estimates (above) and 80% Confidence Intervals (below) for Major Material Classes.

Site	Ceramics	Flaked Tools	Debitage	Nonflaked Tools
5MT23	4,665,745	250,402	3,193,310	41,607
	683,683	38,481	475,747	7,648
5MT2182	876,550	38,557	487,495	9,619
	143,440	5,951	86,918	1,801
5MT4477	161,712	7,224	54,636	1,536
	73,823	3,861	31,429	881
5MT2161	220,768	2,981	72,754	1,969
	170,245	1,991	53,513	1,160
5MT2194	4,275	225	2,558	45
	1,324	105	836	43
5MT2215	24	167	2,107	72
	33	133	1,157	61
5MT2226	23,432	5,020	24,700	384
	6,940	1,800	8,093	245
5MT2336	98,697	16,258	174,168	8,708
	46,507	7,104	80,008	1,082
5MT2854	35,212	3,121	24,866	143
	12,333	1,154	11,495	84
5MT4545	10,680	1,080	13,608	168
	2,417	436	4,579	82
5MT4614	11,195	748	8,316	104
	2,775	214	2,254	46
5MT4650	60,066	2,200	41,739	324
	33,743	1,044	23,628	149
5MT2729	3,636	899	2,145	8
	2,730	441	1,173	3
5MT4751	4,604	9,430	8,410	117
	1,975	333	2,132	87
5MT2242	1,096	839	55,701	350
	371	271	9,795	171
5MT4681	0	59	3,597	40
	0	34	446	45
5MT4789	12	20	384	2
	14	14	257	3
5MT2199	6	197	4,282	0
	7	62	1,297	0

TABLE 3 Correlations Between Populations of Major Artifact Categories. Values shown are Spearman's rank-order correlation coefficients (r_s) between point estimates for major classes of materials for all habitation and seasonal sites ($n = 14$).

	Flaked Tools	Debitage	Nonflaked Tools
Ceramics	.77	.97	.93
Flaked tools		.84	.83
Debitage			.96

MODELING ARTIFACT ACCUMULATION RATES

It has long been recognized that the amount of material accumulated at a habitation site is related to the number of people responsible for that deposition, and the length of time over which the deposition takes place.[4] Despite the fact that both momentary human population and duration of site use remain of great interest to archaeologists, and despite considerable uncertainties surrounding traditional structure-based paleodemographic reconstruction techniques, there have been few attempts to apply arguments based on material depositional rates to paleodemographic problems.[47,50,74,81,94]

The formidable questions that surround arguments about depositional rates of artifacts are essentially the same as those surrounding estimates of population based on accumulations of structures; that is, what is the relationship between the material and a population unit in the systemic context (e.g., how many vessels of what type does a household have?) and how rapidly do the materials enter the archaeological record? Hassan has noted that "the use of artifacts...to determine [human] population size rests, in most cases, on the assumption that the accumulation of...artifacts at a site is proportional to the number of persons and an average, constant rate...of artifact discarding per person."[8]

It is difficult to test these assumptions directly in the archaeological record. However, we can derive certain expectations based on the assumption that there is an average, constant rate of artifact discard per person or per household, recognizing, of course, that these rates may differ considerably among different classes

[8]See Hassan,[33] p.77

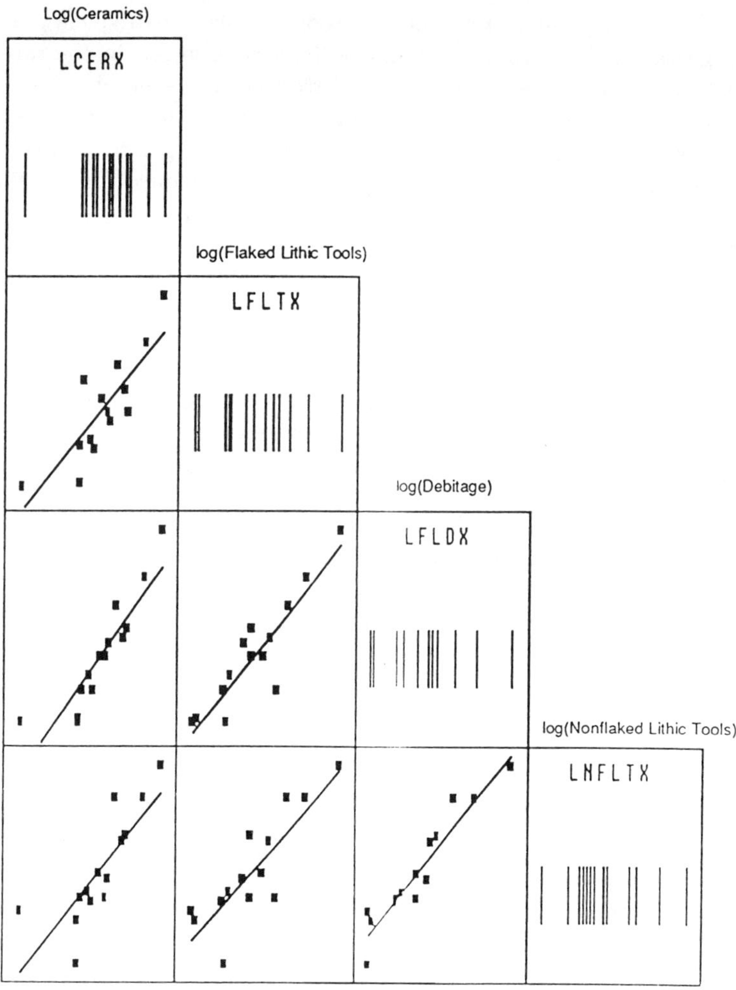

FIGURE 2 Scatterplot matrix for population total estimates. Scatterplots for the relationship between estimates of total quantities of materials in habitation sites and seasonal sites in the DAP area that were sampled probabilistically. Each data point represents one site ($n = 14$ sites). The data are plotted in their logarithms (though the correlations were calculated on raw values) to avoid obscuring the lower end of the distributions. The stripes along the diagonal represent the density of observations across the range of each variable.

of tools, depending on factors such as tool use-life and frequency of use.[86] One such expectation is that there should be high consistency in the rates of deposition of various categories of artifacts relative to other, even functionally unrelated artifacts, over the long term. Specifically, if artifacts from sites are to be useful for paleodemographic inference, there ought to be strong correlations between populations of various artifact categories across sites occupied for varying lengths of time by groups of various sizes, reflecting regularity in the depositional process for various material types.

We can examine this proposition with data from the Dolores Archaeological Project (see Kohler and Gross[48] for discussion of the DAP probability sampling program). Tables 1 and 2 provide the basic data for correlations among various artifact types (Table 3). Limited activity sites are excluded from the analysis. The Spearman rank-order correlation coefficient is used, since it is relatively unaffected by the extreme difference in values for 5MT23 and the smaller sites, and is therefore a more legitimately interpretable measure of the underlying relationship than the Pearsonian coefficient. The smallest r_s is significant at the .005 level of probability. Obviously, the estimates are highly correlated with each other across this sample of 14 habitation and seasonal habitation sites; the strong linearity of the relationships is illustrated by scatterplots (Figure 2).

Such a strong relationship is somewhat surprising for artifact categories which have no apparent functional relationship, such as ceramics and flaked lithic debitage. It is easiest to conclude that there is a third variable linking such categories that is responsible for their relationship. That variable is person-years of occupation, a measure which combines the average momentary population with the duration of occupation.

The expectation of regularities in accumulation across habitation sites in differing materials is upheld; minor exceptions are themselves of interest for identifying extraordinary situations warranting explanation. A question remains, however, as to whether artifact accumulation rates can be calibrated to give direct estimates of human population variables.

Although this is difficult to do from the archaeological record alone, some steps towards a solution are possible. Several of the sites in the DAP probability sample also had enough extensive excavation to allow reasonably accurate estimates of the number of household units, based on architectural data, to be extracted from the excavation reports. These sites are compiled in Table 4 along with estimates of their household-years of occupation, based on the assumption of a 15-year use-life for the architectural manifestations of households. Households themselves are counted within discrete, recognizable episodes of construction called *elements* by DAP staff. Thus, Site 5MT2161 is considered to have 11 household-elements, with one household contributed by Element 1 and 10 households by Element 2.

TABLE 4 Estimated Household-Years of Occupation for Selected DAP Habitation Sites.

Site 5MT#	Household-Elements	Multiplier (years)	Household-Years	Reference
4545	2	x15	30	Brisbin & Varien 1986
2854	2	x15	30	Kuckelman 1986
4614	2	x15	30	Yarnell 1986
2194	1	x15	15	Brown 1986
2226	2	x15	30	Nelson 1985
2161	11	x15	165	Sebastian 1986
2182 Area 1	8	x15	120	Wilshusen 1986
2182 Area 2	16	x15	240	Wilshusen 1986
4477	23	x15	345	Kuckelman 1988

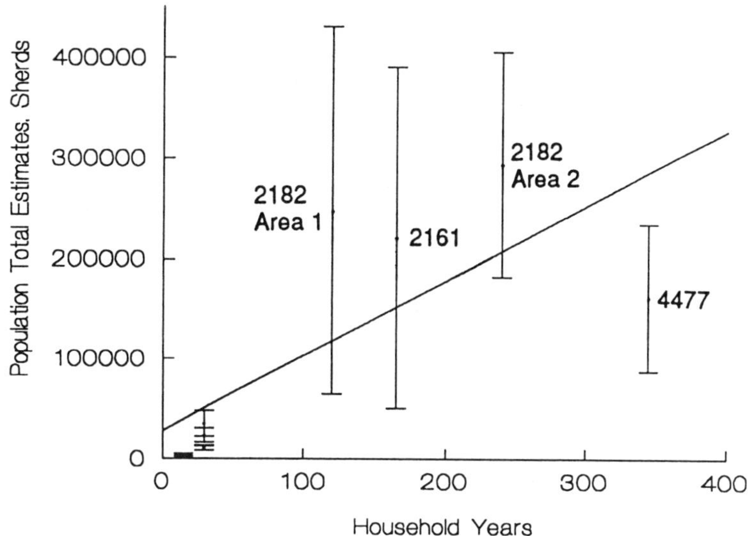

FIGURE 3 Household-years and ceramic populations. Estimated household years of occupation and ceramic population total estimates for the nine residential sites (or areas therein) for which reasonably accurate occupational histories can be inferred from substantial excavation. The error bars mark 80% confidence intervals around the point estimates for the population totals. (See also Table 4 and Eq. (1).)

A straight line is fitted to the relationship between household-years of occupation and the estimates of ceramic populations (from Table 2) in Figure 3. With standard errors in parentheses below the parameter estimates, the least-squares model is:

$$\begin{aligned} \text{Ceramic Population Totals} &= 28{,}395 \;+\; 749\ (\text{Household-years}) \\ &\quad\ (39{,}979) \quad\ (255) \end{aligned} \tag{1}$$

$$\text{with } F_{(1,7)} = 8.66, \quad p > F = .022, \quad \text{and} \quad r_2 = .76.$$

It is perhaps surprising that this relationship is as strong as it is, given the breadth of several of the confidence intervals and the facile assumptions about house use-life used in Table 4.

Interesting structure to the residuals gives us some confidence that we are not just passing a compromise line through very noisy data. Those sites lying below the line are either (in the case of those near the origin) early sites in which the primary occupation was in pit structures, or are, as for 5MT4477, components of villages which contain a late occupation. In the first case, the smaller-than-expected ceramic accumulations at habitations based primarily in pit structures suggests either that such structures have shorter use-lives than later pueblos, or that, as suggested by Gilman[31] and others, pithouses are used seasonally. In the second case, 5MT4477 is part of McPhee Village, a tightly packed series of room blocks in an exceptionally productive portion of the Dolores area. Simulations suggest that competition for nearby agricultural land was intense, and the fieldhouse pattern is well established in that area by this time. Therefore, 5MT4477 may have smaller accumulations of ceramics than would be expected, based on its apparent household-years of occupation, because of intense use of seasonal structures during the growing season, or because it was abandoned before the dwellings had reached their expected use-life. Critical examination of these possibilities using other data is possible but is not appropriate here. Two points should be stressed, however: analyses of this type must control for several types of uncertainties for us to have any confidence in them; and yet, with care, examination of artifact accumulations may eventually provide tools sensitive enough not only to estimate gross parameters of occupation intensity, but even rather subtle differences in seasonality of occupation.

Ignoring the rather large intercept, for which a value of 0 in the population from which this sample is drawn is a possibility, Eq. (1) suggests that some 750 sherds (±255) were deposited, on the average, per household per year, by the Dolores Anasazi.

TABLE 5 Summary of ethnographically observed ceramic inventories and breakage rates.

Group	Location	Vessels Per Household	Breakage Rate (fraction broken per year)	Source
Amahuaca	Amazon Basin		–	De Boer 1974
Fulani	N. Cameroons		.15	David & Hennig 1974
Huichol	W. Mexico	3-25	.16-.67	Weigand 1969
Kalinga	Philippines	8	–	Longacre 1985
Maya (Chuj)	W. Guatemala	57	.82	Nelson 1991
Shipibo-Conibo	Amazon Basin	17	.46	DeBoer & Lathrap 1979
Tarahumara[1]	N. Mexico	13	.30	Pastron 1974
Tarascan	W. Mexico	62	–	Foster 1960
Tunebo	Colombia	24	1.29	Osborn 1979
Wanka	Highland Peru[2]	8.4	.41	Hagstrum 1989

[1] Rough estimates based on the ethnographer's partially quantified descriptions. Only seven or eight pots were in use in any one dwelling; others were stored in alternate houses or in caves.

[2] Some metal containers were in use. Compensating for this, Hagstrum[9] estimates the local pre-Hispanic household inventory at about 15 vessels, with a breakage rate of about .44.

CERAMIC BREAKAGE RATES IN LIVING SOCIETIES

An alternative to reasoning from the archaeological record is to use ethnographically derived parameters. Several ethnoarchaeologists have inventoried household ceramic assemblages and observed or inferred the breakage rates applying to these ceramics (Table 5). Although the estimates for number of pots in use vary considerably, there is more similarity in the total number of vessels broken per year, with most groups averaging two (some of the Huichol ranchos and the Tarahumara) to four pots broken per household per year. This consistency is due to generally higher breakage rates where there are smaller inventories, and vice versa. Foster[29] believes that the 50–75 vessels he observed in modern Tarascan households is higher than would be expected in most prehistoric contexts. Nelson[66] suggests that the large inventories and high breakage rates in the Highland Maya case may be due to perpetual high humidity and other environmental conditions that affect firing conditions and composition of the clay body. Omitting those aberrant cases from consideration, a reasonable estimate would be that the average household in the Southwest broke some three pots per year.

Data bearing on the size of the household ceramic inventory in the DAP area are not strong, but suggest that Dittert's[10] estimate of 18–25 vessels per household for the Cebolleta region may be applicable.[10] Other local evidence points to somewhat smaller numbers. Through ingenious backward calculation of the average number of vessels of various types in use at any one time in a completely excavated Pueblo I hamlet near the Dolores area, Lightfoot[58] estimates that the five or six resident households maintained a total inventory of some 42 vessels. Moreover, there is some local evidence that the number of pots in the household inventory was smaller in the earlier portions of the sequence, from A.D. 600–725, and again after A.D. 910, than in the main portion of the cultural sequence in the valley. Therefore, as is also suggested by Figure 3, a single ceramic breakage rate may not apply to the entire DAP sequence.

However, the middle portions of the sequence, from about A.D. 725–910, may be similar enough to warrant a single calibration. As an example, based on these ethnographic figures, we might assume a household inventory of 20 vessels with an average breakage rate of .15 which, in the long run, would result in an average of three vessels broken per household per year. (The moderately large inventory is used to take into account considerable storage in ceramic vessels, while breakage rate, equal to the lowest of the ethnographic cases, reflects the low usage and handling to which this portion of the inventory is subjected.) With an average vessel weight of about 1000 g (from a series of reconstructible vessels from the Grass Mesa site) and a mean sherd weight of 5.35 g (the average for all ceramics in the probability sample from habitation sites), this results, over the long term, in 3000 g of ceramic breakage and deposition per year, or in terms of sherd count, about 560 sherds/year/household.

This estimate is not too far out of line with the regression-based estimate above; taken together, they suggest that breakage rates of between 500 and 1000 sherds per household are plausible in Dolores during this period. While much better than nothing, such estimates need to be further refined through analysis of the ethnoarchaeological record, to further isolate sources of variability,[60] and analysis of the archaeological record, to learn how to identify those sources of variability.

[10] See Upham,[90] p. 153.

A FURTHER REFINEMENT

So far we have been concerned with the estimating population totals of materials at discrete places—sites—and asking whether it is reasonable to attempt to translate those estimates of materials into some estimate of occupational intensity such as household-years. The answer seems to be a cautious yes. If we can estimate household-years, it further follows that if we can isolate either of its two main components (average momentary population size and duration of occupation), then we can solve for the other. If materials suggest that a site was occupied for 25–50 household-years, and we have independent architectural information that causes us to estimate the average momentary population of households at five, then a reasonable estimate for the continuous duration of occupation is clearly 5–10 years.

While such solutions should prove useful for small, uncomplicated sites, what of large sites with complicated histories of occupation? At least two separate problems present themselves: it will not be very useful to estimate a total figure for household-years of occupation (because it won't help us learn how that population changed in size through time) and we will not normally have reliable architectural estimates for some *average* number of contemporaneously occupied households (because our excavated sample of households will be so small).

In such situations it may be possible to *partition* the total ceramic population into the contribution made by each of several ceramically identifiable periods of occupation using one of the various techniques for the statistical analysis of finite mixture distributions.[88] This family of techniques appears to have been introduced to archaeology by Stahle and Dunn[87] (see also Cowgill[19]); a detailed example relevant to this chapter is analyzed by Kohler and Blinman.[50] The practical requirements include a well-known local chronology, separate estimation of the population totals for each chronologically significant ceramic variant, and a relatively large number of chronologically sensitive ceramic variants (greater than the number of periods to be unmixed). Using linear unmixing, Kohler and Blinman are able to estimate the absolute accumulation of ceramics at the Grass Mesa site in each of four of the five periods into which the main portion of the Dolores occupation had been divided. In this particular case, which is rather unusual for such a large site in that independent architecturally based estimates of numbers of households are also available, these techniques make it possible to argue for a hiatus in the occupation that had previously gone unrecognized, and to suggest an increasingly seasonal use of the Grass Mesa Subphase pithouses of the terminal occupation. In general, even without the information concerning household number available in this case, such analyses gain their power by allowing us to simultaneously exploit both the chronological and the demographic information inherent in ceramic accumulations.

ESTIMATING HOUSEHOLD SIZE FROM VESSEL CAPACITY

One characteristic of the foregoing analyses is that they assume average household size to be constant. Yet this may be unwise, given that completed family size in ethnographically and historically recorded populations ranges widely. Turner and Lofgren[89] raise the possibility that household sizes can be measured from the capacities of household cooking vessels. They go so far as to suggest that absolute househould sizes can be estimated by using the Anasazi ladle as a standard. Their study produces plausible estimates for several phases spanning an 1100-year period.

Nelson[67] evaluates the measure with ethnographic data from the village of San Mateo Ixtatán in the Maya Highlands of Guatemala. Based on observations of 2,907 pots from 51 households, he concludes that cooking-vessel capacity varies with social status and wealth as well as with household size. The wealth-related variation may be explained by reference to the feasting obligations of high-status households, as well as their need to feed planting and harvesting crews on the unusually large landholdings.

It is unclear whether Nelson's study represents an adequate test of the relationship between cooking-vessel capacity and household size. Because the study was conducted within a single community, the sample of vessels and households does not strictly parallel the time-series samples that archaeologists normally analyze. Only 10% of the households have their own potters; perhaps they generally make cooking pots in a size that is appropriate for the perceived "average" family in that community. At least two other settings need to be examined to resolve this problem. The first would be a community in which pottery is being made only for household consumption; potters making vessels for their own households might tailor vessels more precisely fitting their families' requirements. Longacre's[60] study suggests that such is not the case for the Kalinga at least. A yet more definitive test of the measure would be to sample several village in which household size varies significantly and other relevant variables can be controlled.

It is worth noting that Nelson found significant correlation between vessel capacity and household size when controlling for social status and wealth. If we take some Puebloan societies to be more or less egalitarian, it may be this residual variation that is of greatest interest. Do we imagine higher status households in prehistoric pueblos hosting feasts more often or on a grander scale than their lesser contemporaries? Did they hold such large amounts of land that extra vessel capacity would have been needed to feed their field crews? If not, then the Pueblo cases may be more straightforward than that of San Mateo Ixtatán. Further study of ethnographies as well as living socieites may resolve these questions.

CONCLUSION

In this paper we have discussed some problems in estimating prehistoric human population sizes, their implications for various theoretical issues, and some potential solutions. We began with the recognition that ignoring weaknesses in our ability to estimate prehistoric populations is as unsatisfactory as is forcing ourselves to "construct around" those weaknesses by self-consciously limiting ourselves to models that give little or no causal priority to demographic factors. We have explored some of the ramifications of inadequate demographic knowledge for our understanding of Southwestern prehistory in the domains of subsistence change and organization of settleement and society. Finally we have suggested that efforts to improve the weakest link in paleodemographic estimation—our appreciation of site use-life—are best made through building middle-range theory about artifact accumulation rates. By playing off artifact-based, architecture-based, and resource-based estimates against each other, we will begin to arrive at reasonable ranges of absolute population estimates. It should be obvious that if estimates of material populations become an important part of our strategies for estimating human population parameters, we will need to revise field procedures (either through total recovery, or by incorporating some probabilistic sampling) to allow such estimates to be formed.

While we are waiting for such estimates to become available, it should also be obvious that many problems can be fruitfully approached by applying consistent rules to archaeological information in order to reconstruct *relative* population size. Archaeologists do not have to know absolute population size to assess statements such as "local population growth precedes aggregation" or "population-resource ratios reach a high immediately before abandonment of a region." On the other hand, it is essential that we not stop our efforts at this ordinal level, since other problems require absolute measures: Was the adoption of agriculture a cause or a consequence of population change? Are there universal thresholds of population that, once crossed, demand changes in sociopolitical organization?[51] Was the population level at abandonment absolutely larger than could be supported by local production?[64]

If answers to the above questions are at stake, then we can ill afford to ignore the problems and prospects outlined in this paper. We conclude, however, by reflecting critically on our current emphasis on population change as a causal variable. The problem is not just that, as Cowgill[17,18] shows so well, population change may be a consequence rather than a cause of some of the processes that we seek to explain; it is that population change might be causally irrelevant. Wilk[100] reviews explanations of the Maya collapse proposed over several decades and to suggest that the explanations (warfare, environmental degradation, internal strife) are as much a mirror of problems and events in contemporary Western society as they are a product of increasingly satisfying empirical results. This reflexive property of explanation lays open the possibility that demography as cause may eventually become passé in Southwestern archaeology, as it has begun to be in Mesoamerica.[8]

For example, we may eventually replace ecological risk-management models of the emergence of sociopolitical complexity (e.g., Lightfoot[57]) with models that emphasize surplus accumulation and political manipulation as purely social processes (e.g., Bender[5]). We should not allow ourselves to reach such conclusions, however, without first coming to grips with the hard realities of demographic estimation, which we, in the Southwest, have at least as great an opportunity of making precise as any archaeologist working anywhere in the world.

REFERENCES

1. Adler, Michael A. "Population Aggregation and the Anasazi Social Landscape: The View from the Four Corners." Paper presented at the Southwest Symposium, Albuquerque, 1990.
2. Ahlstrom, Richard Van Ness. "The Interpretation of Archaeological Tree-Ring Dates." Ph.D. Dissertation, Department of Anthropology, University of Arizona, 1985.
3. Anyon, Roger, and Steven A. LeBlanc. *The Galaz Ruin: A Prehistoric Mimbres Village in Southwestern New Mexico.* Albuquerque: University of New Mexico Press, 1984.
4. Baumhoff, Martin A., and Robert F. Heizer. "Some Unexploited Possibilities in Ceramic Analysis." *Southwestern J. Anthropology* **15** (1959): 308–316.
5. Bender, Barbara. "Emergent Tribal Formations in the American Midcontinent." *Amer. Antiquity* **50** (1985): 52–62.
6. Binford, Lewis R. "Post-Pleistocene Adaptations." In *New Perspectives in Archaeology*, edited by S. R. Binford and L. R. Binford, 313–342. Chicago: Aldine, 1968.
7. Blake, Michael, Steven A. LeBlanc, and Paul E. Minnis. "Changing Settlement and Population in the Mimbres Valley, Southwest New Mexico." *J. Field Archaeology* **13** (1986): 441–464.
8. Blanton, Richard E., Steve Kowalewski, Gary Feinman, and Jill Appel. *Ancient Mesoamerica: A Comparison of Change in Three Regions.* Cambridge: Cambridge University Press, 1981.
9. Blanton, Richard E., Steve Kowalewski, Gary Feinman, and Eva Fisch. "Regional Evolution in the Valley of Oaxaca, Mexico." *J. Field Archaeology* **6** (1979): 369–390.
10. Blinman, Eric. Personal communication.
11. Brisbin, Joel M., and Mark Varien. "Excavations at Tres Bobos Hamlet (Site SMT4545), a Basketmaker II Habitation." In *Dolores Archaeological Program: Anasazi Communities at Dolores: Early Anasazi Sites in the Sagehen Flats Area*, compiled by A. E. Kane and G. T. Gross, 117–210. Denver: USDI, Bureau of Reclamation, 1986.
12. Brown, Gary A. "Excavations at Casa Bodega Hamlet (Site 5MT2194), a Pueblo I Habitation Site." In *Dolores Archaeological Program: Anasazi Communities at Dolores: Early Anasazi Sites in the Sagehen Flats Area*, compiled by A. E. Kane and G. T. Gross, 507–545. Denver: USDI, Bureau of Reclamation, 1986.
13. Cameron, Catherine M. "The Effect of Varying Estimates of Pit Structure Use-Life on Prehistoric Population Estimates in the American Southwest." *Kiva* **55** (1990): 155–166.

14. Cohen, Mark N. *Food Crisis in Prehistory: Overpopulation and the Origins of Agriculture.* New Haven, CT: Yale University Press, 1977.
15. Cordell, Linda S. *Prehistory of the Southwest.* San Diego: Academic Press, 1984.
16. Cordell, Linda S., and Fred Plog. "Escaping the Confines of Normative Thought: A Reevaluation of Puebloan Prehistory." *Amer. Antiquity* **44** (1979): 405–429.
17. Cowgill, George. "On the Causes and Consequences of Ancient and Modern Population Changes." *Amer. Anthropologist* **77** (1975): 505–525.
18. Cowgill, George. "Population Pressure as Non-Explanation." In *Population Studies in Archaeology and Biological Anthropology*, edited by A. C. Swedlund, 116–126. *Amer. Antiquity* **40** (1975), Pt. 2, Memoir 30.
19. Cowgill, George. "Formal Approaches in Archaeology." In *Archaeological Thought in America*, edited by C. C. Lamberg-Karlovsky, 74–88. Cambridge: Cambridge University Press, 1989.
20. David, Nicholas. "On the Life Span of Pottery, Type Frequencies, and Archaeological Inference." *Amer. Antiquity* **37** (1972): 141–142.
21. David, Nicholas, and Hilke Hennig. *The Ethnography of Pottery: A Fulani Case Seen in Archaeological Perspective.* McCaleb Module in Anthropology 21. Reading, MA: Addison-Wesley, 1972.
22. Deboer, Warren R. "Ceramic Longevity and Archaeological Interpretation: An Example from the Upper Ucayali, Eastern Peru." *Amer. Antiquity* **39** (1974): 335–343.
23. Deboer, Warren R., and Donald W. Lathrap. "The Making and Breaking of Shipibo-Conibo Ceramics." In *Ethnoarchaeology: Implications of Ethnography for Archaeology*, edited by C. Kramer, 102–138. New York: Columbia University Press, 1979.
24. Diehl, Michael. "Architecture as a Material Correlate of Mobility Strategies: Some Implications for Archaeological Interpretation." *Behav. Sci. Resh.* **26** (1992): in press.
25. Euler, Robert C. "Demography and Cultural Dynamics on the Colorado Plateaus." In *Perspectives on Southwestern Prehistory*, edited by G. J. Gumerman, 192–229. Cambridge: Cambridge University Press, 1988.
26. Fish, Suzanne K., and Stephen A. Kowalewski. *The Archaeology of Regions: A Case for Field-Coverage Survey.* Washington: Smithsonian Institution Press, 1990.
27. Flannery, Kent V. "Adaptation, Evolution, and Archaeological Phases." In *Guila Naguitz: Archaic Foraging and Early Agriculture in Oaxaca, Mexico*, 501–508. Orlando: Academic Press, 1986.
28. Flannery, Kent V. "Los origenes de la agricultura en Mexico: las teorfas y las evidencias." In *Historia de la Agricultura. Epoca prehispanica—Siglo XVI*, edited by T. R. Rabiela and W. T. Sanders, 237–266. Mexico City: Coleccion Biblioteca del INAH, Instituto Nacional de Antropologia e Historia, 1989.

29. Foster, George M. "Life-Expectancy of Utilitarian Pottery in Tzintzuntzan, Michoacan, Mexico." *Amer. Antiquity* **25** (1960): 606–609 .
30. Fowler, Andrew P. "Archaeological Clearance Investigation, Acque Chaining and Reseeding Project, Zuni Indian Reservation, McKinley County, New Mexico." ZAP-020-78S. Zuni Archaeology Program, Pueblo of Zuni, 1980.
31. Gilman, Patricia A. "Architecture as Artifact: Pit Structures and Pueblos in the American Southwest." *Amer. Antiquity* **52** (1987): 538–564.
32. Hagstrum, Melissa B. *Technological Continuity and Change: Ceramic Ethnoarchaeology in the Peruvian Andes.* Ann Arbor: University Microfilms International, 1989.
33. Hassan, Fekri A. *Demographic Archaeology.* New York: Academic Press, 1981.
34. Hodder, Ian. *Symbols in Action: Ethnoarchaeological Studies of Material Culture.* Cambridge: Cambridge University Press, 1982.
35. Hodder, Ian. *Archaeology as Long-Term History.* Cambridge: Cambridge Universtiy Press, 1987.
36. Holmes, Barbara E., and Andrew P. Fowler. "The Alternate Dams Survey, An Archaeological Sample Survey and Evaluation of the Burned Timber and Coalmine Dams, Zuni Indian Reservation, McKinley County, New Mexico." Zuni Archaeology Program, Pueblo of Zuni, 1980.
37. Hunter-Anderson, Rosalind D. L. *Prehistoric Adaptation in the American Southwest.* Cambridge: Cambridge University Press, 1986.
38. Hunter-Anderson, Rosalind D. L. "An Archaeological Survey of the Yellowhouse Dam Area." Manuscript on file, Office of Contract Archaeology, University of New Mexico, Albuquerque, 1978.
39. Johnson, Gregory A. "Organizational Structure and Scalar Stress." In *Theory and Explanation in Archaeology: The Southhampton Conference*, edited by C. Renfrew, M. J. Rowlands, and B. A. Segraves, 389–421. New York: Academic Press, 1982.
40. Judge, W. James, and Lynne Sebastian. *Quantifying the Present and Predicting the Past: Theory, Method, and Application of Archaeological Predictive Modeling.* Denver: U.S. Department of the Interior, Bureau of Land Management, 1988.
41. Kane, Allen E. "Prehistory of the Dolores River Valley." In *Dolores Archaeological Program: Final Synthetic Report*, compiled by D. A. Breternitz, C. K. Robinson, and G. T. Gross, 353–435. Denver: USDI, Bureau of Reclamation, 1986.
42. Kane, Allen E. "McPhee Community Cluster Introduction." In *Dolores Archaeological Program: Anasazi Communities at Dolores: McPhee Village*, compiled by A. E. Kane and C. K. Robinson, 1–59. Denver: USDI, Bureau of Reclamation, 1988.
43. Kelley, J. Charles, and Ellen Abbott Kelley. "An Alternative Hypothesis for the Explanation of Anasazi Culture History." In *Collected Papers in Honor*

of Florence Hawley Ellis, edited by T. R. Frisbie, 178–223. Papers of the Archaeological Society of New Mexico No. 2. Norman, OK: Hooper Publishing, 1975.

44. Kintigh, Keith W. "Archaeological Clearance Investigation of the Miller Canyon and Southeastern Boundary Fencing Projects, Zuni Indian Reservation, New Mexico." Zuni Archaeology Program, Pueblo of Zuni, 1980.
45. Kintigh, Keith W. "Final Report: Archaeological Survey Along the Lower Zuni River." Report Submitted to the Arizona State Historic Preservation Office, September 24, 1984.
46. Kintigh, Keith W. "Final Report: Prehistoric Settlement Along the Lower Zuni River." Report Submitted to the Arizona State Historic Preservation Officer, 1987.
47. Kohler, Timothy A. "Ceramic Breakage Rate Simulation: Population Size and the Southeastern Chiefdom." In *Newsletter of Computer Archaeology* **14** (1978): 118.
48. Kohler, Timothy A., and G. Timothy Gross. "Probability Sampling in Excavation: A Program Review." In *Dolores Archaeological Program: Synthetic Report 1978-1981*, prepared under the supervision of D. A. Breternitz, 72–76. Denver: USDI, Bureau of Reclamation, 1984.
49. Kohler, Timothy A., Janet D. Orcutt, Kenneth L. Peterson, and Eric Blinman. "Anasazi Spreadsheets: The Cost of Doing Agricultural Business in Prehistoric Dolores." In *Dolores Archaeological Program: Final Synthetic Report*, compiled by D. A. Breternitz, C. K. Robinson, and G. T. Gross, 525–538. Denver: USDI, Bureau of Reclamation, 1986.
50. Kohler, Timothy A., and Eric Blinman. "Solving Mixture Problems in Archaeology: Analysis of Ceramic Materials for Dating and Demographic Reconstruction." *J. Anthropological Resh.* **6** (1987): 1–28.
51. Kosse, Krisztina. "Group Size and Societal Complexity: Thresholds in Long-Term Memory." *J. Anthropological Archaeology* **9** (1990): 275–303.
52. Kuckelman, Kristin A. "Excavations at Aldea Sierritas (Site 5MT2854), a Basketmaker III/Pueblo I Habitation Site." In *Dolores Archaeological Program: Anasazi Communities at Dolores: Early Anasazi Sites in the Sagehen Flats Area*, compiled by A. E. Kane and G. T. Gross, 285–417. Denver: USDI, Bureau of Reclamation, 1986.
53. Kuckelman, Kristin A. "Excavations at Mesa Negra Pueblo (Site 5MT4477), a Pueblo I/Pueblo II Habitation." In *Dolores Archaeological Program: Anasazi Communities at Dolores: McPhee Village*, compiled by A. E. Kane and C. K. Robinson, 407–558. Denver: USDI, Bureau of Reclamation, 1988.
54. Lekson, Stephen H. "Mimbres Multicomponency." Paper presented at the 55th Annual Meeting of the Society for American Archaeology, Las Vegas, 1990.

55. LeBlanc, Stephen A. "Aspects of Southwestern Prehistory: A.D. 900–1400." In *Ripples in the Chichimec Sea: New Considerations of Southwestern Mesoamerican Interactions*, edited by F. J. Mathien and R. H. McGuire, 105–134. Carbondale: Southern Illinois University Press, 1986.
56. Leone, Mark P., Parker B. Potter, Jr., and Paul A. Shackel. "Towards as Critical Archaeology." *Current Anthropology* **28** (1987): 283–302.
57. Lightfoot, Kent G. *Prehistoric Political Dynamics: A Case from the American Southwest.* DeKalb, Illinois: Northern Illinois University Press, 1984.
58. Lightfoot, Ricky R. "Abandonment Processes in Prehistoric Pueblos." In *The Abandonment of Settlements aml Regions: Ethnoarchaeological and Archaeological Approaches*, edited by C. M. Cameron, and S. Tomka. Cambridge: Cambridge University Press, 1990.
59. Longacre, William A. "Pottery Use-Life Among the Kalinga, Northern Luzon." In *Decoding Prehistoric Ceramics*, edited by B. Nelson, 334–246. Carbondale: Southern Illinois University Press, 1985.
60. Longacre, William A. "Sources of Variation in Kalinga Ceramics." In *Ceramic Ethnoarchaeology*, edited by W. A. Longacre. Tucson: University of Arizona Press, 1991.
61. Matson, R. G. *The Origins of Southwestern Agriculture.* Tucson: University of Arizona Press, 1991.
62. McGuire, Randall H., E. Charles Adams, Ben A. Nelson, and Katherine A. Spielmann. "Drawing the Southwest to Scale: Perspectives in Microregional Relations." Paper Presented at the School of American Research Advance Seminar entitled "The Origin and Evolution of Prehistoric Southwestern Society," Santa Fe, New Mexico.
63. Miller, Daniel and Christopher Tilley. *Ideology, Power, and Prehistory.* Cambridge: Cambridge University Press, 1984.
64. Minnis, Paul E. "Domesticating People and Plants in the Greater Southwest." In *Prehistoric Food Production in North America*, edited by R. I. Ford, 309–340. University of Michigan Anthropological Papers, No. 75. Ann Arbor: Museum of Anthropology, 1985.
65. Muller, Jon. *Archaeology of the Lower Ohio River Valley.* Orlando: Academic Press, 1986.
66. Nelson, Ben A. "Ceramic Frequency and Use Life: A Highland Mayan Case in Cross-Cultural Perspective." In *Ceramic Ethnoarchaeology*, edited by W. A. Longacre, 162–181. Tucson: University of Arizona Press, 1991.
67. Nelson, Ben A. "Ethnoarchaeology and Paleodemography: A Test of Turner and Lofgren's Hypothesis." *J. Anthropological Research* **37** (1981): 107–129.
68. Nelson, Ben A., and Steven A. LeBlanc. *Short-Term Sedentism in the American Southwest: The Mimbres Valley Salado.* Albuquerque: University of New Mexico Press, 1986.
69. Nelson, Charles G. "Dovetail Hamlet (Site SMT2226), a Pueblo I Habitation. In *Dolores Archaeological Program Technical Report DAP-165.* Report

submitted to the Bureau of Reclamation, Upper Colorado Region, Salt Lake City, in compliance with Contract No. 8-07-40-S0562, 1985.

70. Netting, Robert McC. "Population, Permanent Agriculture, and Polities: Unpacking the Evolutionary Portmaneau." In *The Evolution of Political Systems*, edited by Steadman Upham, 21–61. Cambridge: Cambridge University Press.

71. Orcutt, Janet D., Eric Blinman, and Timothy A. Kohler. "Explanations of Population Aggregation in the Mesa Verde Region Prior to A.D. 900." In *Perspectives on Southwestern. Prehistory*, edited by P. E. Minnis and C. L. Redman, 196–212. Boulder, CO: Westview Press, 1990.

72. Osborn, Ann. *La Ceramica de los Tunebos: Un Estudio Ethnografico*. Bogota: Fundacion de Investigaciones Arqueologicos Nacionales, Banco de la Republica, 1979.

73. Pastron, Allen G. "Preliminary Ethnoarchaeological Investigations Among the Tarahumara." In *Ethnoarchaeology*, edited by C. B. Donnan and C. W Clewlow, Jr., 93–114. Archaeological Survey, Institute of Archaeology. Monograph 4. Los Angeles: University of California, 1974.

74. Pauketat, Timothy R. "Monitoring Mississippian Homestead Occupation Span and Economy Using Ceramic Refuse." *Amer. Antiquity* **54** (1989): 288–310.

75. Pertersen, W. *Population*. 2nd edition. New York: MacMillan.

76. Plog, Fred. "Exchange, Tribes, and Alliances: The Northern Southwest. " *Amer. Archaeology* **4** (1984): 217-223.

77. Powell, Shirley. "Anasazi Demographic Patterns and Organizational Responses: Assumptions and Interpretive Difficulties." In *The Anasazi in a Changing Environment*, edited by G. J. Gumerman, 192–229. Cambridge: Cambridge University Press, 1988.

78. Reid, J. Jefferson, Michael B. Schiffer, Stephanie M. Whittlesey, Madeleine J. Hinkes, Alan P. Sullivan III, Christian Downum, William A. Longacre, and H. David Tuggle. "Perception and Interpretation in Contemporary Southwestern Archaeology: Comments on Cordell, Upham, and Brock." *Amer. Antiquity* **54** (1989): 802–814.

79. Reynolds, Robert G. "An Adaptive Computer Model for the Evolution of Plant Collecting and Early Agriculture in the Eastern Valley of Oaxaca." In *Guila Naguitz: Archaic Foraging and Early Agriculture in Oaxaca, Mexico*, 439–500. Orlando: Academic Press, 1986.

80. Rindos, David. *The Origins of Agriculture: An Evolutionary Perspective*. Orlando: Academic Press, 1984.

81. Schiffer, Michael B. "The Effects of Occupation Span on Site Content." In *The Cache River Archaeological Project: An Experiment in Contract Archaeology*, assembled by M. B. Schiffer and J. H. House. *Arkansas Archaeological Survey Research Series* **8** (1975): 265–269.

82. Schlanger, Sarah. *Prehistoric Population Dynamics in the Dolores Area, Southwestern Colorado.* Ann Arbor: University Microfilms International, 1985.
83. Schlanger, Sarah. "Population Studies." In *Dolores Archaeological Program: Final Synthetic Report*, compiled by D. A. Breternitz, C. K. Robinson, and G. T. Gross, 493–524. Denver: USDI, Bureau of Reclamation, 1986.
84. Schlanger, Sarah. "Patterns of Population Movement and Long-Term Population Growth in Southwestern Colorado." *Amer. Antiquity* **53** (1988): 773–793.
85. Sebastian, Lynne. "Excavations at Prince Hamlet (Site SMT2161), a Pueblo I Habitation Site." In *Dolores Archaeological Program: Anasazi Communities at Dolores: Early Anasazi Sites in the Sagehen Flats Area*, compiled by A. E. Kane and G. T. Gross, 333–498. Denver: USDI, Bureau of Reclamation, 1986.
86. Shott, Michael J. "On Tool-Class Use Lives and the Formation of Archaeological Assemblages." *Amer. Antiquity* **54** (1989): 9–30.
87. Stahle and Dunn. "Ana Analysis and Application of the Size Distribution of Waste Flakes from the Manufacture of Bifacial Stone Tools." *World Archaeology* **14** (1982): 84–97.
88. Titterington, D. F., A. F. M. Smith, and U. E. Markov. *Statistical Analysis of Finite Mixture.* Somerset: John Wiley & Sons, 1986.
89. Turner II, C. G. and L. Lofgren. "Household Size of Prehistoric Western Pueblo Indians." *Southwestern J. Anthropology* **22** (1966): 117–32.
90. Upham, Steadman. *Polities and Power: An Economic and Political History of the Western Pueblo.* New York: Academic Press, 1982.
91. Upham, Steadman. "Archaeological Visibility and the Underclass of Southwestern Prehistory." *Amer. Antiquity* **53** (1988):245–261.
92. Upham, Steadman, and Gail M. Bockley. "The Chronologies of Nuvakwewtaga: Implications for Social Processes." In *The Sociopolitical Structure of Prehistoric Southwestern Societies*, edited by S. Upham, K. G. Lightfoot, and R. Jewett, 447–490. Boulder, CO: Westview Press, 1989.
93. Van West, Carla. "Modeling Prehistoric Climatic Variability and Agricultural Production in Southwestern Colorado: A GIS Approach." Ph.D. Dissertation, Department of Anthropology, Washington State University, Pullman, 1990.
94. Varien, Mark D. "Measuring Site Use-Life and Seasonality of Occupation: Accumulation Rate Studies." Paper Presented at the 55th Annual Meeting of the Society for American Archaeology, Las Vegas.
95. Watson, Patty Jo, Steven A. LeBlanc, and Charles L. Redman. "Aspects of Zuni Prehistory: Preliminary Report on Excavation and Survey in the El Morro Valley of New Mexico." *J. Field Archaeology* **7** (1980): 201–218.
96. Weigand, Phil C. *Modern Huichol Ceramics.* Mesoamerican Studies, No. 3. Carbondale: Southern Illinois University Press, 1969.

97. Weigand, Phil C., Garmen Harbottle, and Edward V. Sayre. "Turquoise Sources and Source Analysis: Mesoamerican and the Southwestern U.S.A." In *Exchange Systems in Prehistory*, edited by T. K. Earle and J. E. Ericson, 15–34. New York: Academic Press.
98. Whittle, Alasdair. *Problems in Neolithic Archaeology.* Cambridge: Cambridge University Press, 1988.
99. Whittlesey, Stephanie. "Uses and Abuses of Mogollon Mortuary Data." In *Recent Research in Mogollon Archaeology*, edited by S. Upham, F. Plog, D. Batcho, and B. Kauffman, 285–293. The University Museum Occasional Papers No. 10. Las Cruces: New Mexico State University.
100. Wilk, Richard R. "The Ancient Maya and the Political Present." *J. Anthropological Resh.* **41** (1985): 307–326.
101. Wills, W. H. "Early Agriculture and Sedentism in the American Southwest: Evidence and Interpretations." *J. World Prehistory* **2** (1988): 445–488.
102. Wilshusen, Richard H. "Excavations at Rio Vista Village." In *Dolores Archaeological Program: Anasazi Communities at Dolores: Middle Canyon Area*, compiled by A. E. Kane and C. K. Robinson, 211–660. Denver: USDI, Bureau of Reclamation, 1986.
103. Wilshusen, Richard H., and Eric Blinman. "Pueblo I Village Formation: A Reevaluation of Sites Recorded by Earl Morrin on Ute Mountain Ute Tribal Lands." *Kiva* **57** (1992): 251–269.
104. Yarnell, Richard W. "Excavations at Prairie Dog Hamlet (Site 5MT4614), a Basketmaker III/Pueblo I Site." In *Dolores Archaeological Program: Anasazf Communities at Dolores: Early Anasazi Sites in the Sagehen Flats Area*, compiled by A. E. Kane and G. T. Gross, 419–506. Denver: USDI, Bureau of Reclamation, 1986.

Offering Explanations

Linda S. Cordell
University Museum, University of Colorado, Boulder, CO 80309

The Nature of Explanation in Archaeology: A Position Statement

In September 1989, a group of southwestern archaeologists met at the School of American Research in Santa Fe where they discussed processes and events that seem to have occurred over the entire Southwest in prehistoric times. Those events took place in locations viewed as differing environmentally (plateaus, mountains and low deserts) and as containing somewhat separate cultural traditions (labelled Anasazi, Mogollon, and Hohokam). The events and processes that were the focus of discussion included the adoption and spread of horticulture and sedentarism, the appearance of large domestic sites presumably reflecting the aggregation of population, the formation of regional alliances, the appearance of social hierarchy, regional abandonments, conflict, etc. In outlining these events and general processes, aside from reconfirming the notion that the Southwest is indeed a single culture area with a long tradition of cultural interaction, there were two underlying and recurrent themes that were not examined in detail. The first was that although having agreed that some similar events and processes had occurred over the whole area, sometimes synchronously, these events and processes had not been explained. The 1990 Santa Fe Institute Workshop on "The Organization and Evolution of Prehistoric Southwestern Society" was convened in order to begin to provide those explanations and explore their implications for general theory about how societies evolve. The second theme concerned disagreement among those archaeologists present about the nature of explanation in archaeology.[1] There were extended discussions about historical versus nomological explanations and one attempt to develop an argument based on a version of coherence theory.[35] All of this seemed confusing to non-archaeologists present and to some of the archaeologists as well. This paper attempts to provide a context within which those disagreements may be understood. It is addressed particularly to non-archaeologists and makes no assumptions about prior knowledge of archaeological method and theory. This paper is a position statement for further discussion and for the longer and more comprehensive treatment of theoretical issues in the chapter by Cordell, Kelley, Kintigh, Lekson, and Sinclair that follows .

Prehistoric archaeology[2] in Western Europe, England, Canada, and the United States is currently undergoing a period of difficulty and indecision. This is reflected in "debate concerning the ultimate goal of archaeological research"[3] and in diverse positions on method, theory, and explanation.[13,20,33] The heterogeneity and divergence in theoretical direction appear as a lack of consensus about central questions worthy of focused research efforts. At the level of concepts, there are few agreed upon frameworks. The positions archaeologists take and the intellectual affiliations they support are as diverse as the labels they carry: ahistorical, behavioral, cognitive, critical, ecological, empirical, evolutionary, feminist, functional, historical, Marxist, processual, post-processual, and structuralist. At the very least, this diversity is confusing to the sympathetic and interested outsider. At worst, the polemical nature of some archaeological discussions is tedious and incomprehensible to the uninitiated.

There are times within any science when "understanding of nature's behavior builds in a crescendo," as Neher[24] (p. 6) characterized the search for cosmic rays in the 1930s. At such times, there is great enthusiasm among researchers and the pace at which important discoveries are made increases. At other times, work is slower and activities lack the excitement of times of crescendo. Archaeology also shows these periodic shifts in tempo. In southwestern archaeology, the 1930s, which followed the development of dendrochronology as a precise method for determining

[1]In 1968, A.C. Spaulding's paper "Explanation in Archaeology" appeared in *New Perspectives in Archaeology*, edited by Binford and Binford. My statement is not meant to be substitute for or an updated version of Spaulding's remarks. Spaulding's paper is a landmark document to which I owe a great deal. It deserves rereading by most archaeologists.

[2]Archaeology is not a unified discipline with a commonly accepted body of theory or commonly adopted frameworks of analysis and interpretation. Archaeology may operate within the context of classics, biblical studies, or art history. In the Americas, the material remains with which archaeologists work are predominantly those of societies without writing and separate from either classical or biblical traditions. In this sense they are "prehistoric." In the United States and Canada, archaeology is most commonly affiliated with anthropology, at least in part because the material of ethnographic anthropology also deals with societies without writing and, until recently, outside the framework of western civilization. In Europe, prehistoric archaeology is commonly affiliated with geology or with institutes of prehistory.

[3]See Trigger,[33] p. 370.

the age of pre-Spanish ruins, was a time of excitement and discovery. A similar time of crescendo occurred in the early 1970s as research focused on exploring prehistoric social organization.[4] The current diversity of perspectives being advocated in archaeology at least suggests "diminuendo" (*sensu* Neher,[24] p. 6).

In addition to times of more or less focus and excitement, archaeology periodically questions its affiliation with science rather than with history, general humanities, and art (see Binford[4,5,6]; Spaulding[28]; Steward and Setzler[30]; Taylor[31]; and Willey and Phillips[36]). This vacillation is not characteristic of most sciences, and I suspect that it is not recognized or understood in archaeological writing when encountered by those outside the field. Much current controversy, particularly between those who would espouse ahistorical vs. historical explanation, is a manifestation of the science vs. humanity problem. For that reason alone, it is worthwhile to explore the dilemma as it occurs in archaeology. In so doing, I hope to elucidate the differences between the kinds of inferences with which archaeologists are comfortable versus those with which they are not. I also compare prehistoric archaeology with paleontology because those fields share a concern for establishing context and developing contingent explanations. I then provide a very brief analytic statement about archaeological explanations for a major development in southwestern prehistory, the eleventh-century Chacoan manifestation in the San Juan Basin, as an example of explanation in archaeology. Finally, I discuss some future needs and suggest one kind of assistance from the physical sciences which would be very useful to southwestern archaeology.

Archaeologists study past human behavior but cannot observe that behavior directly. Archaeologists infer past behavior by observing the remains of that behavior in the present as archaeological sites, landscapes, artifacts, and features. The present configuration of all of these must be some combination of past human behavior, natural geomorphological and other non-cultural processes, and the behaviors of archaeological recovery (generally excavation but also survey and mapping of surface finds and recording of all kinds). Michael Schiffer[27] provides the most extensive discussion of processes that "transform" (Schiffer's term), or act upon, the archaeological record. Sorting out the various processes that are responsible for a particular archaeological situation depends upon understanding physical laws and their interaction with models of human behavior. Archaeologists will be most comfortable with interpretations of archaeological occurrences that derive from non-cultural transformation process transformations and physical laws because knowledge of physical laws is more firmly established than is knowledge of human behavior.

For example, the date of the last use of a hearth in a room in a site in Chaco Canyon if derived from archaeomagnetism will be considered more reliable than one derived from the pottery in the fill of the hearth. Similarly, the date of construction of a structure in Chaco Canyon that has been derived from tree rings will be

[4]The seminal papers on prehistoric southwestern social organization were those of James N. Hill[16] and William A. Longacre.[21] A volume from a School of American Research advanced seminar on prehistoric southwestern social organization[22] is a classic.

accepted more readily than one based on pottery types found in construction debris or in the walls. The archaeomagnetic dates and the tree-ring dates depend primarily on non-anthropogenic processes, despite the fact that the use of the hearth for heating or the tree for building material is, of course, referable to cultural behavior. The pottery types found in the debris at Chaco Canyon are defined on stylistic and technological grounds. Changes in pottery style and technology are not known to occur at a regular rate. Pottery types are assigned dates through cross dating. In cross dating the type is dated by its occurrence on and in archaeological sites dated by some other method (generally tree rings but also radiocarbon dating). When the type is found elsewhere on a previously undated site, the new site is given the same date as the pottery type. Cross dating is a common practice fraught with problems such as failure to know and account for the distance of the site to be dated from the source of manufacture of the pottery, the "social distance" of the inhabitants of the site to be dated to the source of the pottery, different rates of stylistic change that depend on undefined communication networks, etc. Obviously, a key factor in the reluctance of archaeologists to agree upon inferences based on cultural behavior rather than those derived from physical and natural sciences, is their acute awareness of the variability in human behavior over time and over space and a concomitant inability to observe much of that behavior in the modern world.

The questions of most interest to archaeology are those about specific acts of human behavior at one extreme and those relating to broad processes of cultural evolution at the other. We may be vitally interested in particular human activities that occurred in A.D. 1150 in Chaco Canyon and also in why some kinds of societies appear to be more stable over time than others. The archaeological record itself may lack the appropriate precision to resolve the first kind of question because the archaeological remains and their context are a product of longer term non-cultural and cultural processes in which A.D. 1150 may not be separable to the desired degree of detail. Questions related to long-term processes of cultural evolution and change are difficult to address because observations of today's world may not be appropriate for the past we need to understand.

Normal processes of scientific evaluation are ahistorical and depend upon constancy in the "laws of nature." The laws of nature are generally covering laws, admitting no exceptions. Obviously those aspects of any particular archaeological situation that can be explained by reference to covering laws will be. The problems arise because archaeology also involves cultural behavior. Whether or not there are laws of cultural behavior and what form those laws may take are questions that have long been key issues in cultural anthropology. In my own conservative archaeological perspective, some observations seem clear. Those archaeologists who work with relatively recent periods (the past 25,000 years) can hold the human body and mental capacity constant. This luxury is not available to those prehistorians dealing with the remains of behavior of earlier forms of humans. There is no *a priori* reason to conclude that *Homo habilis* killed other animals for food, shared food of any kind, maintained a "home base," or carried out any other "universal," human activities.

When archaeologists can hold the human body and mental capacity constant, as can all Americanists, there is still the difficulty of knowing the range of cultural behavior in a world in which the boundary conditions are no longer available for scrutiny. For example, there are numerous (though tremendously diminished) populations of hunter-gatherers living in a variety of environments in today's world. Some of these people were contacted and studied before their traditional life ways were much altered by the encroachments of colonial or industrial nations. These are the classic anthropological examples of hunter-gatherers, from the Inuit to the Yahgan, that are case studies in all introductory anthropology courses. Yet, no one knows what the cultural behavior of hunter-gatherers was like when there were only hunter-gatherers in the world, when hunter-gatherers inhabited the most productive environments on earth and competed for resources only with each other .

The ethnographic record will forever be an incomplete, and distorted, view of all possible past societies. Although archaeologists commonly call upon the ethnographic cases to interpret archaeological remains, they do so at risk. The risk is greatest, and the interpretations weakest, when the archaeological inference relates to some aspect of culture known to vary as a result of the interaction of the society with modern western culture or as a result of interference resulting from such interaction or contact.[5] For example, Martin Wobst[38] carried out some early, and still exemplary and useful, simulation experiments in archaeology that attempt to develop knowledge about hunter-gatherer group size. The simulations were based on assumptions made about a necessarily tiny fraction of the world's hunter-gatherers whose populations and group sizes had been dramatically affected by the encroachment of colonial societies. The inferences regarding hunter-gatherer group size may or may not be valid at all for any preconquest North American population. In general, inferences about prehistoric hunter-gatherer group size, kinship structure, and territoriality may not be as persuasive as those concerning behavior that is less constrained by the cultural environment, such as methods of stone tool manufacture given human musculature and certain kinds of rock or, possibly, basic caloric and nutritional requirements given as a range of values for people of comparable body size and activity levels inhabiting similar natural environments.

Although, when dealing with the archaeological record of the past 25,000 years, human mental capacity can be held constant, this does not mean that introspection is an appropriate method for inferring prehistoric patterns of thought or behavior. This admonition may seem crude and unnecessary. On the other hand, in 1966, Luis Alvarez obtained what must certainly have been a very large sum of money, from the Atomic Energy Commission, and the cooperation of an international team of scientists and several U.S. Federal agencies, in order to locate previously undiscovered chambers in the Egyptian Pyramid of Chephren. Alvarez described the rationale behind his experiments as follows:

[5] No society is or has been a closed system. Thus, one should also consider interactions of prehistoric hunter-gatherer groups and prehistoric agriculturalists long before contact with a colonial or industrial world.

> That Chephren would have built no chambers seemed contrary to everything I know about human behavior. Since Chephren's grandfather had put two chambers in his pyramid and Chephren's father had three, I reasoned that Chephren would more logically have put four chambers in his pyramid! So, the Pyramid Project, as it was to be known, was based on my simple guess concerning the way people behave (see Alvarez,[1] p. 181).

In fact, as he readily admitted, Alvarez's guess was wrong. It is significant though that many anthropologists follow similar paths. Those are most commonly revealed in inferred universals of human behavior. For example, humans are described as inherently and universally aggressive killer-apes,[9,23] territorial,[3] or peaceful.[32] They are described as universally motivated by individual quests for power or dominance[11,19] or as universally manipulated in class struggles.[37] While there certainly must be very abstract or broad universals of human thought and action based on a common structural design, cultural behavior is highly variable. It is not appropriate to explain highly variable behavior in terms of constant inherent features. Pronouncements about inherent universals reveal more about scientific culture and the beliefs of scientists (including the themata discussed by Holton[18]) than they do about human culture in general.

It is unlikely that cultural behavior, with all its complexity and variability, will be explicable deductively (the covering law admitting of no exception) from nomological and complete (all relevant variables taken into account) statements. Nevertheless, a great many archaeologists believe that anthropological explanations will be probabilistic-statistical and partial (rather than deductive and complete). Spaulding[28] and Trigger[33] (pp. 372–383) provide extended discussion of these points. Trigger[33] (p. 375) finds the "statistical relevance" or "how possibly" explanations (of Salmon[25]; Salmon[26]; and Dray,[10] respectively) particularly useful. He states:

> These [explanations] should not be viewed as an alternative to deductive explanations since both extensively employ arguments of this sort. An important characteristic of "how possibly" explanations is the reconstruction of a chain of events, accompanied by an effort to account for these events and the sequence in which they occurred. Explanations ideally should be based on well-established laws of human behavior, although common sense must often be used as a "filler" because of the lack of such theory. Many answers to questions that arise as part of "how possibly" explanations take the form of additional data that eliminate one or more alternative possibilities...The concept of archaeological explanations taking the form of alternative possibilities, some of which eventually may be eliminated by new data, is a corollary of this approach...In due course, new archaeological findings or research in other fields also may help to provide generalizations that can replace common-sense or empirical solutions to problems (see Trigger,[33] p. 376).

Thus, Trigger's view, which I find compatible with my own and what I see as much of archaeological practice, admits that past events and recurrent processes provide the context and contingencies of the particular events one wishes to explain. In "reconstructing" the contingencies, an argument may be framed in terms of "how possibly" statements which are accompanied by attempts to account for the contingencies and the order of their occurrence by reference to general laws, or at least empirical generalizations. "How possibly" statements are not equally valid scenarios or narratives about the past. They are not equally plausible sketches of what might have happened. They are not alternative histories. In the sense that establishing contingency relationships—and evaluating them—provides a history of events, archaeological explanation may be termed "historical." It must always be recalled, however, that there is an equally tortuous path between the reconstructed contingencies and the archaeological record that is also a product of non-cultural factors and archaeological recovery.

A number of archaeologists have a quite different view of "historical" explanation. According to Hodder,[17] Wylie,[39,40] Conkey,[7] and others, most or all archaeological inferences, interpretations and "straightforward observational experiences" (see Wylie,[39] p. 73) are biased by Western middle class values, gender and ethnic prejudices, and imperialist political concerns. At least one of these scholars (see Hodder,[17] p. 30) has encouraged the development of multiple narratives of the past, all of which are considered equally valid or invalid, a position which several have pointed out is philosophically nihilistic.[8,33]

Aside from the issue of nihilism, the fact that scientific instruments of observation may be biased does not preclude the utility of the method of science. Science does not prevent bias. Rather, science does provide a method for exposing and eliminating bias once that bias is suspected (see, for example, Spencer-Wood[29]). Certainly, there does not seem to be any indication that history, literature, art or Marxist philosophy has eliminated bias in interpretation. I personally contend that although science, like all other human activity, is imperfect, it is not only the only game in town but a good one with which to explore the past. "Historical" explanations that provide context and contingencies must be evaluated according to the procedures and methods of scientific confirmation. Until that is accomplished, these "histories" are just-so stories. My problem with them is that their authors seem indifferent to any need to attempt to confirm or disconfirm them.

In my view, prehistoric archaeology is a science that is quite similar to paleontology and is related to anthropology in the same way that paleontology is related to geology and biology. Archaeology is concerned with prehistoric societies and paleontology is concerned with prehistoric organisms. Both sciences are restricted to present observations and must therefore be concerned with evaluating the representativeness of the modern record and the way in which it was formed. In paleontology, this takes the form of taphonomic studies and, in archaeology, of studies of formation processes and the representativeness of the archaeological record. In both fields, aspects of interpretation can depend upon the physical sciences, and when that is the case, the interpretations are widely accepted. Both sciences are also concerned about the general character of evolution on the one hand and the specific events of the past on the other. In both cases, politics and other aspects of world view have biased observations (in paleontology toward always interpreting "progress" in the fossil record). Both depend on science for the tools to counter bias when it is suspected. Both fields are also incomplete with respect to general theory and most contentious in regard to issues that involve general theory. Obviously, the theory of evolution by means of natural selection puts paleontology on much firmer ground than archaeology. Even in paleontology, however, there are problems regarding the inclusiveness of natural selection and, more important, the character of evolution as evidenced by arguments between saltationists and gradualists and debates about macroevolution.

In arguing for the importance of historical explanations in paleontology, Gould[15] makes two points that are particularly relevant to prehistoric archaeology. First, he notes that historical explanations are subject to scientific evaluation in which "the documentation of evidence, and probability of truth by disproof of alternatives, may be every bit as conclusive as for any explanation in traditional science" (see Gould,[15] p. 284). Second, he argues the importance of general laws of nature:

> Am I really arguing that nothing about life's history could be predicted, or might follow directly from general laws of nature? Of course not; the question that we face is one of scale, or level of focus. Life exhibits a structure obedient to physical principles. We do not live amidst a chaos of historical circumstance unaffected by anything accessible to the "scientific method" as traditionally conceived. ...nvariant laws of nature impact the general forms and functions of organisms; they set the channels in which organic design must evolve. But the channels are so broad relative to the details that fascinate us (see Gould,[15] p. 289)!

My view is that archaeology is indeed a science that depends upon physical and natural sciences and whichever of the social sciences are concerned with developing general laws of human behavior (anthropology, of course, but also human geography, economics, psychology, etc.). Largely because there are presently a lack of well-established general laws of human behavior, and because anthropologists are acutely aware of human cultural diversity, archaeologists will feel most comfortable with interpretations of aspects of the archaeological record that depend on the general laws of the physical sciences (cf. Dunnell 1992). Archaeologists also develop historical scenarios that provide plausible sketches of what might have happened that would explain particular archaeological situations. While these should also refer to general laws (or at least empirical generalizations), in fact and unfortunately, they rarely do. Further, archaeologists seem to lag behind their paleontological and other colleagues in meeting their obligation to evaluate and test aspects of the historical sketches they propose. Devising appropriate tests could be one of the more valuable contributions any of us could make to the field. A final note in this regard concerns a matter of words. Some archaeologists, perhaps wishing to emphasize the importance of historical context, use the phrase "historical process." Process, however, refers to quite abstract and repeatable events rather than to particular historic contingencies. The phrase "historical process" can only refer to general statements about the course of history, such as the Hegelian or Marixist dialectic. In my opinion, those historical processes are not laws of human behavior or empirical generalizations about how the world works. There are no historical processes. (For a similar conclusion based on slightly different argument, see Flannery[14] and Spaulding.[28]) The descriptions of so-called historical processes are merely plausible scenarios of historic events that must be fully evaluated.

An important test of the health of any science, I think, is whether or not it allows us to learn anything new, particularly something unexpected and even unpopular. Since archaeology seems so dependent upon an ethnographic record that we know must be an incomplete view of all possible societies that have ever existed, the question about learning anything new is an especially relevant one. In order to gain perspective on this issue, I tried to evaluate such an occurrence in a related field, again paleontology. The example I chose is the proposition that the major extinctions that occurred between the Cretaceous and the Tertiary (including the dinosaurs) had been the result of asteroid impact.[6] In reviewing the example, I was curious about how the explanation had been developed and whether or not a similar scenario might occur in archaeology and, if so, how.

The Alvarez view of Cretaceous-Tertiary (K-T) extinctions is interesting in that in paleontology, catastrophic explanations are particularly unpopular as, of course, is recourse to extraterrestrial events. The discovery and its acceptance had at least some fortuitous aspects. First, the initial question raised by Walter Alvarez was whether or not the K-T extinctions had occurred rapidly or slowly. Walter Alvarez had sampled a clay deposit from near Gubbio, Italy that separated a Cretaceous limestone deposit from a Tertiary one. His reasonable question was how might it be possible to discover the period of time over which the clay deposit had formed. Fortuitously, he discussed this with his father, who happened to be Luis Alvarez. While this indeed seems lucky chance, it does raise the issue of the fact that the international community of physical scientists, and to some extent physical and natural scientists, is both larger and has much more frequent and open communication than does the international community of archaeologists. Problem areas in archaeology are not widely known and discussed in international scientific circles. Also, scholarly meetings in archaeology provide occasions for presenting conclusions rather than discussing problems.

Luis Alvarez suggested that measuring iridium would be most appropriate to determine the time that it took the clay layer to form. The earth's crust is depleted of iridium and most of the iridium present in the crust is assumed to be of extraterrestrial origin from meteoritic debris that is uniformly and continuously deposited

[6] In part, my own interest in looking into this event concerned an evaluation of whether or not the "great man" theory of history was crucial. In other words, I was curious about how important Luis Alvarez was to the discovery and its acceptance by the scientific community.[2]

over the earth's surface. With an estimated sedimentation rate 10 millimeters per 1000 years, the clay should have contained 0.004 ppb of iridium. Measuring iridium in that concentration required neutron activation analysis, in this case carried out with the Lawrence Berkeley Laboratory reactor. Again, it may have been a lucky accident that Luis Alvarez suggested using iridium as a clock (as opposed perhaps to measuring decay rates of organics in the clay), but it was not an unusual choice. In fact, iridium was found in concentrations of 9 ppb, which required an extraterrestrial source of catastrophic proportions.

Importantly, the explanation depended on the laws of physics. It did not depend on hypothesizing something quirky about the mating behavior of dinosaurs. Further, iridium of extraterrestrial origin might have been derived from sources other than a meteor (such as a supernova or nebular cloud), but in hypothesizing meteoritic origin, a number of hypotheses could be evaluated and further measurements were required to do so. All involved laws of physics and not behavioral peculiarities of extinct species. Further, the meteor hypothesis, if confirmed, does not provide a complete explanation. There must be additional statements about the effects of such a meteor on the earth's climate and vegetation and how those relate to the extinctions. All of these statements require assessment. Yet, in proposing meteoritic origin of iridium, the context of the extinctions is dramatically changed.

In asking myself whether or not a similarly unpopular, surprising explanation might be developed in archaeology, and using the K-T extinctions as a model, I believe that the answer is yes, it could. While I can think of no instance of this having already happened in southwestern archeology, I believe that our understanding of a an archaeological situation to be explained has proceeded along similar lines which need to be pushed further. This has occurred, I think, in the case of Chaco Canyon and the Chacoan system.

Gwinn Vivian[34] argues that before the 1970s archaeologists working in Chaco Canyon generally tried to explain (or account for) the Chaco developments by reference to the archeology of the canyon itself. Although outlier communities[7] had been known for years and some had been excavated, the importance of the outliers to considerations of the system were not appreciated. Various how possibly explanation and just-so stories had been advanced to account for the Chacoan great houses and other formal features. While none of these has really ever been adequately evaluated, they became largely irrelevant in the late 1970s and 1980s. The change occurred because a number of investigators documented road systems linking outliers to Chaco Canyon itself; tree-ring dates established the contemporaneity of Chacoan great houses and small sites, and tree-ring dates and architectural studies were beginning to provide data on how very quickly the major sites in the canyon and the outliers were constructed. Thus, the nature of the system to be explained changed dramatically, and most previous explanatory models were found to be inadequate.

[7] Chacoan outliers consist of buildings that are built in Chaco style and have one or more of the "unusual" Chacoan features (such as a great kiva, tower kiva, or both), but are outside Chaco Canyon proper. There are currently some 120 known outliers.

At a number of national meetings archaeologists working in the San Juan Basin gave symposia on new understanding of the Chaco system. I believe I served as a discussant at three of these over a period of probably five years. The explanations offered for the newly defined Chaco system included inferring that the system served as a redistributive center for food, evening out unpredictable crop yields in various parts of the San Juan Basin. Another scenario envisioned the system as functioning as a predominantly vacant ceremonial center in Chaco Canyon, surrounded by outlier communities whose inhabitants participated in periodic ritual pilgrimages to the center, where food stuffs were consumed. In yet another view, the outliers were seen as recipients of excess population from the canyon. In outlier communities, these people joined local groups, all of whom adopted a Chacoan religious organization. These "historical sketches" are not exhaustive of the ones presented nor are they the more outlandish of them. They seem to me typical of archaeological explanations in two respects. First, they are primarily concerned with the functioning of past cultural systems. They are therefore analogous to developing scenarios for the K-T extinctions based on the sex lives of dinosaurs. Second, and more critical for me, over the five or so years during which I heard these explanations offered, no one had attempted to disconfirm or confirm any of them.

What we need here is an honest attempt to evaluate alternatives, and an attempt that depends on firmer ground than alternate reconstructions of extinct social systems. If food were being exchanged or moved over long distances, would it be possible to develop trace element analyses that would let us discriminate different areas where corn was grown? If the outliers were founded by excess population from the canyon, should not the outliers, in general, be more recent than sites in the canyon? What broad studies of dating of outliers has been done?

The lack of evaluation does remind us that accumulating data in archaeology is a time-consuming process. It is also often very expensive. In thinking about the need for precision dating of ceramic types, outliers, and the like, I am reminded that accelerator dating of radiocarbon specimens was announced to the archaeological community in 1976. Yet, this technique is not only still considered problematical but currently costs more than $500 per sample! It would take all the connections Luis Alvarez was able to mobilize for us to do the kind of work we most need with this technique and others. Perhaps archaeologists could enlist the assistance of our hard science colleagues in the development and application of techniques of measurement and data capture that are most appropriate for the kinds of testing and evaluation we need.

REFERENCES

1. Alvarez, L. W. "Using Cosmic Rays in the Search for Hidden Chambers in the Pyramids." In *Discovering Alvarez*, edited by W. P. Trower, 181–185. Chicago: University of Chicago Press, 1987.
2. Alvarez, L. W., W. Alvarez, F. Asaro, and H. V. Michael. "Extraterrestrial Cause for the Cretaceous-Tertiary Extinction." *Science* **208** (1980): 1095–1108.
3. Ardrey, R. *African Genesis; A Personal Investigation into the Animal Origins and Nature of Man.* London: Collins, 1961.
4. Binford, L. R. "Archaeological Systematics and the Study of Cultural Process." *Amer. Antiquity* **31** (1965): 302–210.
5. Binford, L. R. "Archaeological Perspectives." In *New Perspectives in Archaeology*, edited by S. R. Binford and L. R. Binford, 5–32. Chicago: Aldine, 1968.
6. Binford, L. R. "Objectivity-Explanation-Archaeology 1981." In *Theory and Explanation in Archaeology*, edited by M. J. Rowlands, C. Renfrew, and B .Abbott-Segraves, 125–138. London: Academic Press, 1982.
7. Conkey, M. "Open Comments About Archaeological Issues Confronting the 1990s." Presented at the 54th Annual Meeting of the Society for American Archaeology, Atlanta: April, 1989.
8. Cordell, L. S. and V. J. Yannie. "Ethnicity, Ethnogenesis and the Individual: A Processual Approach Toward Dialogue." In *Processual and Postprocessual Archaeologies: Multiple Ways of Knowing the Past*, edited by R. W. Preucel, 96–107. Occasional Paper No. 10, Center for Archaeological Investigations, Southern Illinois University, 1991.
9. Dart, R. *Adventures with the Missing Link.* Philadelphia: The Institutes Press, 1967.
10. Dray, W. *Laws and Explanation in History.* Oxford: Oxford University Press, 1957.
11. Dumont, L. *Homo Hierarchicus; An Essay on the Caste System.* Translated by M. Sainsbury. Chicago: University of Chicago Press, 1970.
12. Dunnell, R. C. "Archaeology and Evolutionary Science." In *Quandaries and Quests, Visions of Archaeology's Future*, edited by LuAnn Wandsnider, 209–224. Occasional Paper No. 20, Center for Archaeological Investigations, Southern Illinois University, Carbondale, Illinois, 1992.
13. Earle, T. K. and R. W. Preucel. "Processual Archaeology and the Radical Critique." *Current Anthropology* **28** (1987): 501–538.
14. Flannery, K. V. "Culture History v. Culture Process: A Debate in American Archaeology." *Sci. Amer.* **217** (1967): 119–122.
15. Gould, S. J. *Wonderful Life.* New York - London: Norton, 1989.
16. Hill, J. N. "Broken K. Pueblo: Prehistoric Social Organization in the American Southwest." Anthropological Papers 18, the University of Arizona, 1970.

17. Hodder, I. "Archaeology in 1984." *Antiquity* **58** (1984): 25–32.
18. Holton, G. "On the Role of Themata in Scientific Thought." *Science* **188** (1975): 328–334.
19. Leach, E. R. *Political Systems of Highland Burma; A Study of Kachin Social Structure.* Cambridge: Harvard University Press, 1954.
20. Leone, M. P., P. B. Potter, and P. A. Shackel. "Toward a Critical Archaeology." *Current Anthropology* **28** (1987): 283–302.
21. Longacre, W. A. "Changing Patterns of Social Integration: A Prehistoric Example from the American Southwest." *Amer. Anthropologist* **68** (1966): 94–102.
22. Longacre, W. A. *Reconstructing Prehistoric Pueblo Societies.* Albuquerque: School of American Research and University of New Mexico Press, 1970.
23. Morris, D. *The Naked Ape; A Zoologist's Study of the Human Animal.* [1st American ed.]. New York: McGraw-Hill, 1967.
24. Neher, H. V. "Some Early Days of Cosmic Rays." In *Discovering Alvarez,* edited by W. P. Trower, 6–8. Chicago: University of Chicago Press, 1987.
25. Salmon, M. H. *Philosophy and Archaeology.* New York: Academic Press, 1982.
26. Salmon, W. C. *The Foundations of Scientific Inference.* Pittsburgh: University of Pittsburgh Press, 1967.
27. Schiffer, M. B. *Formation Processes of the Archaeological Record.* Albuquerque: University of New Mexico, 1987.
28. Spaulding, A. C. "Explanation in Archaeology." In *New Perspectives in Archarology,* edited by S. R. Binford and L. R. Binford, 33–39. Chicago: Aldine, 1968.
29. Spencer-Wood, S. "A Feminist Program for Nonsexist Archaeology." In *Quandaries and Quests, Visions of Archaeology's Future,* edited by LuAnn Wandsnider, 98–114. Occasional Paper No. 20, Center for Archaeological Investigations, Southern Illinois University, Carbondale, Illinois, 1992
30. Steward, J. H. and F. M. Setzler. "Function and Configuration in Archaeology." *Amer. Antiquity* **4** (1938): 4–10.
31. Taylor, W. W., Jr. *A Study of Archaeology.* Vol. 69. Memoir Series of the American Anthropological Association. Menasha: American Anthropological Assoc., 1948.
32. Thomas, E. Marshall. *The Harmless People.* New York: Knopf, 1959.
33. Trigger, B. G. *A History of Archaeological Thought.* Cambridge: Cambridge University Press, 1989.
34. Vivian, R. G. "The Chacoan Prehistory of the San Juan Basin. In *New World Archaeological Record,* edited by J. B.Griffin. San Diego: Academic Press, 1990.
35. Wilcox, D. R. Conference discussion, 1989.
36. Willey, G. R. and P. Phillips. *Method and Theory in American Archaeology.* Chicago: University of Chicago Press, 1958.
37. Wilmsen, E. N. *Land Filled with Flies; A Political Economy of the Kalahari.* Chicago: The University of Chicago Press, 1989.

38. Wobst, H. M. "Boundry Conditions for Paleolithic Social Systems: A Simulation Approach." *Amer. Antiquity* **39** (1974): 147–178.
39. Wylie, A. "The Reaction Against Analogy." *Advances in Archaeological Methods and Theory* **8** (1985): 63–111.
40. Wylie, A. "Gender Theory and the Archaeological Record: Why is There no Archaeology of Gender?" Presented at the 54st Annual Meeting of the Society for American Archaeology, Atlanta, 1989.

Linda S. Cordell,* Jane H. Kelley,† Keith W. Kintigh, Stephen H. Lekson,‡ and Rolf M. Sinclair*****
*University Museum, University of Colorado, Boulder, CO 80309
†Department of Archaeology, University of Calgary, Calgary, AB T2N IN4
**Department of Anthropology, Arizona State University, Tempe, AZ 85287
‡Crow Canyon Archaeological Center, 1777 S. Harrison Street, Denver, CO 80210 and Laboratory of Anthropology, Museum of New Mexico, P.O. Box 2087, Santa Fe, NM 87504
***Division of Physics, National Science Foundation, Washington, DC 20550

Toward Increasing our Knowledge of the Past: A Discussion

INTRODUCTION

Archaeologists today are reevaluating the theoretical and epistemological bases of their research (e.g., Earle and Preucel[16]; Hodder[25]; and Preucel[38]). This activity appears to be part of what Willey describes as three decades of asking:

> (1) What are the legitimate and desired goals in the interpretation of archaeological data? (2) What are the limitations and constraints to be placed upon such interpretations? (3) And how, in our search for objectivity, may we best become aware of the subjective notions and biases that we all, in our various ways, bring to the archaeological enterprise (Willey,[56] p. 106)?

Throughout its history, archaeology has moved between periods during which its practitioners view themselves principally as humanistic scholars and times when they see themselves primarily as scientists. This uncertainty, as Trigger[51] argues, is likely conditioned by the broader social and political milieux in which archaeologists live and work. This paper does not examine the background or history of changing archaeological perspectives. Rather, it addresses issues related to how archaeologists might more effectively use diverse perspectives within the framework of science to increase our understanding of the prehistory of the Southwest.

Understanding Complexity in the Prehistoric Southwest,
Eds. G. Gumerman and M. Gell-Mann, SFI Studies in the Sciences of Complexity, Proc. Vol. XVI, Addison-Wesley, 1994

The archaeologist authors of this paper normally do archaeology rather than write about its philosophical bases. We are not comfortable in the role of pundits and do not intend to prescribe to our peers. Yet, the workshop focus on modeling and explaining the organization and evolution of past societies requires some information about the epistemological context of the data with which we work. There is no seamless narrative of the events composing southwestern prehistory. There are a variety of interpretations of those events. It is our intention to assist in clarifying how we, as archaeologists, can use these interpretations to gain more information about the past we wish to understand.

In what follows, we first outline our philosophical perspective and then discuss how we might use various models and partial models of southwestern prehistory to further increase our knowledge about the past. The Santa Fe Institute provides a setting in which novel, controversial, even unpopular, ideas can be discussed openly. Encouraged by this atmosphere, we offer a number of recommendations for future work. These suggestions are unusually broad. They range from a program to develop new sources of comparative data to the usually taboo subject of remedying problems in the social organization and professional culture of archaeology. These thoughts are proffered in a spirit of constructive advice that we hope readers might find of interest.

PHILOSOPHICAL ISSUES: THE POST-PROCESSUALIST CRITIQUE AND SCIENTIFIC METHOD

The authors of this chapter share the belief that the goal of archaeology is the contribution of knowledge to objective understandings of human behavior. While we believe that progress can be made toward this scientific goal, we recognize that the nature of these understandings and the strength of our beliefs in them have not only to do with scientific evaluation of relevant archaeological data but with theoretical and sociological issues. In this section, we discuss the philosophical context of our position and indicate ways in which we think progress toward that goal can be enhanced.

Over the last two decades, much of the influential philosophically oriented literature in archaeology has focused either on the Hempelian program advanced by the New Archaeology or, more recently, its post-processualist critiques. However, we suspect that few American archaeologists would swear undying allegiance to either a Hempelian covering-law model or to one of the hard-core post-processualist programs. Even fewer probably behave, i.e., practice archaeology, fully in accord with one set of beliefs or the other. While healthy instincts have led most archaeologists to reject these extreme programs, we hope that an explicit discussion of our own position may be productive.

Some post-processualist critique[25,26,33] is directed at the use of the deductive nomological (DN) model of scientific explanation in archaeology.[4,21,22,29,42,55] The DN model includes a hierarchy of concepts (including theory, law, hypothesis, and test implication) linked through formal, deductive arguments. It requires that explanations specify all relevant variables and admit of no exceptions. In the DN model, the truth of the premises logically guarantees the truth of the conclusions. Among other things, the evident complexity of most human behavior leads post-processual critics (e.g., Hodder[25] and Saitta[41]), as well as the authors, to reject the DN model as the only explanatory template.

Another aspect of recent criticism is directed at the reliance of new archaeologists on economic and environmental variables (which seemed more accessible to the available methods) to the exclusion of historical and symbolic components of human behavior. Critics assert the importance of understanding gender roles, social classes, religion, and more generalized belief systems, a cluster of constructs, which, for convenience, we label "ideology." Again, much of this criticism seems well placed. It seems clear to us that ideological and historical factors are crucial to our understanding of much human behavior. However, we disagree with the position that identifying a plausibly relevant past event does, in and of itself, explain anything. Reconstructing the historically based, contingent conditions of the past and understanding the operative processes must be done in a rigorous, scientific manner.

We also part company with those critics who reject the desirability or possibility of science. In extreme post-processual formulations, any narrative account of the past is considered as valid as any other. Fortunately, this notion seems to be losing favor. We recognize that our theoretical perspectives and interpretations are influenced by our own social and political contexts. We see that as a reason for caution and skepticism in evaluating arguments, not as a cause for complete abandonment of the scientific enterprise (see Trigger[52] for a more complete exposition of this argument).

Thus, we retain a commitment to the scientific method, a position which we believe is crucial to advancing archaeology. However, we believe that as long as knowledge claims can be scientifically assessed, a variety of interpretive frameworks may be appropriate vehicles to carry archaeological explanations. As Spaulding[46] has pointed out, there is nothing inherent in a scientific approach that precludes the use of ideological or historical factors. Indeed, in the Southwest, these factors must be considered more often and more seriously than they usually have been. We hope that acceptance of a broad range of relevant arguments may place theoretical discussion on more realistic levels for archaeological problems and data bases.

Part of the glue that holds our discipline together is what is loosely called the scientific method. At the most basic, this means that we subscribe to the idea that we can make judgments concerning better and worse fits of data to ideas. We execute the evaluation process through assessment of the logical structure of arguments, the plausibility of assumptions, the exclusion of alternative positions, the appropriateness of data collection techniques, and the strength of the support provided by the data. Although we can evaluate knowledge claims, we may sometimes wind up with a hung jury. In some cases we just can't say whether x is more likely than y given the data base or the degree of development of the hypotheses or theories. Lack of resolution of such questions is not indicative of a weakness in the approach; rather these unresolved questions can be used to point the way to needed research.

Unfortunately, evaluation, which may take the form of testing, is usually presented as a binary decision made at some particular day of reckoning. Too often, archaeological testing of ideas has been tied to single investigators or single projects. One tests this hypothesis and either it does or doesn't work. Given the complexities of human behavior (of both scientists and their subjects), it is most unlikely that one individual or a single project would be able to perform a complete evaluation of all of the underlying assumptions and multiple levels of linking arguments involved in testing a particular hypothesis.

In cases where so many variables are potentially implicated, as in archaeology, and where potentially useful data bases are so diverse and enormous, we need to shift our views about evaluation of a hypothesis, and indirectly, a theory, from a binary, single-operation model to a much broader one. The consideration of multiple lines of evidence is a well-recognized but under-utilized way to detect errors, expose bias, and strengthen our ideas. Also needed are multiple critical evaluations of important ideas from different perspectives, using different data sets. The value of this suggestion is perhaps best illustrated by the many considerations of the key hypotheses advanced by Longacre[32] and Hill[24] (see also Stanislawski[47] and Plog[35]).

As Cowgill[11] has argued, we need to transform our thinking (statistical and otherwise) away from simply accepting or rejecting hypotheses to evaluating the degree of fit between alternative hypotheses and data. An assessment of contrasting or competing alternatives may allow us to exclude certain possibilities from further consideration while at the same time pointing to one or a few preferred alternatives. Far too often, testing, in the narrower sense, is too constrained by subjective ties between hypothesis and data for the test to be as impartial and objective as we might like to think.

THEORETICAL PREDISPOSITIONS AND PROBLEM SELECTION

Few archaeologists think very much about how and why they select the problems they investigate in the field. Yet, much of what archaeologists do is structured by their starting point. Because of subtle but pervasive influence the issue of problem selection has on the way in which archaeology is practiced, we should pay more attention to the choice of substantive problems, methodologies, models, and theoretical frameworks.

Traditional philosophy of science, especially positivism, is focused on justification and, therefore, excludes discovery from consideration.[22,23] The conventional wisdom is that discovery and justification must be kept separate. Recognizing that this boundary should be maintained does not mean that we should not examine both sides of the discovery and justification fence. Sociological approaches to the philosophy of science, notably those of Kuhn[30] and Popper,[36] are concerned with where theories come from in the dynamic process by which scientific knowledge is acquired.

That archaeology should be so particularly shaped by theory choice is related to the degree to which data underdetermine theory (Quine's "Duhem thesis"; closely related to the problem of equifinality, discussed below). Simply stated, many possible theories could equally well account for the data that are available to us.[13]

The high priority placed on novelty and originality, which often means creative importing of ideas from other disciplines, also has an important impact on the field.[29] Borrowed models carry attendant problems of transferability. For example, the World Systems approach was formulated to model the rise of capitalism. New, or nearly new, frameworks such as the concept of strong and weak patterns[50] carry underlying assumptions, and their fit to previous classification systems and approaches are generally not fully worked out. Some recommendations hinging on sociological considerations are presented in more detail in the final section of this chapter.

IDEOLOGY AND HISTORY

Ideological and historical factors, as broadly defined and discussed above, are obviously important in the operation of cultural systems, but they are difficult to identify and evaluate archaeologically. Despite their importance, archaeologists, post-processualist or otherwise, have not developed general methods that are adequate for identifying ideological and historical influences on the archaeological record, evaluating their relative effects on human systems, or incorporating them in interpretations. In some cases, ethnographic accounts, particularly those from modern southwestern peoples, provide insights into the importance of ideology in their cultures and, through analogy, may suggest ways to incorporate ideology in our archaeological interpretations. In other cases appropriate direct analogies may be lacking.

Some recent studies in southwestern archaeology specifically include inquiry into or interpretation of ideology. These suggest models of ways in which the ideological aspects of culture may be approached archaeologically and ways in which archaeological methods of Salado polychromes to be an excellent example of ideologically based interpretation. The logical structure of her argument and the evaluative process are those of normal scientific archaeology. Crown developed three possible interpretations to account for the timing, production, and distribution of Salado polychromes. The data best fit the interpretation that the introduction of Salado polychromes reflects the introduction of a cult.

We also examined ideological issues with respect to interpretations of the Chaco system in the San Juan Basin of northwest New Mexico. Three current alternative models of the Chaco "phenomenon" might be termed the redistribution, pilgrimage, and tribute models. The redistribution model (as proposed by Judge[27]; Judge and others[28]; and further elaborated by Powers and others[37]; and Tainter[49]) is an economic, risk-sharing model in which food from all over the San Juan basin was pooled by leaders in Chaco Canyon for redistribution to outlying areas during times of local shortage. In contrast, the pilgrimage model, suggested later by Judge and others[28] when expectations of the redistribution model were not supported, has a strong ideological basis. According to this argument, Chaco Canyon was a place of great religious importance to which Anasazi pilgrims traveled (over the roads) and brought food for participation in periodic rituals. Others (Akins and Schelberg[2] and Schelberg[44]) have argued that Chaco Canyon was the locus of strong political power, and that tribute was exacted from the Chacoan hinterlands by way of secondary centers, the so-called Chacoan outliers.

Each of these models depends to some extent on ideological factors, thus it should be possible to develop methods of evaluating them. In fact, some contrasting expectations for these models can be developed, but no conclusive tests have been formulated or executed. It is likely that none of these models is stated with such precision that it would be possible to contrast them directly without making assumptions with which their supporters would disagree. Nevertheless, exploring ways in which the models might be contrasted, evaluated, and tested provides an indication of the kinds of methods and data that would be required for further research.

One informative test could involve the use of trace-element analyses of prehistoric corn, which might be used to identify the location in which it was grown. If the Chacoan system were primarily redistributional, we would expect substantial quantities of corn from several source areas to have been dispersed (through redistribution) throughout the San Juan Basin (via Chacoan outliers or whatever). Thus, finding corn that was grown at the Salmon Ruin, along the San Juan River north of the canyon, was then deposited both within the canyon at Pueblo Bonito and east of the canyon at Pueblo Pintado would lend support to the redistributional model as opposed to the pilgrimage model. Finding corn exclusively where it was grown and within the canyon would tend to contradict the redistributive model. However, in neither case would the source of corn provide a crucial test of either model, in part because there are a variety of behaviors that might be associated with redistributive systems and pilgrimages that have not been made explicit. For example, pilgrims might well transport corn, for their own use or as offerings, from one portion of the system to another.

Carrying through with this line of discussion, however, examining the intensity of road use might allow us to distinguish a tribute from a pilgrimage model. With the pilgrimage model we would expect a high intensity of use by large numbers of people, possibly indicated by the density and distribution of prehistoric road-side debris. A marked absence of such debris would tend to disconfirm the pilgrimage model in favor of a redistribution or tribute model. Both redistribution and tribute models would require demonstrating extensive storage facilities in Chaco Canyon. These suggestions have not indicated a crucial test of one or more of the models, and they may themselves be somewhat simplistic. On the other hand, they help to clarify the need for precision in explicating models and their underlying assumptions and the need for creativity in developing evaluative tests.

At a more general level, one method for identifying ideological components of prehistoric systems may be to find multiple examples of strong patterns in the archaeological record that do not appear susceptible to economic/ecological explanations. For example, these might include consistent alignments or many consistent spatial arrangements of features, as well as the consistent appearance of "ritual artifacts." In this context, one way to assess the relative importance of the suggested ideological elements in the archaeological system is by estimating the amount of energy that was invested in producing them. This approach, however productive it might be, is open to the criticism of treating ideological matters as residual factors, those without largely utilitarian value. From an anthropological perspective, it must be recognized that to appeal to ideological or historical factors only as residuals or as a sort of last resort does not do them justice.

FAILURES TO BE EXPLICIT, CRUCIAL TESTS AND EQUIFINALITY

The discussion of alternative models for the operation of the Chacoan system raises the problem of our inability to accept or reject arguments in southwestern archaeology. An exploration of the factors that facilitate or impede the evaluation of argument, included the facts that archaeologists typically fail to be explicit about their assumptions, definitions, variables, and linking arguments. There is also a failure to specify the scope or boundary conditions for arguments. In the context of historical reconstructions, this is extreme, but the problem is also prevalent in scientific archaeology.

Clearly we need to be moving toward critical tests in which one or more alternatives can be eliminated. Yet, even when archaeologists have been fairly explicit, this is difficult to do. The various scenarios for Chaco, as discussed above, are good examples. Even if we demonstrated that corn grown at an outlier were stored at Pueblo Bonito, no crucial test of the current models would have been performed.

In general, the issue of explicitness was raised, in part, by processual archaeology in the context of espousing a particular style of explanation, that associated with the DN model. In the context of this discussion, we find a number of explanatory styles appropriate, but we reiterate that being explicit about assumptions and definitions remains crucial. Explicitness may have fallen aside because of an initial confusion between scientific justification and the logic of discovery. When the new archaeology failed to achieve all of its stated goals, the discipline moved away from procedures that should not have been discarded. Clearly, being explicit about assumptions and procedures—at all levels—should be reinstated into whatever kind of archaeology is conducted.

A good example of the lack of specificity is the confusion over the term complexity. Some archaeological investigators refer to managerial and or hereditary elites. Others use complexity to refer to diverse situations of social differentiation. Test implications derived from each of these perspectives would be completely different from one another. Other examples can be found in more seemingly basic issues, such as chronometric dating. Two different investigators taking the same set of tree-ring dates and the same contexts may well infer the length of occupation of a site totally differently. In this case the disparity is probably procedural. Investigators use different procedures to determine clusters of dates from which to infer construction and remodeling. Unless they are explicit about which procedures they use, their different interpretations cannot be evaluated. We are all aware of various procedures used to evaluate radiocarbon dates that lead to dramatically different conclusions. At a minimum, investigators need to make their procedures explicit. It is preferable, of course, that they also make their data available to others for reanalysis through publication and curation. Until we know whether or not new data or more refined methods of analyses are required by understanding the procedures already used, there can be little progress.

As the above discussion of Chaco suggests, it is often difficult to develop crucial tests because models and their underlying assumptions may not be made adequately explicit. In dealing with complex systems, however, we should also anticipate problems of equifinality. Equifinality refers to situations in which ambiguous inferences cannot be resolved, identical referents are subject to different and contradictory interpretations, or when the same products or results derive from two or more processes (Benson,[3] p.182 and Sullivan,[48] p.191). For the scientist who is interested in isolating the cause of a particular phenomenon, situations of apparent equifinality are particularly important because they may indicate where our models, theories, or observations are currently inadequate. For example, initially, the conflicting interpretations of the Grasshopper burial data appeared to be such a situation. In this case, contrasting models of social organization were inferred from the same burial data. Attempting to isolate the locus of disagreement, whether based in procedures used in data recovery and analysis or in inference, brings out deficiencies in our current methods and theory. Thus, Cordell and others[10] believed a key problem for the Grasshopper burials could be attributed to data recovery and analysis and was the result of excavation procedures that produced a biased sample of the burial population and subsequent improper application of life-table data. While this may still be the case, peeling away the layers of the argument have been highly rewarding. We have learned that life-table data currently used (i.e., Coale and Demeny[9]) probably are not appropriate for prehistoric southwestern societies, and in on-going research others at this conference and elsewhere[34] are developing new models.

BENEFICIAL EVALUATION, EXPLANATORY STYLE, AND ISSUES OF SCALE OF PROBLEM

While specific conclusions concerning their inferences of descent and residence have been largely discounted, Longacre's[32] and Hill's[24] pioneering studies are recognized as singularly important in stimulating productive research in the Southwest (and elsewhere). Among other things, their work resulted in a raft of advances in ethnoarchaeology, studies of style, and experimental studies, as well as refined and more explicit archaeological field research. It seems useful to inquire why these studies by Longacre and Hill were so extraordinarily influential and successful in generating productive research.

These studies were not only novel in their attempts to access cultural anthropological variables, they were unusually explicit in laying out and pursuing their research designs, the value of which we discuss above. Longacre and Hill also defined research questions that were relatively accessible to the scientific methods of archaeology as they were understood at the time. This leads us to the observation that different classes of research problems are susceptible to different modes of explanation or understanding. For example, actualistic studies of ceramic performance may be pursued using straightforward experimental procedures (despite the fact that failure to adequately specify the boundary conditions for the results often severely limits the relevance of these studies). Questions of residence rule, while considerably more difficult and requiring many more assumptions and more complex linking arguments between hypotheses and data, still seem tractable under standard modes of scientific explanation.

The difficulties are extreme, however, when dealing with the big questions, such as the development and operation of the Chaco or Colonial Hohokam systems, the adoption of agriculture, or the development of increased social and political complexity. The greatly increased difficulty of dealing with these classes of problems seems to have to do with the fact that they are concerned with the understanding of cultural processes. Additionally, these problems are not conceivably solved by individual researchers working at individual sites. Furthermore because of the focus on process, these problems do not have fixed data sets; instead the relevant data bases are large in scale and constantly expanding.

It seems unreasonable to us to expect that the explanations of these large-scale problems will have the same form or structure as the smaller-scale problems. Generally "explanations" of large-scale, complex processes take the form of essays in which questions, hypotheses, data, and linking arguments are woven into arguments that are seen as more or less plausible by those who read the arguments. Binford's[6] "Post-Pleistocene Adaptations" is a good example of the form such essays take. Unfortunately, the degree of plausibility of the arguments often depends in substantial portion on the predispositions of the audience.

Yet we believe that improvements can be made. Such essays or explanatory sketches need, in one way or another, to speak in several voices, to clearly separate the assumptions, hypotheses, and strong, data-based inferences from the more speculative and perhaps anecdotally supported portions of the argument. In addition to more clearly separating the components of the arguments, one simple improvement, now excluded by our major journal, would be the use of footnotes to amplify critical assumptions and aspects of the arguments. We applaud and encourage more use of data appendices to further allow the evaluation of arguments by independent researchers.

In archaeology, we seem to benefit when a particular inference or problem attracts the interest and imagination of multiple archaeologists who bring different perspectives and expertise to bear on a common problem or inference over a fairly extended period of scrutiny. Kelley and Hanen[29] (p. 346) classed this a "second level of testing," and the Longacre and Hill cases exemplify the benefits to the field that can emerge from the application of multiple perspectives from multiple positions involving overlapping but differentiated areas of expertise and experience. The problem is that relatively few cases seem to attract a spectrum of sustained archaeological scrutiny that exposes ambiguities, uncertainties, and errors in the entire constructed edifice from its foundations to its assumptions and theoretical coherence. This is at least partly due to the rewards for originality and novelty in archaeology. Also, archaeologists tend to be rather individualistic, and even those who accept myriad constraints on their activities as a result of, for example, agency or contract guidelines, will not take kindly to being told that they have to evaluate this and that theory. And who would tell them?

Another evaluation tactic, and one that is closely related to second-level testing, concerns the selection of the best among competing hypotheses, a process sometimes called "inference to the best explanation."[19] The comparative method of anthropology can be seen as predisposing us toward extended scrutiny of hypotheses and competitive evaluation of several. A third evaluation strategy that pays dividends in archaeology is the marshalling of support from multiple lines of evidence that are as independent as possible. For example, if evidence from environmental reconstruction, sourcing information, dating techniques, etc. seems mutually supportive of an idea, it is stronger than if all supporting information is in one basket. The interpretation or hypothesis could still be wrong, of course, and it is difficult for a single archaeologist or even a research program to provide the necessary consideration of plausible alternatives, or in-depth evaluation of underlying assumptions.

We agree that one of our goals is to understand human variability, but it seems unlikely that we will have non-trivial laws of human behavior. We acknowledge that as yet, we do not seem to have any such laws of human behavior. In suggesting that we broaden the basis of explanation, we propose that there is room for inductive statistical forms of explanation, such as Salmon's,[43] although we don't yet know the scale limitations of the SR (statistical relevance) model. In order to develop the context and contingencies relevant to the events or processes we wish to understand, we must do historical reconstructions. Archaeologists might then follow some of the procedures historians use in these pursuits, such as Dray's[15] "how possibly" questions. We suspect that archaeological explanations will likely include how possibly questions, explanatory sketches, and some empirical generalizations. We will most likely argue both inductively and deductively, although not at the same time.

Given all this potential variability in explanatory style, we run the risk of being criticized for advocating a philosophy of "anything goes." It is therefore imperative that we maintain evaluative techniques and place far more emphasis on evaluation and confirmation. We can and do evaluate knowledge claims, data-acquisition techniques, and the appropriateness of methods of analysis, etc. We must continue to develop techniques of confirmation such as those discussed above. Archaeological data are malleable and subject to a variety of interpretations. Yet, even archaeological data are not endlessly malleable. We do need to remember that much of what we do know is often well supported. For example, with respect to the San Juan Basin, it is just no longer possible to view sites of the relevant period without reference to the scale of the Chacoan system.

EVOLUTION AS APPROACH AND PROBLEM

The chief concern of this workshop is modeling the evolution of society in the Southwest. In part for this reason, it is useful to examine some general concerns about evolutionist approaches in archaeology. Although the vocabulary of evolution and evolutionary theory is pervasive in archaeology, there may be considerable confusion about evolutionary thinking and the utility of evolutionary models. In anthropology in general and in archaeology, cultural evolutionary writing encompasses everything from the nineteenth century schemes of Tyler and Morgan through classical Marxism, to the materialist neo-evolutionary approach of Leslie White, the ecological neo-evolutionary position of Julian Steward and the selectionist Darwinian evolution writing of David Rindos and of Robert Dunnell (see Trigger,[51] pp. 289–341 for an excellent discussion and abundant references). As Trigger notes, the major differences among these lie in what is perceived as the cause of cultural change. Inquiry into the cause of cultural change is important and on-going, with no single view clearly prevailing. Trigger's discussion suggests that at a minimum, archaeologists should be precise about their view of evolution. In our view, productive research may derive from more than one evolutionary perspective. It is critical then that in order to fairly evaluate and test ideas about cultural evolution, archaeologists must be as explicit as possible about what they see as the cause of evolutionary change and the mechanisms of evolution they envision are operating. It will also be valuable if archaeologists work toward delineating change that is of evolutionary character from that which is not. It is to be expected that at least some change occurs because of some historical accident or contingency and then is evolutionarily neutral or not selected against.

Cultural evolutionary thinking, at its narrowest, assumes that all cultural phenomena are adaptive or otherwise explicable in evolutionary terms; that everything has a function; that the function the observer has isolated as important is in fact working as assumed; or, teleologically, that a present function is the function originally served and selectively retained. Whether or not a narrow evolutionist program is adopted, we suggest that cultural evolution be pursued as a research question rather than as an assumption.

We would argue that over the length of human evolution, the proportion of human behavior that links directly to explication in terms of Darwinian models of reproductive advantage appears to decrease. Yet, cultural behavior may still be addressed through different or more broadly phrased evolutionary models. Over the long-time period that concerns archaeologists, human behavior changes, and culture itself develops and becomes more complex. Models of evolution that include learning and the ability of the species to change behavior rapidly become more appropriate for the later ends of the period of interest. Theories of human behavior may not be universal. Some theories should be more appropriate to understanding variability among archaeological remains of *Homo erectus* populations than variability among Neolithic communities of *Homo sapiens.* Similarly, theories of microeconomics may be perfectly useful for some aspects of modern economies but not to others or to the economies of tribes. It is important then, at a minimum, for archaeologists to specify the boundary conditions for which the theory with which they are working appears to be most useful, and to be open to moving to different theoretical perspectives as they become relevant.

When we address selective advantage and adaptation, we must be concerned with matters of scale. Scale is particularly important in evolutionary research that is directed toward understanding adaptations. An adaptation presupposes some sort of selective advantage. As such, an adaptation may or may not be synonymous with economy. In the Southwest, we are accustomed to thinking of the sum of household economies as an adaptation. Consequently, the scale of most evolutionary thinking in the Southwest is rather small, encompassing the site and its sustaining area or a drainage or, at best, a phase or focus. Indeed evolutionary processes that operate on spatial and temporal scales above the scale of economy are more appropriately considered selectively adapted. For example, long-term, multigenerational land use, involving cycles of settlement, resource depletion, movement, and resource recovery have been suggested for both the Anasazi and the Mogollon areas. These are difficult to envision as economic, because they do not result from conscious economic decisions. Yet, these phenomena are of the appropriate scale to be studied as adaptations. Again, focusing research on matters of what is apt to be an evolutionary adaptation versus what is not, is a worthy endeavor.

In evolutionary thinking, boundary conditions are as important as scale. We can view things in an evolutionary framework without seeing them as part of a universal or unilinear scheme of evolution. We can have an evolutionary (selected) trajectory in the Southwest that may or may not be relevant to another (global) trajectory. If we believe the Southwest is relevant to another situation, we would have to demonstrate that the boundary conditions are the same. For example, Norman Yoffee,[58] maintains that the sequence in the Southwest is irrelevant to understanding evolution of complex societies as it occurred in Mesopotamia. That is, the Southwest does not represent a truncated part of that overall trajectory. If we believe that the Southwest is relevant, then we would have to demonstrate that the Southwest sequence shares the boundary conditions as well as the selective pressures of portions of the Mesopotamian sequence.

Yoffee's statement reminds us that we cannot assume that a particular evolutionary trajectory occurred in the Southwest based on some general notions of unilinear evolution. Further, we know that in order to understand the context of the changes that did occur in the Southwest, looking at ethnographic analogs may be inappropriate. Our ethnographic data derive from contexts in which state-level societies dominate the world. The archaeological record of the Southwest was accumulated largely in a context of the absence of states. Thus, it may be appropriate to suggest the use of a comparative archaeological data base for pursuing broadly evolutionary questions. The use of comparative archaeological data, described below, must, we think, be in addition to the continued development of actualistic and ethnographically—based studies for middle range issues.

TOWARD A COMPARATIVE ARCHAEOLOGY

A comparative archaeology would be suitable for examining issues for which archaeological analogs are more appropriate than ethnographic cases. A comparative archaeology would be concerned with both synchronic patterns and long-term diachronic sequences; with the nature of transitions and cumulative patterns that result from the combination of behavior and transformation processes that determine how the record looks. Two examples hint at the potential productivity of a comparative archaeology as a research strategy; Flannery's[17] discussion of the village as settlement type in the Near East and Mesoamerica and a study by Carol Raish[40] on the relative stability of pre-state farming communities.

Flannery's work focused on house form and the arrangement of dwellings in Neolithic communities in general. He noted that in the Near East there had been a transition from villages composed of circular huts arranged in compounds to villages composed of rectangular houses. In Mesoamerica, the prevailing form was also one of villages with rectangular dwellings but with weak evidence suggesting a prior use of circular hut compounds. Whether or not there is support for Flannery's model of the Neolithic processes that underlie the two different patterns, it is the comparative approach of archaeological cases that is useful. Raish was interested in apparent differences in the duration of the pre-state farming period in Old and New World sequences. She identified a pattern of shorter sequences in the Americas, with the exception of Peru, and proposed that the presence and economic importance of domesticated animals might be a factor conditioning the differences.

A comparative archaeology, then, focuses on the identification and analysis of purely archaeological patterns. Currently, we may see comparisons of two or three regions with which an individual scholar is familiar. While these can be very useful, they may also fail as elementary sampling exercises. Comparative archaeology, minimally, will help calibrate archaeological conventions used by archaeologists working in different regions. What is a ceremonial center? a village? a regional system? Comparison on this level at least indicates discrepancies between regional usages and conventions. Ultimately, it may lead to longer-term growth of knowledge, lifting post-Pleistocene studies out of severe provincialism and making archaeology cross-cultural.

Comparative frameworks for data may be considered at three scales or levels with which we commonly work: the artifact, the site, and the landscape or region. Much of the observational language appropriate to the artifact level is common to archaeologists world-wide. This is a result of archaeology's historical association with art history and museums, which provide an existing observational structure; the foundation of observational languages in the physical, chemical, and mechanical properties of artifacts; and the relative simplicity of the unit (artifact) compared to units at larger scales (i.e., regions). Obviously, the language of formal and functional artifact types is less universal than that of the physical properties of artifacts. Yet, even here, there is a degree of universal usage (an arrowhead is probably quite consistently identified as an arrowhead).

As the scale and complexity of the level of phenomena examined increases from artifact to site to landscape, the complexity and richness of our observational language decreases. Thus, observational language for sites is limited primarily to simple dimensions such as measurements of size. An observational language for sites with architecture that encompasses both form, perhaps, as measured through connectivity, and condition (as ruin) is only beginning to be developed and has not been applied to any broad comparative sample. On a larger scale, it would be illuminating to assemble a broad sample of archaeologists who believe they are working with Neolithic state-level societies, and compare their concepts of the landscapes appropriate to such societies and the landscapes they have observed.

Comparative frameworks usually encompass reconstructions of past societies through some form of cross-cultural analysis. We seldom attempt very broad comparisons of purely archaeological patterning, in either physical remains or at very basic levels of analysis. Examples of potentially interesting patterning for comparison include such peculiarly archaeological phenomenon as multi-componency, horizon styles, and particular cultural landscapes defined by non-site distributions of artifacts. These are just a few of many familiar archaeological phenomena for which a comparative base could be useful. We know in principal what archaeologists mean when they use the terms such as multi-component or horizon style, but we do not know their patterning or associations. We have no basis from which to evaluate whether or not the particular archaeological phenomenon we are concerned with is unique, rare, or common in similar contexts world-wide basis. Such knowledge is potentially crucial to our understanding.

The greater potential value of a comparative archaeology should be reached by comparing purely archaeological patterns both synchronously and diachronically. For example, at the level of synchronic patterns, we might want to look at the nature of storage technology and its scale, the sizes of settlements, and settlement layouts. At the level of regions, we might look at the geographic extent of horizon styles in different world regions, holding site size, subsistence base, or other factors equal. Within the context of diachronic patterns we might be concerned with settlement longevity in different areas of the world, the longevity of regional patterns, such as the length of time populations appear to have been mobile in areas where sedentism eventually prevailed. We might also be concerned with the nature of transitions, the length of time over which a particular change occurs, or whether or not there are intermediary steps, or from one world region to another, or how widespread within a past society a specific transition is. Patterns of a cumulative nature are those derived from both behavior and site formation processes. Comparative archaeology would allow us to examine those situations that lead to palimpsests versus those that produce continuous smears and those that produce tells.

Within the context of a comparative archaeology, we would be concerned both with the collected data and, more importantly, their boundary conditions. Thus inquiry is directed toward similar patterns in different contexts. With this, one would begin to compare these with respect to the nature of boundary conditions. The nature of the boundary conditions when compared could allow us to make inferences about how they contribute to the similarities and differences in the patterns. Obviously, there must be construction of linking and warranting arguments between boundary conditions and the observed patterns.

We see the development of a comparative archaeology as analogous to what we already do when we compare assemblages and sites on a very local level. We would be taking methods we routinely use and applying them on a different, very much broader, scale and to combinations of data bases that we do not traditionally put together. What may be critical, is that for the larger-scale problems this could be particularly important because there are no comparative examples on the local level. For example, there is only one Chaco Canyon/San Juan Basin system in the Southwest and only one Hohokam/ball-court system in the Southwest but similar entities may occur in a world wide sample.

Thus, we recommend serious, systematic comparative analysis of direct archaeological patterns as well as higher-level social reconstructions based on those patterns. We propose a comparative archaeology, in part, as a remedy for the provincialism of southwestern archaeology. Southwestern archaeology in the 1990s must break out of severely localized analytical frameworks, and a comparative approach will assist in accomplishing this goal. Beyond that, however, we are confident that a comparative approach will also benefit the broader field of anthropological archaeology.

In working out a comparative archaeology, we suggest that it is the strength of anthropology and it brings us back to what anthropology is all about. As archaeologists, we appreciate and hope to take into account the problems in dealing with cross-cultural comparisons that our ethnographic colleagues have learned and mostly resolved. For example, establishing the relevance of the boundary conditions and, in fact, the relevance of the comparison at all need to be established. We anticipate problems of comparability of measurement and observations and comparability of terminology. We still think the endeavor is worthwhile despite problems and abuse. We think Yoffee's[58] approach provides valuable insights and is a good start.

Within this context, some or all of the following might be useful: the development of a Human Relations Area File (HRAF) of archaeology; the development of a metalanguage with which to discuss and compare dimensions; or the development of a new kind of conference with cross-cultural archaeologists coming together. If a comparative archaeology were to be developed, we might look at institutions like arid land institutes and the *Journal of Comparative History* for models. We also recognize that such a comparative archaeology would have major implications for graduate student training and curricula.

All of this must proceed with the continuous development of middle range theory so that the patterns we hope to discover are interpretable either in terms of "laws of nature," "laws of physics," or meaningful/understandable behavioral patterns. For example, there are patterns of demography derived from burial samples that may show up time and again around the world. We need to know if the pattern is a result of a consistent process of mortality (age and sex death) in human populations or if they are a consistent bias of the archaeologist. Inferences may require additional mathematical modeling. Clearly, the use of a comparative archaeology as a method must also follow all the normal scientific procedures of confirmation.

CHECKING THE FACTS, A CONTINUING NEED

We have discussed broadening the base of explanation in archaeology and called for more attention to evaluation and testing. We are also keenly aware that various historical scenarios are being developed and discussed that relate to the specific areas of complexity addressed by this conference. Among them are the Chacoan and Hohokam regional systems and the distribution of polychrome ceramics in the late prehistoric Anasazi and Casas Grandes areas. Although, obviously, we do have considerable data on each of these, it is also true that some very basic information is either non-existent or poorly supported. Even where the data are quite good, there may be a lack of models that have been subject to even minimum confirmation procedures. We mention just a few of these here in order to call attention to them. The hope is to avoid expending effort developing explanations for past situations that we cannot really demonstrate existed. Rather, we hope that effort would be devoted to clarifying these issues.

With respect to the Chacoan regional tradition, we note that there are too few cutting tree-ring date for the Hosta Butte sites. In his synthesis, Vivian[54] lists only one! Although Hosta Butte and Bonito style sites are most likely contemporary, much better chronological control is required. Dating the spans of occupation of the Hosta Butte sites and the Chacoan outliers also must be undertaken before scenarios are developed that relate to the contemporaneity of sites that may not, in fact, be strictly contemporary. Further information is also needed on the Chacoan roads. The full extent of the roads needs to be traced and located. There may be differences in the roads within versus those outside the canyon beyond the frequently noted differences in width.

Many of the models that deal with Chaco are concerned with trade and exchange and their role in buffering agricultural short-falls. These should remind us that we need to develop better models to assess productivity and success of Chacoan agriculture. The simulations done by Lynne Sebastian,[45] and Carla Van West's[53] study that combines geographic information systems with paleoenvironmental reconstructions are potential guides here. We also need sourcing studies of trachyte (sanidine basalt) temper and other Chacoan ceramic constituents, the timbers used for construction, and of foodstuffs, if possible. Finally, in regard to the basic facts about Chaco, at a minimum we think we must reevaluate the basis on which outlier sites have been identified and determine the range of sites that has been considered Chacoan.

Questions relating to a Hohokam regional system are, at present, even more pressing. We need studies of the regional distributions, dating and contexts of ceramics, ball courts, stone palettes, clay figurines, and the irrigation systems. Until now, the Hohokam regional system has been defined primarily on the basis of the distribution of ball courts. Ways of describing this regional system that are independent of the distribution of ball courts are needed now. Whether or not there were exchange systems at the scale of ball-court communities and larger units needs to be determined, as does the relationship between the Hohokam system and the Mimbres Classic that is reflected in ceramic styles and possibly also in clay figurines. In this regard, there is a need for some very basic information, such as basic survey data from the San Simon area. Another quite basic query is for information about plants used in the Hohokam area for fuel for heating, cooking, and for ceramic production. If it is proposed that some long-distance trade brought ceramics to Chaco, at least in part, because the canyon lacked adequate wood to fire pottery, then the situation for the Hohokam should have been even more extreme. What solutions did the Hohokam develop?

In order to describe the systems that are represented by the distributions of polychrome pottery in parts of the Southwest after A.D. 1250, some basic observational data are needed. These include the loci of production of the polychromes, especially demonstration of the extent of local and non-local production and much more precise chronological placement of small versus large sites. Most of the questions we have about this period, however, relate to the need to develop more and better models. We need to explicate how we make the connections we do between site size, ceramic distributions, and proposed socio-political alliances. We need to determine additional specific organizational features associated with alliances that might be recognized archaeologically. Similarly, we need to address how we recognize elites. What are the models that allow us to infer the presence of elites in the archaeological record from distributions generated by alternative organizational forms? We need to model and understand regional exchange and elite exchange differentiated explicitly from non-administered exchange. The Rio Grande provides an important contrasting area for study of the proposed late prehistoric regional systems. We need to know whether or not we have sufficient data to reject alliances based on the lack of large-scale exchange in that area. It will be important to contrast Rio Grande glaze-ware production and distributions to those of the western Anasazi.

The case of Casas Grandes requires development on a variety of levels. We continue to need systematic survey around Casas for settlements, agricultural land, and roads to add to the impressionistic windshield surveys. We know very little about what led to Casas; we don't know what precedes it. There are a variety of mound sites surrounding Casas that need to be dated. We need to establish the existence of the trade system and the scale of the trade on both a local and regional basis. Our eventual understanding of Casas will require better culture history provided through survey, dating, and sourcing of ceramics, copper, foodstuffs, and wood. We will need to go back and reanalyze the functions of previously excavated rooms. Although we may accept interpreting some rooms as workshops, we should examine the functions of other rooms in terms of architecture, artifacts, and features. Using the same methods Breitburg[7] has used for turkeys, it should be possible to source the macaws from Casas and determine whether or not these are the same genetic population as those from Grasshopper. The kinds of information obtained from the Casas Grandes area should allow us to compare the Casas system with what we know of Chaco in terms of scale and organization rather than looking at what may or may not have been transmitted from one to the other.

There are additional classes of information that will be useful for the whole Southwest. It should be helpful to update the classic study Breternitz[8] did on tree-ring dated pottery types and also include ceramic assemblages. We should also be able to extend tree-ring series by deriving probabilities of fits within the tree-ring sequence for samples where there are currently apparently an insufficient numbers of rings. We also need to compare models of chronological placement of abandonments that use archaeomagnetic dates versus those that are based on tree-rings. It is possible that such comparison might revise our understanding of regional population dynamics.

There are a variety of issues related to model building in general, that arose out of discussion of the cases of complexity with which we are dealing but are not tied to those cases. In most of these instances the need seems to be for actualistic studies, simulations, or broader ethnographic comparisons. Thus, we need methods to identify functional differentiates in space use in small residential sites, as well as in areas of public architecture. These would include making broad ethnographic comparisons, an assessment of features, and artifact distributions. We need to develop measures of intensity of use for different kinds of functionally differentiated features.

We also suggest examining recognized, non-state regional systems in order to see how such societies pursue a range of options in transporting goods and how these, in turn, may relate to different modes of regional organization. Similarly, we need a broad ethnographic view of just how many different kinds and scales of organizational structures can be accommodated within kinship-based organizations. We need to know the importance of different scales of kinship structures within different organizational forms, for example, the possibility of groups with village economies to forming alliances. Our wish list includes wanting better models of and ways to infer essentially local exchange and interaction from archaeological data and more cross-cultural data on labor as a measurable item. We need to understand how time is allocated and used among societies at different levels of complexity.

We need to develop better models of sustainable surplus achievable in different southwestern environments under different technologies. We need to compare and refine models of productivity and storage such as those developed by Lynne Sebastian,[45] Carla Van West,[53] and Michelle Hegmon.[20] And finally, we would like to prepare simulations of different types of exchange varying the numbers of people and objects involved.

SOME SOCIOLOGICAL RECOMMENDATIONS

There are a variety of factors which impede the development of Southwest archaeology that relate to the sociology of the science of archaeology rather than to particular issues in the philosophy of science. We address some of these here and make some suggestions for their remediation.

In order to achieve the strongest inferences possible within the present context of disciplinary resources and problem orientations, some attention must be paid to non-logical, socio-political factors that influence the implementation of scientific goals.[18,30] Among the factors that emerge as significant in southwestern archaeology (as is the case more generally in North American archaeology) are the critical split between data acquisition and the integration of new evidence into received archaeological views, the low incidence of thorough and multi-faceted evaluation of proposed hypotheses, the failure to make the implications of models and hypotheses sufficiently explicit to allow adequate evaluation, and a scarcity of periodic review mechanisms. Implicated are certain aspects of training archaeologists, the reward systems, the socio-political context within which archaeology is practiced, and the organizational structure of the discipline.

It would seem that, as social scientists, we should be able to devise schemes that will lessen the liabilities of such factors and narrow the division between the various archaeological venues so that the discipline can advance upon a reasonably common front toward shared objectives. As we continue to debate which strategies are the most appropriate for our discipline, we should remember that archaeology is not alone in facing a range of methodological issues, and indeed most academic disciplines in the social sciences are currently undergoing rigorous self-scrutiny.

Although there is growing sensitivity to the impact of sociological issues on the doing of archaeology, the role of organizational structure on the science side of archaeology is still underappreciated. Becoming aware of organizational constraints leads us to consider ways to make the socio-political aspects of science a more explicit part of the the awareness of archaeologists. What we want to do is make conscious decisions on the socio-political side of archaeology to maximize the science side.

Southwestern archaeology is conducted within academic (university and museum) and CRM (cultural resources management) contexts. In fact, the vast majority of field work today is done within CRM. Except for very large, long-term CRM programs, the dichotomy generally works against pursuing scientific goals in the most effective ways and presents archaeology with some severe problems. We need to ask if there are measures that might be implemented within current realities which might act to lessen the gap between the two. Several possibilities come to mind. Universities should encourage graduate students and undergraduates to use CRM collections and documentation for theses and honors papers. NSF might set aside a certain amount of its budget for research that combines CRM and academic data and perspectives. Federal and state agencies might consider ways to make the archaeological information under their control more accessible to the discipline through internal reallocation of resources or through some sort of incentive plan to get contractors to prepare and submit journal articles. Perhaps subventions could be given for publications making archaeological contracting results available for peer review and incorporation into broader archaeological formulations. It could also be useful to exchange personnel between academic and heritage management contexts, or to give cross-over experiences in other ways such as encouraging CRM managers to seek research grants with salary stipends.

The review and evaluation mechanisms now in place for archaeology, in general, tend to operate at the locus of individuals and single projects. Within academia, there is a low reward for research that replicates or partially replicates previous research. Rather, the highest rewards are for innovative or novel research. While repetition of entire experiments may not be a common method of advancing the field in hard sciences, it is nonetheless the case that new information may be gained when previous results are obtained as well as when they are not.

What is weakly developed, and what has a very weak position in the organizational structure of the discipline, is a higher-order platform from which broader and longer-term concerns can be assessed. Attention to the larger scale of research needs is perhaps rendered less attractive in part because archaeologists tend to regard themselves as a group of individualists, and individualism is fostered by many aspects of field work as well as by the reward system. We can see a place for regional assessments of research needs that could be made available to concerned parties such as granting agencies.

Vital evaluation work is not getting done in southwestern archaeology. If we want to encourage firming up hypotheses, double checking results, and getting multiple perspectives on the strengths and weaknesses of an idea or model, then the rewards for individuals and projects need to be revamped to acknowledge the contributions of research that is not necessarily new and novel but that is, perhaps, a careful reevaluation of a body of data or an old idea. Not only will we have to value such research and give positive recognition to those who do it, but agencies funding research will need to acknowledge its value as well.

Part of the problem in conducting thorough evaluation procedures is situated in the preferred disciplinary form of citation and writing. We are now trained to use internal citations, and the use of footnotes has become much less common than it once was, or still is in some other disciplines. Our major journals use internal citations routinely. We encourage the short-circuiting of the scientific process through our failure to insist on clear differentiation of assumptions, enabling arguments, narrative, and explication of the reasons certain citations are included. Thus, one step we can take in our efforts to improve evaluation is to petition the editors of archaeological journals to accept and, indeed, encourage clarity through various structural means including the use of footnotes.

For philosophical reasons we have argued that the use of multiple working hypotheses and multiple lines of evidence is a useful strategy in archaeology. One way in which these multiple perspectives can be facilitated in archaeology is through rewarding publication of papers with multiple authors and co-authors. In archaeology, co-authorship is much rarer than in other disciplines in which this publishing style is effectively employed. Co-authors and co-editors may not receive appropriate recognition at tenure and promotion time, and papers with multiple authors are not considered equally weighty on C.V.'s. Within CRM, publication may not be rewarded at all and may be actively discouraged by compensating personnel only for field work, lab work, and the generation of reports that will be retained only as "gray" or submerged literature that is seldom open to peer review. With regard to the "gray" literature, the southwestern archaeological informational base is growing so large, and the hypotheses advanced so numerous, that we need to seek mechanisms to insure routine and non-extraordinary access to this rather vast data base, as well as to expose more of that data base to peer review.

When an idea is in the literature, there seems to be no easy way of eradicating it once it has been proved unworthy and/or has been repudiated by its proposer. One thinks of the concept of the Desert Culture (Aikens,[1] pp. 200–202). Interestingly enough, that phrase has tended to become useful in a more heuristic way as it has come to stand as a shorthand for an arid lands kind of adaptation. There is, however, still apparent occasional confusion over what is meant by references to the Desert Culture. Or, to take another example, the revised view of the chronology of Bat Cave,[57] and the implications of this for the history of agriculture in the Southwest, has not yet been fully internalized by the archaeological community. Such examples seem to be largely a communication and information flow problem brought about by the enormous amount of literature being generated, as well as its subsequent distribution and internalization by relevant practitioners. Some lag is certainly inevitable. The point would seem to be to shorten the lag time as much as possible.

A number of review/evaluative/problem-specific mechanisms are now employed in southwestern archaeology. We simply feel that there should be more of these and that attention should be paid to broadening the representation of archaeologists in the kinds of seminars and workshops that are sometimes seen as elitist. Some aspects of the communications problems might be addressed within the context of the periodic critical reviews which we advocate for the Southwest, again in the spirit of wanting more of these reviews rather than suggesting anything earthshaking and novel. In using the phrase critical review, we wish to emphasize both healthy skepticism and critical evaluation as well as the communications aspect. Small working conferences focused on a common theme and the southwestern overviews commissioned by the Forest Service and Bureau of Land Management are useful mechanisms for higher-order reviews. Such efforts are sometimes one-shot fixes, and the problem of up-dating the overviews, for example, or texts on a regular basis persists. However, some of the existing overviews are in the process of being updated, and this is an important precedent. Review mechanisms help keep members of the profession informed about concepts that are holding up and those that have less utility or are downright wrong.

Not surprisingly, many of these sociological problems are perpetuated in the context of training students, including, for example, the dissemination of accepted writing and citation styles. We tend to train students to be specialists in an area, and in a part of that area. There seem to be few people who are equally at home on both sides of the New Mexico/Arizona line, and so the number of people who have an in-depth knowledge of the entire Southwest are few in number. Regional or areal expertise combinations (Southwest and South Plains, Mesoamerica and the Celebes) as well as many other kinds of combinations of expertise acquired by archaeologists influence problem orientation, methodological preferences, and how likely they are to see relationships/similarities/commonalities of process in different places. Combinations of expertise affect what people are willing to consider as plausible or worth studying as well as capabilities for evaluating hypotheses.

Although the authors do not regard archaeological training as the equivalent of acquiring an intellectual straight jacket (Kuhn,[30] p. 5 and Kelley and Hanen,[29] pp. 103, 131–133), we do see training as an essentially conservative endeavor that excludes even as it opens doors. Which doors are opened and which doors are closed have sociological components as well as more logical ones. We should be careful not to confuse the inevitable limitations of training with more fundamental issues about what we accept as plausible problems and solutions in scientific archaeology.

An important part of the evaluation process is understanding the ways in which alternative models proposed for the same phenomena are actually overlapping, and ways in which they are genuinely competitive. To do this one has to dissect the models, peeling them back layer by layer. It would seem that the competing models for Chaco or hypotheses could be peeled back in this manner. The authors considered ways to separate out the component parts of the proposed alternatives in the Chacoan case only to find that the models actually overlapped in their test implications to the point that we were unable to offer suggestions for critical tests that would let us decide between them. Not only should professional archaeologists do more peeling back of models and hypotheses, but we should train our students to do it as a necessary part of formulating discriminating evaluation procedures.

We have said little about the socio-political context in which archaeology is practiced in the Southwest, or implications of this complex situation for the kind of archaeology that is done. This topic opens a new area of inquiry that is beyond our mandate in this paper. We note only that major players include universities, a multitude of federal and state agencies and programs, contracting firms, the general public and, importantly, Native Americans. This complex socio-political context has important implications for the scale and allocation of resources, access to basic data (including archaeological data that are still in the ground) research orientation, use of certain methods, and disposition of the results of archaeological investigations.

No single person, no single project, no single period of science, can ask all relevant questions and collect all relevant information and offer all relevant solutions. Selections must be made. Other questions can always be asked and other information gathered. The important thing is to be clear about the effects of making one selection and thereby eliminating many others. Consider what our views of Casas Grandes, Chihuahua might be today if the massive Joint Casas Grandes Project had NOT been directed by Charlie Di Peso (see Di Peso[14]). Di Peso was at home in Mexico, completely bilingual, knowledgeable of Mesoamerica, and committed to the proposition that people moved on the landscape at certain times and places. His vision of the past was on a larger scale than that customarily held by southwestern archaeologists. He interpreted Paquimé as a trading center which played a vital role in dynamic Southwest/Mesoamerican relationships. It is hard to think of another southwestern archaeologist who, at that time, would have adopted the perspectives Di Peso championed. Our knowledge of Casas Grandes is, to the present day, largely dominated by Di Peso's account, although new perspectives are now emerging. As long as we must say that our knowledge about the past would be different if one problem orientation was used in a project rather than another one (or that our knowledge would be different if some other archaeologist had dug Paquimé a quarter of a century ago), then we must acknowledge that sociological factors are among the many variables that play a significant role in the practice of archaeology.

While actual sociological studies of southwestern archaeology would be very interesting and valuable, our position in this paper is simply to stress the need for archaeologists to understand some of these factors in order to move from a state of muddling along to one of more conscious decision making in order to improve the level of scientific archaeology that we can accomplish with our current state of knowledge and resources.

CONCLUSIONS

We espouse the position that the goal of archaeology is to increase our knowledge about the past. We suggest that archaeology may benefit by broadening the frameworks of interpretation that are used but at the same time retaining a commitment to science and the scientific method. We propose that archaeologists pay more attention to the range of issues that is involved in the selection of problems to undertake in field work and analysis. Attention was deflected from problem selection by the philosophical position of positivism. Although we agree that the act of discovery should be kept separate from justification, the context of problem selection requires study on its own.

We hope that archaeologists will include historical and ideological factors in their research. At the same time, we recognize that these areas are particularly vulnerable to a lack of explicitness that is required for thorough confirmation. There seems to be a variety of factors related to archaeological methods and to the sociology of professional archaeology that have hindered the development of research focused on evaluation, testing, and confirmation. We have suggested several strategies, such as the use of multiple lines of evidence and working to clarify apparent cases of equifinality, that might be used to improve our record in these areas.

In order to advance our knowledge of the evolution of complexity in the prehistory of the Southwest, we advocate developing questions directed toward identifying and discriminating evolutionary change and modifying the scale at which evolutionary questions are posed. In addition to continuing middle-range research required to understand archaeological patterning, we also propose the utility of comparing archaeological data from the Southwest with a broad sample of archaeological data from around the world. Among other things, such a comparative archaeology would allow better resolution of which patterns are uniquely southwestern and which are not. Despite the high quality of the data we control and the strength of some of our inferences and models, we have indicated a series of questions that are important to the focus of this conference and which require quite straightforward observations and on-going analyses.

We have also looked at some of the broad sociological factors that influence our discipline. Some of these factors may impede advances in knowledge, and we suggest ways in which these may be mitigated or changed. A great deal of research seems to be required so that less of what we know about the past can be attributed to the personality and originality of particular archaeologists.

REFERENCES

1. Aikens, M. C. *Hogup Cave.* University of Utah Anthropological Papers No. 93. Salt Lake City: University of Utah Press, 1970.
2. Akins, N., and J. Schelberg. "Evidence for Organizational Complexity as Seen from the Mortuary Practices at Chaco Canyon." In *Recent Research on Chaco Prehistory,* edited by W. Judge and J. Schelberg, 89–102. Albuquerque: Reports of the Chaco Center, No. 8. Division of Cultural Research, National Park Service, 1984.
3. Benson, C. L. "Anasazi Social Organization and Social Structure: Evaluation of the Evidence." In *Status, Structure and Stratification, Current Archaeological Reconstructions,* edited by M. Thompson, M. T. Garcia, and F. J. Kense, 181–189. Proceedings of the Sixteenth Annual Conference. Calgary: The University of Calgary, 1985.
4. Binford, L. R. "A Consideration of Archaeological Research Design." *Amer. Antiquity* **29** (1964): 425–441.
5. Binford, L. R. "Archaeological Systematics and the Study of Cultural Process." *Amer. Antiquity* **31** (1965): 302–210.
6. Binford, Lewis R. "Post-Pleistocene Adaptations." In *New Perspectives in Archeology,* edited by S. R. Binford and L. R. Binford, 313–343. Chicago: Aldine, 1968.
7. Breitburg, E. "Prehistoric New World Turkey Domestication: Origins, Developments and Consequences." Ph.D Dissertation, Southern Illinois University, 1988.
8. Breternitz, D. "An Appraisal of Tree-Ring Dated Pottery in the Southwest." Anthropology Papers of the University of Arizona, No. 10, Tucson, 1966.
9. Coale, A. J., and P. Demeny. *Regional Model Life Tables and Stable Populations,* 2nd ed. New York: Academic Press, 1983.
10. Cordell, L. S., S. Upham, and S. L. Brock. "Obscuring Cultural Patterns in the Archaeological Record: A Discussion from Southwestern Archaeology." *Amer. Antiquity* **52** (1987): 556–577.
11. Cowgill, G. "The Trouble With Significance Tests and What We Can Do About It." *Amer. Antiquity* **42** (1977): 350–368.
12. Crown, P. "The Production of the Salado Polychromes in the American Southwest." Paper delivered at the 56th Annual Meeting of the Society for American Archaeology, New Orleans, April 1991.
13. Dancy, J. *Introduction to Contempory Epistemology.* Oxford: Basil Blackwell, 1985.
14. Di Peso, C. *Casas Grandes: A Fallen Trading Center of the Gran Chichimeca,* 13 vols. Flagstaff: Amerind Foundation, Dragoon, and Northland Press, 1974.
15. Dray, W. *Laws and Explanation in History.* Oxford: University Press, 1957.
16. Earle, T. and R. Preucel "Processual Archaeology and the Radical Critique." *Current Anthropology* **28** (1987): 501–538.

17. Flannery, Kent. "The Origins of the Village as a Settlement Type in Mesoamerica and the Near East: A Comparative Study." In *Man, Settlement and Urbanism*, edited by P. Ucko, R. Tringham and G. W. Dimbleby, 23–55. Cambridge: Schenkman, 1972.
18. Gero, J. M., D. M. Lacy, and M. L. Blakey, ed. *The Scio-Politics of Archaeology.* Research Reports Number 23. Amherst: Department of Anthropology, University of Massachusetts, 1983.
19. Harman, G. "Inference to the Best Explanation." *Phil. Rev.* **74** (1965): 88–95.
20. Hegmon, Michelle. "Social Integration and Architecture." In *The Architecture of Social Integration in Prehistoric Pueblos*, edited by W. D. Lipe and M. Hegmon, 5–15. Cortez, Colorado: Occasional Papers of the Crow Canyon Archaeological Center, 1989.
21. Hempel, C., and P. Oppenheim. "Studies in the Logic of Explanation." *Phil. Sci.* **15** (1948): 135–175.
22. Hempel, K. G. *Aspects of Scientific Explanation and Other Essays in the Philosophy of Science.* New York: Free Press, 1965.
23. Hempel, K. G. *The Philosophy of Natural Science.* Englewood Cliffs: Prentice-Hall, 1966.
24. Hill, J N. "Broken K. Pueblo: Prehistoric Social Organization in the American Southwest." Anthropological Papers of the University of Arizona **18** (1970): Tucson.
25. Hodder, I. "Theoretical Archaeology: A Reactionary View." In *Symbolic and Structural Archaeology*, edited by Ian Hodder, 1–16. Cambridge: Cambridge University Press, 1982.
26. Hodder, I. "Archaeology in 1984." *Antiquity* **58** (1984): 25–32.
27. Judge, W. "The Development of a Complex Cultural Ecosystem in the Chaco Basin, New Mexico." In *Proceedings of the First Conference on Scientific Research in the National Parks*, edited by R. Linn, 901–905. National Park Service Transactions and Proceedings, Series 5, 1979
28. Judge, W. J., W. B. Gillespie, S. H. Lekson, and W. H. Toll. "Tenth Century Developments in Chaco Canyon." In *Collected Papers in Honor of Erik Kellerman Reed*, edited by A. Schroeder, 65–98. Archaeological Society of New Mexico, Anthropological Papers No. 6, 1981.
29. Kelley, J. H., and M. P. Hanen. *Archaeology and the Methodology of Science.* Albuquerque: University of New Mexico Press, 1988.
30. Kuhn, T. J. *The Structure of Scientific Revolutions.* Chicago: University of Chicago Press, 1970. (First edition 1962.)
31. Lekson, S. H., T. C. Windes, J. R. Stein, and W. J. Judge. "The Chaco Canyon Community." *Sci. Amer.* **259** (1988): 100–109.
32. Longacre, William. "Archaeology as Anthropology: A Case Study." *Anthropology Papers of the University of Arizona* **17** (1970).
33. Miller, D., and Christopher Tilley. *Ideology, Power, and Prehistory.* Cambridge: University Press, 1984.

34. Paine, Richard P. "Model Life Tables as a Measure of Bias in the Grasshopper Pueblo Skeletal Series." *Amer. Antiqueity* **54(4)** (1989): 820–824.
35. Plog, S. E. "The Inference of Prehistoric Social Organization from Ceramic Design Variability." *Michigan Discussions in Anthropology* **1** (1976): 1–47.
36. Popper, K. *The Open Society and Its Enemies.* London: Routledge, 1966.
37. Powers, R., and W. B. Gillespie, and S. H. Lekson. "The Outlier Survey: A Regional View of Settlement in the San Juan Basin." In *Reports of the Chaco Center No. 3.*, Cultural Research Division National Park Service, Albuquerque, 1983.
38. Preucel, R., ed. *Processual and Postprocessual Archaeologies: Multiple Ways of Knowing the Past Center for Archaeological Investigations.* Occasional Paper No. 10. Carbondale: Southern Illinois University, 1991.
39. Quine, W. V. D. *Ontological Relativity.* New York: Columbia University Press, 1960.
40. Raish, C. "Domestic Animals and Stability in Pre-State Farming Societies." Ph.D. Dissertation, University of New Mexico, Albuquerque 1988.
41. Saitta, Dean. "The Poverty of Philosophy in Archaeology." In *Archaeological Hammers and Theories*, edited by J. A. Moore and A. S. Keene, 300–305. New York: Academic Press, 1983.
42. Salmon, M. H. *Philosophy and Archaeology.* New York: Academic Press, 1982.
43. Salmon, W. C. *The Foundations of Scientific Inference.* Pittsburgh: University of Pittsburgh Press, 1967.
44. Schelberg, J. "Analogy, Complexity, and Regionally-based Perspectives." In *Recent Research on Chaco Prehistory*, edited by W. J. Judge and J. D. Schelberg, 5–21. Reports of the Chaco Center, No. 8. Albuquerque: Division of Cultural Research, National Park Service, 1984.
45. Sebastian, L. "Leadership, Power, and Productive Potential: A Political Model of the Chaco System." Ph.D. Dissertation, University of New Mexico, Albuquerque, 1988.
46. Spaulding, A. "Distinguished Lecture: Archeology and Anthropology." *Amer. Anthropologist* **90** (1988): 263–272.
47. Stanislawski, M. B. "Review of Archaeology as Anthropology: A Case Study." *Amer. Antiquity* **38** (1973): 117–121.
48. Sullivan, A. P. "Inference and Evidence in Archaeology: A Division of the Conceptual Problems." In *Advances in Archeaological Method and Theory*, Vol.1, edited by Michael B. Schiffer, 183–222. New York: Academic Press, 1978.
49. Tainter, J. *The Collapse of Complex Societies.* Cambridge: University Press, 1988.
50. Tainter, J. and F. Plog "Writing and Rewriting Southwestern Prehistory." Position paper presented at the Santa Fe Institute Workshop on "The Organization and Evolution of Prehistoric Southwestern Society," Santa Fe, 1990.
51. Trigger, B. G. *A History of Archaeological Thought.* Cambridge: Cambridge University Press, 1989.

52. Trigger, B. G. "Hyperrelativism, Responsibility, and the Social Sciences." *Canadian Review of Sociology and Anthropology* **25** (1989): 776–797.
53. Van West, C. "Modeling Prehistoric Climatic Variability and Agricultural Production in Southwestern Colorado: A GIS Approach." Paper delivered at the 56th Annual Meeting of the Society for American Archaeology, New Orleans, April 24–28, 1991.
54. Vivian, R. G. "The Chacoan Prehistory of the San Juan Basin." San Diego: Academic Press, 1990.
55. Watson, P. J., S. A. LeBlanc, and C. L. Redman. *Explanation in Archaeology: An Explicitly Scientific Approach.* New York: Columbia University Press, 1971.
56. Willey, G. R. "A History of Archaeological Thought, by Bruce Trigger." *J. Field Archaeology* **18** (Spring 1991): 106–109.
57. Wills, W. H. *Early Prehistoric Agriculture in the American Southwest.* Santa Fe: School of American Research Press, 1988.
58. Yoffee, Norman. "Too Many Chiefs? or Safe Texts for the '90s." In *Archaeological Theory—Who Sets the Agenda?*, edited by A. Sherratt and N. Yoffee. Cambridge: Cambridge University Press, 1990.

Randall H. McGuire
Department of Anthropology, SUNY, Binghamton, NY 13902-6000

Historical Process and Southwestern Prehistory: A Position Paper

The feast day of San Lorenzo dawned in Picuris Pueblo on August 10, 1990, just as it had for over 350 years since Franciscan Friars built and dedicated a church there to Saint Lorenzo. By noon on this August 10, a crowd of several hundred had gathered in the bare plaza of the pueblo to observe the dances and rituals with which the people of Picuris celebrate the fiesta of their patron saint. The sun was high and the day hot. At the west end of a roped-off area stood a two-story high pole, topped with cross bars from which hung a sheep, a watermelon, and loaves of bread bound up in brightly colored swaths of cloth. At the east end of this area stood a brush arbor sheltering the image of San Lorenzo.

The crowd that had gathered was a mixed lot. Anglo tourists wandered the plaza, cameras around their necks, children in tow, constantly asking each other and anyone else "when are things going to start." Dark complected Hispano men in cowboy shirts, boots, blue jeans, and gimme caps critically discussed the reconstruction of the church in an archaic dialect of Spanish. A smattering of European tourists, German and French to judge by their speech, moved through the larger crowd. A bare handful of Indian people skirted the plaza sitting in the shade of houses and pickup trucks.

The pueblo officials carefully orchestrated the movements and actions of the crowd. A loud speaker periodically announced instructions in English, to "have a camera in the plaza you must purchase a camera permit" and "no photographs

Understanding Complexity in the Prehistoric Southwest,
Eds. G. Gumerman and M. Gell-Mann, SFI Studies in the Sciences of
Complexity, Proc. Vol. XVI, Addison-Wesley, 1994

are allowed of the sacred clowns." Several men of the pueblo wandered through the crowd dressed in plains Indian-style clothing and prominently displaying police badges.

Then a surprise. A small group of Huichol Indians, visiting Santa Fe from Mexico, came into the plaza to dance for the crowd. They were dressed in elaborately embroidered white garments, the men wore straw hats circled with turkey feathers, and they danced to the sound of a homemade violin. The governor of the pueblo asked them to dance for the saint and they did so, each kneeling in veneration of the image when they had finished.

At the west end of the plaza the crowd parted and the corn dancers entered. The men dressed in embroidered kilts, with two eagle feathers in their hair danced in one line and the women dressed in dark blue wool dresses and ankle-high white moccasins danced in the other. All carried and wore fir boughs. The audience's attention focused on the youngest dancers, barely four or five years old, at the end of the lines. The tourists jostled each other to photograph these youngest dancers to a chorus of "how cute."

The dancers wove their lines in and out, and back and forth. They seemed ready to drop, by the time the water clowns entered the plaza. Each clown wore a leather hat with two horns on top and had painted their bodies in black and white stripes. They gamboled with the dancers and each other, and occasionally paused to tell a joke in English. They exited following the two rows of dancers. In a few minutes they returned with toy bows and arrows, hunting sheep. When their toy weapons failed to bring down the ewe at the top of the pole, they engaged in a routine of pratt falls and slap stick attempting to climb the pole. One of the younger clowns moved through the crowd passing the hat. Here a gaggle of high school friends, mainly girls, recognized him. For a brief flirtatious moment, the sacred clown became an American high school kid, with a gold earring, tennis shoes, and a Bart Simpson hair cut. The moment passed and he moved on. Finally, one of the clowns climbed the pole and lowered the watermelon, sheep, bread, and cloth. The public performance complete the crowd dispersed.

Nothing particularly special or momentous happened in Picuris on this San Lorenzo day, but the activities of that day manifest the richness of culture and experience that is the Southwest. In that plaza gathered the triad of culture, Indian, Spanish, and Anglo, that we associate with the Southwest. The events, activities, and relations, that played themselves out on that day embody the conflicts and contradictions between these cultures that shaped and created the Southwest as a cultural entity. The corn dance before the image of San Lorenzo is rooted in the synergy of catholicism and native religion built with much blood and suffering nearly 300 years ago. The pueblo of Picuris existed for many years in relative isolation from a larger world and, in this time, it shared a church, life, and conflicts with the surrounding Hispano community. The critical interest of modern Hispanos in the reconstruction of the pueblo church reflects this past. For the last 150 years, Picuris has been part of the United States and under the rule of a dominant Anglo culture. For almost 100 of those years, this culture sought to erase all that was unique to Picuris, but more recently, it has turned the uniqueness and richness of Indian cultures, including the feast day of San Lorenzo at Picuris, into a marketable commodity. The Anglo spectators must be controlled by camera permits and tribal deputies, and directed by loud speaker announcements because they do not know how to properly behave and will assume they have the right to behave as they please.

We cannot make sense of these cultural relations or their role in shaping the Southwest and the events in Picuris without reference to historical processes. That is the specific series of interconnected changes through time that beget the social relations manifest on that day in August. The understanding of these processes requires that we consider particular examples of lived human experience in a larger spatial and temporal context. This context does not begin with Spanish documents or end at the borders of New Mexico. It extends back into prehistory, across the entire Southwest and beyond.

We could approach this history like Ralph Linton[1] at his breakfast table.[2] The pole climb is a custom that originates in Mesoamerica, and the items that the clowns recover from the top of the pole are all old-world domesticates brought to the Southwest by the Spanish. The plains dress of the tribal deputies is a style that developed on the great plains as the Indian groups there borrowed items of dress from their neighbors and the Americans pushing in on them from the east. The Huichols used a European-derived instrument, the violin, in their dance and decorated their hats with the feathers of a bird domesticated in the Southwest. And so on.

This is the approach to history that the New Archaeology critiqued and the stereotype of an historical approach that many archaeologists carry. As the critiques pointed out twenty years ago, it is not a very satisfying approach to understanding archaeology or culture. Diffusion leaves open the crucial questions of why some traits are accepted and others not, and how do these traits affect the societies that accept them? History in the sense of Linton is not something that explains the past or present but is, instead, something that requires explanation.

Another contemporary anthropological approach to history treats history as cultural difference.[2,11] This approach equates history with culture or structure. Study focuses on either a context of symbols[2] or a structure,[11] which is historically derived and to which human actors attach meaning. Human behavior is always meaningful behavior so that action can only occur in the context of these symbols or structure. From this vantage point the historically created culture that must precede any human action severely channels such action.

[1]Ralph Linton,[4] pp. 326–327.

[2]Ralph Linton wrote a text book on anthropology which was the most commonly used text in introductory classes from the 1940s until the 1960s. In this text he takes an "all American" breakfast of bacon, waffles, orange juice, coffee, and eggs, and shows that most of these items originated somewhere else in the world and diffused to the United States. It is one of the most generally known expositions of the cultural diffusion approach.

These contexts of symbols and structures doggedy resist change, so that shifts in culture are relatively gradual over time. Geertz[3] speaks of historical change as a continuous cultural process with few, if any, sharp breaks. A sequential account of what people did does not capture this slow process of change in meaning or allow one to finger exactly when and where change occurred. Sahlins[11] does not share Geertz's emphasis on meaning and action but focuses instead on the conceptual scheme or structure of history. He speaks of the relation and interaction of structure and event, whereby any event in history depends upon the contingent circumstances of structure which mediates how people will behave. People can creatively reconsider their conceptual schemes and they are most likely to do this when confronted by phenomena that do not fit these schemes, especially in the form of people that bear different conceptual schemes.

If we look at history as cultural difference, then the dynamic of history derives from the fact that not all groups of people share the same culture (context of symbols or structure). These symbols and structure may overlap and differ in complex ways. At Picuris, the Anglo spectators, by and large, did not seem to understand why the Huichols were asked to dance for the santo or even the significance of the brush arbor and the statue within it. They had come to see an "Indian dance." The culture of the Huichol and the Picuris overlapped in the veneration of the saint, but none of the Huichol knew the context of symbols to pull off the transformation from sacred clown to high school kid that I witnessed on the edge of the crowd. These differences and overlaps were historically created. Both the Huichol and the Picuris were subject to the conceptual scheme of Spanish catholicism and incorporated it into their own pre-existing culture. Only the Picuris have had to reconcile their culture directly with the culture of Anglo-America, while the Huichol have confronted the culture of a Mexican nation.

The events of history, like the events of San Lorenzo day at Picuris, take the form that they do because of the incomplete sharing and differences in conceptual schemes of their participants. To view history as cultural difference necessitates that we regard these patterns of symbols and conceptual schemes as hard bounded entities that have histories of their own, in some sense separable from a larger history.

The events of the San Lorenzo day celebration at Picuris are very much about boundedness, yet we also know that this boundedness is a historical product. The ability of the Huichol, Picuris, and Hispanos to maintain a distinct culture, conceptual scheme, and pattern of symbols, depends in large part on their ability to maintain a cultural boundary between them and the larger dominant culture. Yet, this larger dominant culture has not always been present and once both the Huichols and the Picuris were not under the sway of another culture. It seems likely that the boundedness we see is more a consequence of this history than simply a manifestation of clearly separable cultures, each with their own history.

[3] Geertz,[2] p. 5.

History as cultural difference also raises severe methodological problems for archaeology. It requires us to examine the contact and conflict between different conceptual schemes and patterns of symbols. This is something we are not well prepared to do. The theory lacks any significant material referent and gives us few clues on how to approach its study in prehistory, with only material culture at hand.

There does exist a third way to view historical process, that is, to look at history as a material social process.[7,10,13] A process that involves not only changes in the relations between cultural groups but also the transformation of whole social orders. To understand the events of San Lorenzo day at Picuris, we would focus on the connections between the cultural groups that were represented there. We would ask about their commonalities and differences and the larger history in which these commonalities and differences emerged. The focus on the connections between cultural contexts examines how these connections develop within a larger, unevenly developing but unified social process. Boundedness becomes, in this theory, a product of history and a phenomena that requires a study of the history of the larger whole in order to understand the boundaries that we see at any given time.

The way in which each cultural group behaved on that San Lorenzo day is a consequence of how they are connected in a modern social order. This social order is the result of a larger history of uneven development. The boundaries of this history, in both time and space, extend far beyond what could be seen on the plaza at Picuris. The presence of the Huichol and the European tourists only hints at the extent of these boundaries. This history is a story of both tradition and transformation. For example, the northern Tiwa of Picuris lived in their mountain valley at least 300 years before any of the other cultural actors of San Lorenzo day 1990 appeared on the scene. Picuris identity, ritual, and symbolism has been made and remade over the centuries since, but the process began with the aboriginal condition. The synergy of Tiwa and catholic belief reflects the conflict and contact between that aboriginal past and the beliefs, policies, and structure of Spanish culture born in Europe and itself transformed in Mesoamerica. That synergy also reflects more modern connections. The water clown society lapsed at Picuris in the 1940s after nearly 100 years of U.S. pressure on the Picuris to assimilate to Anglo-American culture. The people of Picuris revived the society in the late 1960s, at a time when the politics, economics, and nature of cultural difference, were being transformed in the U.S. social order. In both of these processes, the people of Picuris struggled to shape their history on their own terms (by their own tradition) but under larger sets of conditions that transformed what that history could be.

The historical processes that define and redefine cultural groups and the connections between them did not begin with the Spanish entrada or the beginnings of a documentary record in the Southwest. They had been going on here for thousands of years before.

Archaeologists have long known that development in the prehistoric world was uneven. They have spoken of key and dependent areas,[8] cores and buffers,[9] and heartlands and hinterlands.[1] The existence of such unevenness provides the dynamic for historical processes of change. Uneven development creates social groups within a social order that have different interests. As they attempt to act in accordance with those interests, the contradictions and ambiguities of the unevenness lead to conflict and change.

When speaking of uneven development, there is a tendency to assume that the unevenness will spring from an imbalance in economics and power which favors one center over other peripheries, much as it does in the modern world. Centrality, however, may be a product of a variety of factors or a combination there of. A social group may be central because of its position in a web of religious, economic, or political relations. One group may be the center for one set of relations, e.g., religious, while a different group is the center for another set of relations, e.g., economic.

We also tend to assume that all groups and relations can be ranked, a questionable assumption. A great number of contrasts can be made between social groups based on linguistics, culture, adaptation, religion, etc. and these distinctions may be ranked or not.[4]

How we place a social group, as central or not, depends in part on the scale at which we examine the web of social relations and what aspects of the social world we choose to look at. In the context of Southwestern prehistory, we may wish to speak of Chaco Canyon as a center, but, in terms of the Southwest and Mesoamerica, the entire Southwest must be thought of as a periphery. In the Hohokam Classic Period, the Phoenix basin was not a center for stylistic innovation, but it did have a more intensive agricultural system and greater social differentiation than surrounding areas.[6]

Marquardt and Crumley[5] speak of the "effective scale" of research, that being "any scale at which pattern may be recognized or meaning inferred." As we change the effective scale of our analysis, we frame a different web of relations. The unevenness in these relations will disappear at a different scale as a new pattern of unevenness appears. Social groups also live and act in a world of varying scales and their position viz-à-viz others changes as their scale of reference changes. Our choice of an effective scale, therefore, brackets an area for study allowing us to view a particular set of social relations while denying us access to sets visible at other scales. Also, we will find that some theoretical models are more informative at one scale and others at a different scale, so that our choice of models in part also depends on the scale of our analysis. The prehistoric world we wish to understand was a complex product of the intersection of all these scales. Thus, our studies of prehistory need to be multi-scaler.

The study of historical process necessarily involves a concern for the detail and richness of particular human experiences, such as the feast day of San Lorenzo. Such a concern frustrates the making of predictive generalizations from these particulars. It does not, however, preclude the search for pattern in historical process in order

[4] Marquardt and Crumley,[5] p. 11.

[5] Marquardt and Crumley,[5] p. 2.

to make generalizations beyond a particular case. In this effort it is probably wise to keep in mind Mark Twain's observation that "history does not repeat itself but sometimes it rhymes."[6] Four issues are raised if we discuss historical processes and the making of abstractions: (1) authority, (2) determination, (3) generalization, and (4) contingency.

AUTHORITY. We as scholars do not live outside of history. We write from a position within a social order, and like the observers in the plaza at Picuris, each of us sees the past differently and attaches different significance to the past depending upon that position. If all of us had been at Picuris on that day, we could agree on a description of the events that happened that day, the number of corn dancers, the time of day, which clown climbed the pole, etc. Any explanation for why these things happened or the meaning of these events would have to fit this description, but such a fit would not be an adequate test of the explanation because many possible explanations could fit the description. We need to also examine why we as scholars might prefer one such explanation over another.

DETERMINATION. Historical determination lies in conditioned action. Human action is cultural action and requires the prior existence of conceptual schemes, social structures, and material necessities (food, tools, technical knowledge, raw materials, etc). These prior conditions constrain how people may act and limit the possible consequences of this action but they are themselves the consequences of prior activity and thought.[7] The starting point is always conditioned action. The prior conditions of cultural life channel historical change by ruling out a large range of possible actions and consequences. These channels are, however, broad and there always remains a wide range of possible actions and consequences, some of which neither the actors or the scholars who study history can imagine.

GENERALIZATION. Generalizations about historical process can examine the pattern of process in a set of cases to determine how best to study change in other similar cases. These generalizations suggest what prior conditions, actions, and consequences we should examine to understand change in a particular case. For example, Wolf[8] observed that in societies bound together by kinship, kinship may be a means of mobilizing labor or a means of controlling access to resources. In the former case, control over the reproductive powers of women will be a key aspect to understanding change because such control grants rights to social labor. In the latter, parentage and lineage will be key because these define membership in the kin group and therefore access to resources.

Such generalization may lead to prognosis rooted in the realities of particular cases. A prognosis is *strictu sensu* a premiss as to the likely course of future events

[6] Salomon,[12] p. 208.

[7] Roseberry,[10] p. 54.

[8] Wolf,[13] p. 90.

which should be based on an analysis of the mechanisms and conditions present in the case but cannot be arrived at by simple deduction. Prognoses are derived from evidence in the real world and they are constantly modified with evidence from the real world. They cannot be reduced to timeless and spaceless generalities.

CONTINGENCY. The notion of contingency in historical process has probably been best recently expressed by Gould[3] when he notes the awesome power of the seemingly insignificant in history." The circumstances that condition human action leave broad channels and much room for actions and consequences that cannot be predicted or known in advance. Small changes in events can cumulatively have dramatic consequences over the course of history. Contingency warns us away from the fallacy of assuming that just because something happened a certain way in the past that it had to happen that way.

Developmental changes similar to historical process that we observe in other contexts did occur in the Southwest. Societies got larger, political structures became more complex, people worked harder to produce more, people learned to harness more energy, and the domination of one individual over another became institutionalized. That these changes occurred in a number of historical trajectories around the world cannot be questioned. Archaeologists usually attempt to explain such changes in terms of evolutionary abstractions. Calling this change evolution suggests that each instance of such change is simply a specific instance of a universal process of change. The commonalities that evolutionary abstraction stress obscures the differences between historical sequences which define those sequences and gives us a fragmented and delusive understanding of the past. Developmental change can only be understood in the context of real historical sequences where non-developmental change, digression, and diversity are as important to the understanding of change as the regularities and abstractions that evolutionary studies seek.

We study historical process to understand real-lived human experience and build abstractions to serve that purpose. This contrasts with other approaches current in archaeology which study particular cases to generate timeless and spaceless abstractions. History is made by the actions of real people not abstractions. People are not free to make history anyway they want. Their actions are conditioned by the material, ideological, and social circumstances of their existence, even as their actions transform these circumstances. These processes do not just manifest themselves in the great events of history, like the late thirteenth-century abandonment of the San Juan basin or the Coronado entrada, but also in the day-to-day, and year-to-year lives of people and in events like the feast day of San Lorenzo at Picuris.

REFERENCES

1. Adams, Robert McC. *The Land Behind Bayhdad.* Chicago: University of Chicago Press, 1965.
2. Geertz, Clifford. *Negara: The Theater State in Nineteenth-Century Bali.* Princeton: Princeton University Press, 1980.
3. Gould, Stephen J. *Wonderful Life: Burgess Shale and the Nature of History.* New York: Norton, 1989.
4. Linton, Ralph. *The Study of Man.* New York: Appleton Century Crofts, 1936.
5. Marquardt, William, and Carole L. Crumley. "Theoretical Issues in the Analysis of Spatial Patterning." In *Regional Dynamics: Burgundian Landscapes in Historical Perspective*, edited by C. L. Crumley and W. H. Marquardt, 1–18. Orlando: Academic Press, 1987.
6. McGuire, Randall H. "From the Outside Looking In: The Concept of Periphery in Hohokam Archaeology." In *Exploring the Hohokam: Prehistoric Desert Dwellers of the Southwest*, edited by G. J. Gumerman. Albuquerque, NM: University of New Mexico Press, 1991.
7. Mintz, Sidney. *Sweetness and Power: The Place of Sugar in Modern History.* New York: Viking, 1985.
8. Palerm, Angel, and Eric R. Wolf. "Ecological Potential and Cultural Development in Mesoamerica." In *Studies in Human Ecology*, 1–37. Anthropological Society of Washington D.C., 1957
9. Rathje, William. "The Origin and Development of Lowland Classic Maya Civilization." *Amer. Antiquity* **36** (1971): 275–285.
10. Roseberry, William. *Anthropologies and Histories.* New Brunswick: Rutgers University Press, 1989.
11. Sahlins, Marshall. *Islands of History.* Chicago: University of Chicago Press, 1985.
12. Salomon, Roger B. *Twain and the Image of History.* New Haven, CT: Yale University Press, 1961.
13. Wolf, Eric. *Europe and the People Without History.* Berkeley, CA: University of California Press, 1982.

Jonathan Haas,* Edmund J. Ladd,† Jerrold E. Levy, Randall H. McGuire,‡ and Norman Yoffee*****
*Collections and Research, Field Museum of Natural History, Chicago, IL 60605
†Museum of New Mexico, Laboratory of Anthropology, 708 Camino Lejo, Santa Fe, NM 87504
**Department of Anthropology, University of Arizona, Tucson, AZ 85721
‡Department of Anthropology, SUNY, Birnghamton, NY 13902-6000
***Museum of Anthropology, University of Michigan, Ann Arbor, MI 48103

Historical Processes in the Prehistoric Southwest

Despite the fact that anthropology took the reconstruction of the history of humankind as its primary goal at the time of its emergence as a discipline, it has embraced approaches that are essentially ahistorical. Evolutionary models of the late nineteenth century sorted contemporary nonliterate societies into categories thought to represent stages of development. Diffusionist theories attempted to make "reasonable" inferences about the past from spatial distributions of synchronic cultural data.

Archaeology developed as almost the only means to gain historical depth despite the limited nature of its data. A reaction against the whole task of historical reconstruction, however, was first voiced by cultural anthropologists of the British functionalist school. Resigned to the use of ahistorical data they turned their attention instead to a search for cultural "laws," viewing human societies as analogous to living organisms adapting to their environments. Not only was history deemed unimportant, but the vast realm of ideology that Radcliffe-Brown called "sentiment" was consigned to a back seat with little or no role to play in the shaping of social structures. Although American ethnologists as Murdock and Eggan sought to combine historical with functional approaches, the interest in history steadily declined among American cultural anthropologists almost until the present time. Now, of course, there has been an explosion of interest in various historical schools and trends: Annales approaches, "world system" and dependency theory, and ethnohistory of all sorts. Works by Wolf,[64] Sahlins,[45] and Roseberry[44] represent this rapprochement between history and cultural anthropology.

The parallel reaction against historical reconstruction took longer to emerge among American archaeologists due, in large part, to the fact that archaeological data is historical. As is well known, the "New Archaeology" flourished in the last three decades with a rhetorical bias against history. Binford and others contrasted what they called "culture history" with "culture process," the former having a pejorative connotation. Briefly, new archaeologists criticized history for being "particularistic" and not "nomothetic." This interest in laws of human behavior included a rejection of "diffusion," that is to say, roughly, that contact among peoples produced social change.[47] This position was clearly a reaction to the previous dominant trend whereby major material changes in the archaeological record were regularly ascribed to migrations of new peoples.[1] New archaeologists (if we may conflate a number of views into this generalization) posited "laws" so that social organization was regarded as an adaptation to environmental conditions and social change was viewed principally as a response to environmental change. Many of these new archaeologists were Southwest specialists, they saw these "laws" as universal but tended to perceive the world through Southwestern lenses.[31,48] The factors found to be important in shaping Southwestern developments were assumed to be equally important in other parts of the world.

Although this is not the place to write an intellectual history of archaeological theory, it should at least be pointed out that many processualist archaeologists hostile to "historical processes" were poorly informed about what historians do, how historical explanations are formed, and especially that some historians (e.g., Braudel) had even a more profound respect for the environment than any new archaeologist. These archaeologists equated history with a particularistic series of haphazard changes, lacking any sense of interdependence or process. Contrary to this characterization, we use the expression historical processes to refer simply to the specific series of changes through time that occurred in a particular historical sequence. These changes form a process because they were interdependent and contingent, each change being constrained by what had happened before and each change then constraining what could happen next. Historical process is historical because it deals with real temporal changes in specific contexts; it is only collaterally concerned with universal, abstract principles or laws that are not linked temporally in a real sequence of change. This notion of history is not necessarily new to Southwestern archaeology and, despite the anti-historical rethoric of the last

[1]A summary of Southwestern prehistory of this scope and brevity can only be controversial. In order to alert the reader to some of this controversy and to where, and about what, we are guessing, we asked the other participants in the Santa Fe Institute symposium to contribute comments. Five individuals were gracious enough to do so: Ben Nelson, Linda Cordell, Kieth Kintigh, George Gumerman, and Steven Lekson. We have incorporated their comments into the text as footnotes with attribution to the authors.

twenty years, it may be argued that most Southwestern archaeologists have been studying historical process for decades.[2]

New archaeologists and others in the last decades who were engaged in working out social evolutionary principles and trajectories tended to be ahistorical. The story of evolutionary concern in cultural anthropology and thence to archaeology, following the debates of White, Steward, and others, need not be recapitulated here. White's[60,61] general evolutio pointedly ignored historical processes, characterizing them as biographies of great men; Steward's[51] specific evolution was based on environmentally specific adaptations[3]; and Sahlins' and Service's[46] compromise position, which included specific lines within generalized evolutionary stages, still diminished historical processes as unimportant to evolutionary investigations.

One result of this rhetoric was that many Southwest prehistorians, who are drawn to study the geological, topographic, climatological, and botanical "facts" about social organizations, tended to reduce societies and modes of change to nothing more than adaptive responses to the environment. Influenced by long-standing characterizations of the Southwest as an environmentally marginal zone, in which prehistoric inhabitants had to scratch out their subsistence, agricultural societies were usually modeled as "tribal," implying little economic differentiation and no political centralization.

In more recent times, notions of environmental limitations have been greatly amended or rejected entirely. Some archaeologists find politics "phenomenally" emerging in Chaco Canyon (Irwin-Williams[25][4]), and some argue that alliances formed either to buffer the environment, to connect elites, and/or to wage war more effectively.[40,53] With these studies a new openness to consideration of "historical processes" is apparent. The point of the forgoing, of course, is not that we agree with any or all of the above new interpretations of Southwestern prehistory. It is simply that with these new views, which have broken from the strictures of what

[2] Linda Cordell: While I understand that the authors cannot present a history of American Archaeology, their view of New Archaeology is idiosyncratic to say the least. For the past 20 years, new archaeologists have accepted a systemic definition of culture, generally that of Leslie White. White describes culture as a system composed of ideological, sociological, and technological subsystems. As with any system, a change in one component will effect the other components. Social organization is not regarded as an adaptation to environmental conditions. Rather, a change in the technological subsystem will have an effect on social organization (and on ideology). New archaeology also did not reject diffusionism. It did reject diffusion as a sufficient condition to explain culture change. The authors confuse New Archaeology and environmental determinism. In fact, most early new archaeology, particularly in the Southwest, was tremendously concerned with trying to derive information about social organization from the archaeological record. Finally, the reaction of New Archaeology was explicitly against Boasian culture history, sometimes referred to as historical particularism by ethnologists as well as by archaeologists.

[3] George Gumerman: But Steward did recognize the role of historical factors in shaping the "character" of a culture.

[4] Linda Cordell: This entirely misrepresents Irwin-Williams who used the term phenomenon in order to extend the discussion to the Chacoan Outliers and roads rather than just the ruins in Chaco Canyon proper.[6]

seems to be environmental determinism, adaptive aspects of Southwestern societies can be placed in both a new historical and a new evolutionary perspective.[5]

No longer must Southwest societies be modeled by analogy to "egalitarian" societies. Neither is it necessary that they be seen as ascriptive, "chiefly" societies nor as nascent states. Whether complex Southwestern societies should be seen as stages in the development of states is now open to examination. It is through consideration of historical processes that these new evolutionary possibilities may be evaluated.

New interpretations of historical processes must be based firmly on empirical findings. Interpretations of "what happened" in Southwestern prehistory have substantially changed in the past several years, and will change again in the future. In some instances the past changes as new and better data are recovered and as better analyses are performed on data long known. One example of such a change in "what happened" is the new dates for the introduction of agriculture in the Southwest.[63] Archaeologists are also systematically rethinking the traditional nomenclature used to label Southwestern prehistoric areas. As assumptions of what these "cultures" are thought to mean—alternatively ethnic groups, ecological adaptations, spheres of interaction—have changed, so have these regional labels been called into question.[50] Yet another example involves the undoubted Mesoamerican contacts and the extent and nature of their influence on Southwestern prehistory (see papers in Mathien and McGuire[32] and McGuire et al.[35]). Previous generations of Southwestern archaeologists tended to regard such contacts as economic in nature. The transmission of elements of Mesoamerican ideology was seen as an aspect of this economic activity. While this analysis seems to have been conditioned by old evolutionary assumptions that economic and ideological institutions were systemically interconnected, new interpretations offer little reason to believe that much long-distance trade took place between the two areas.[26,35,41,43] Empirical evidence for the transmission of Mesoamerican values and beliefs to the Southwest where they were reinterpreted suggests that ideological institutions can be diffused without being ineluctably tied to economic imperatives.[6]

[5]Stephen Lekson: The marginality of the Southwest (and it surely is "marginal" compared to most other places in North America where people tried to live by farming) explains the success of environmental "deterministic" approaches. The Southwestern environment has a lot to say about what preindustrial people can and can't do here, and what they actually did. Thus New Archaeology's ecological approaches worked and continue to work.[19] Environmental "determinism" (or causal primacy), like most of our theories, is probably time/space/contingency specific and naturally—so to speak—comes to grief on a global, longitudinal sample. That doesn't mean it's everywhere fruitless: take the Southwest, for example, where the explanatory mix is heavily weighted to environment.

[6]Linda Cordell: If we wait long enough, someone will claim that the wheel is a brand new invention. In 1965, Albert H. Schroeder published a pair of articles in *American Antiquity* on what he termed "unregulated," "regulated," and "pattern" diffusion from Mesoamerica into the U.S. Southwest that examines this issue in detail.

If every generation is destined to interpret the past anew, a commitment to seeking historical processes will not liberate the theory-laden observer from his or her own place in modern time and space. Our interpretations of historical processes in the Southwest must fit the empirical findings of Southwestern archaeology; thus not all interpretations we may wish to make about this past will be equally valid. The empirical finding of Southwestern archaeology will always, however, fit many different stories; just as all interpretations will not be acceptable, we will rarely, if ever, find a single true or even best prehistory. This means that multiple histories may be written for Southwestern prehistory and that not all of these histories will be written by archaeologists. The native peoples of the region interpret the prehistory of the Southwest in a spiritual sense very differently from the analytical views of archaeologists. Native American notions of the past, their reasons for talking about the past, and the cultural significance of the interpretations they make are qualitatively different from our interests as archaeologists.[3] Native American interpretations of their pasts can come into conflict or agreement with archaeological interpretations as they both attempt to interpret the same places or series of events in the past. Archaeologists have much to learn from these histories, both in the spiritual understanding that they give and in the critique that they invite about our own ideology of the past.[7]

Finally, and perhaps most importantly, an allegiance to historical processes implies a concern with "contingent explanation." Human action is contingent because the range of possible actions that people can consider are both limited and enabled by the consequences of prior human action.[34,44] The methods that people can use to make a living in an environment, their religious beliefs, their social organization, and so on, depend upon pre-existing structures of social relations, beliefs, and technologies. These social relations, beliefs, and technologies are themselves the products of previous human action. Thus, when people act to change these things, they enable new actions and structures while they constrain others. The effects of this process can be cumulative over time. So, seemingly minor events can have dramatic consequences in the course of history. Contingent change is both determined and unpredictable. Contingent explanations, therefore, must recognize historical processes of change while keeping firmly in mind that just because history happened in a certain way does not mean that it had to happen that way.

If, as new archaeologists claimed, explanations of culture change cannot be reduced to a sequence of chronological events, neither can the past be explained without some reference to itself. Historical processes thus are not themselves explanations, but they are the raw stuff from which explanations can be formulated. This focus on historical processes has significant impact on the interpretation of material culture. In this context, formally similar distributions of artifacts may represent different social formations. The distribution of Kana-a black-on-white, for example, is as widespread as that of Jeddito yellow ware. It would be fallacious, however, to

[7]George Gumerman: Spiritual understandings of the past are derived from belief systems and such systems are, by their very nature, not subject to confirmation by testing.

infer that they represent similar kinds of "alliances." Kana-a black-on-white was manufactured throughout the area in which it is found. By contrast, Jeddito ware was manufactured in a central location and exchanged in the surrounding region. Thus, while Kana-a black-on-white reflects a widespread ceramic style used by diverse social groups, Jeddito yellow denotes an early stage in the development of Hopi ethnicity. Furthermore, Kana-a style is found on a variety of ceramic wares, but Jeddito style only occurs on Jeddito ware types.

Archaeologists now see Southwestern prehistory as a complex interweaving of environmental adaptations and social choices, of interregional contacts and regional spheres of interaction. There are patterns of sedentism, regionalization, population aggregation, systemic collapse, and demographic reorganization. All of these historical processes require "contingent explanations" and attention to the scale of history. If the Kayenta region has a common history shared by Kiet Seel and Betatakin,[21] each of these villages also has its own history.[10] Thus, both the regional and the village levels of analysis are important and interesting.

Historical processes are a growth industry in archaeology world wide[52] and surely deserve consideration in Southwestern studies. The "prehistory" of the Southwest is the history of the region and the peoples living in it before there are written records. We must be less concerned with the "typical" and more with particular sequences and variations which make the Southwest as much a unique development in world history as it is a laboratory for the study of social change.

THE ORGANIZATION OF THIS PAPER

The choice of scale and the basic periodization scheme we use in our discussion was given to us by the conference organizers. We have chosen to follow these dictates because in some ways it is best that we make our arguments for a historical analysis at this broad, pan-Southwestern scale. It is at the broadest scales that universal generalizations seem most appropriate. They tend to groan and break under the weight of specifics and the details of how people lived. It is at this broadest scale that we most need to show the importance of history.

Having chosen this scale we must beg the reader to recognize what we are painting with a very broad brush and that we have had to make many educated guesses about what was going on. We have also had to choose frequently between conflicting interpretations of the events of Southwestern prehistory. We have not fully developed our discussions of these choices and assumptions, because to do so would require a monograph. What we have tried to do, however, is to demonstrate the importance of incorporating history into broad archaeological analyses, and we have done so in the context of specific cultural events and sequences.

We have identified several general processes, or patterns of development in Southwestern prehistory. There were qualitative shifts in the development of that history, so that we are not dealing with a smooth trajectatory of growth. We do not think the quantitative shifts we identify represent all that occurred, but they are some of the major ones. The times between these qualitative shifts were not long periods of stasis. Quantitative change occurred within these periods and this change was important to understanding the transformations that followed. The process of change in Southwestern prehistory was, therefore, neither a smooth evolution nor a stair step of phases but some complex intermixture of both these processes. The seeds for the qualitative transformations occur within the variation that proceeds them. In each period there is a dominant trend that characterizes the period, but there was also much variation from that trend. In the succeeding period the new dominant trend is created out of this diversity, often on the margins of the earlier central development.

DEVELOPMENT OF REGIONAL TRADITIONS: A.D. 300–900

If we look across the Southwest at the beginning of this time period, no pattern of distinct regional traditions emerges; rather we see the end of a millennium of a pan-regional Southwestern tradition. This is the time, when on the broadest scale, there seems to be a common historical foundation for the Southwest. It is, at least to some extent, this historical foundation in the Archaic that gives a sense of common unity to the Southwest in subsequent eras and serves to differentiate the Southwest from other broad cultural traditions in North and Middle America.[24] The beginnings of plainware utilitarian ceramics and maize agriculture distinguish the Southwest from neighboring areas to the north, east, and west. To the south we do not have enough data to determine if a boundary existed.

Population is relatively low in the Southwest at this time, and in spite of some use of maize, the people are still fairly mobile foragers. Extrapolating back from modern patterns of social organization, it is likely that these early Southwestern peoples had bilateral descent organization, agamous settlements, and Steward's "family level of social integration."[51][8] To some extent the mobility and loose social organization combine to give the common historical tradition to the Southwest. Although there were local differences in the availability of different kinds of resources and in environmental conditions, no fully sedentary occupations occur. People cross multiple zones, and a family level of organization links each small family group with a changing broad network of similar groups. Few local differences can be observed among such groups.

These similarities do not, however, mean there were no differences. In ceramics, for example, the beginnings of regional traditions are to be found in the period from

[8] Linda Cordell: In this case, I actually agree with the authors, but they should be aware that there is a large body of ethnographic and ethnological literature debating the existence of Steward's "family level of socio-cultural integration."[49]

A.D. 300 to 600. People used local clays and minerals in the production of ceramics, and differences in local availability laid the groundwork for the regionally specific wares, e.g., buff wares in southern Arizona, white wares and red wares farther north. Thus, even when there is a common Southwest history on the supraregional scale, there are the beginnings of different histories at the regional scale. After A.D. 300, we begin to see the emergence of regionally distinct traditions, with the differentiation of the Anasazi in the north occurring first, then with the Hohokam and Mogollon traditions diverging shortly thereafter.[9] By no later than A.D. 500 we can now see the manifestations of multiple Southwest histories. The three broad traditions have been identified most clearly on the basis of the three classic ceramic traditions: the black-on-white wares of the Anasazi, the red-on-brown wares of the Mogollon, and the red-on-buff wares of the Hohokam.[10]

Why do these regional traditions diversify out of the earlier pan-regional archaic tradition? There is the obvious fact that the geographic/environmental conditions of the three regions are quite distinct, with the deserts of the Hohokam, the mountains of the Mogollon, and the plateaus of the Anasazi. However, it must be noted that there are Hohokam and Anasazi in the mountains, and Mogollon in plateau and desert environments. Geography also accounts for cultural differences between regions, as, for example, the Mogollon Rim tended to mark a cultural separation between northern and southern Arizona.[11] There were differences in the social milieu of the three regions as well. In particular, the Hohokam were closer to the northern borders of Mesoamerica, whereas the Anasazi and Mogollon were a large step removed from this much more complex cultural system.

In pointing out such localized conditions, we are not saying that there were no general processes operating in this pattern of diversification. Specifically, population continued to grow and continued to provide an impetus for the intensification of production of maize. This increased reliance on agriculture across the Southwest, as manifested, for example, in the appearance of trough metates, in turn resulted in increased sedentization and the "closing off" of the familial-type social networks of the previous period. Regional differentiation therefore can be attributed to much more sedentary village life, increased interaction within regions, and decreased interaction between regions. These common conditions, in addition to inducing regional differentiation at one level-resulted in significant broad similarities crosscutting the regions. Across the Southwest during the period from A.D. 300 to 600, in addition to the continued reliance on maize agriculture and sedentization, we see proliferation and elaboration of ceramics, more permanent residences, and various manifestations of subsurface architecture. Village size is relatively small, with rarely more than a few families living together in irregular clusters. There are no indications

[9] Linda Cordell: A.D. 500 is a much better date.

[10] Stephen Lekson: Ceramic "cultures" are pernicious, particularly those based largely on clay geology plus-or-minus a little technology (e.g., gray, buff, and brown wares). It is a real eye-opener to think about the past without any reference to traditional ceramic ("cultural") divisions.

[11] Linda Cordell: Is this the group who decry environmental determinism?

of centralization or hierarchy in any of the three regions, but there are some signs of communal ceremonial structures that would indicate greater integration at the local level (ball courts in the Hohokam, and great kivas among the Mogollon and Anasazi). On a broad scale, these commonalties indicate that the people of the Southwest were facing similar kinds of general environmental/demographic technological conditions and were responding to those conditions in similar ways.

Nevertheless, the different histories were set in motion, and the local manifestations of adaptive strategies were regionally specific. As mentioned above, the regions are characterized by distinct ceramic traditions, and these ceramic traditions are directly derivative from the earlier resource-locality-dependent plainware assemblages. There are also differences in the kind of designs that appear in the first decorated ceramics in each area. In the Mogollon and Hohokam areas, there are quartered designs that are quite similar to ceramics found further to the south and into Mesoamerica.[12] In the Anasazi a complex of relatively simple designs are largely unique to the northern whiteware complexes.[5,30] While difficult to interpret, these specific differences do suggest that early in the histories of the various traditions, ceramics may be assuming different social meanings and cultural roles. Stepping above the level of ceramics, the villages of this period are different in the three regions, with distinctive regional types of pithouses and ceremonial structures. Hohokam, for example, have roundish or "kidney bean"-shaped "houses-in-pits," while the Mogollon have square or rectangular "pithouses" with long ramp entries.[59] There are some indications that these differences can be attributed to residential longevity, environmental conditions, and/or simply human idiosyncrasies. While there has been some work to demonstrate that there are functional advantages to rectangular versus round residences in terms of storage and expendability, these advantages do not really seem to apply to pithouse architecture.[37] At the same time, it should be recognized that the appearance of rectangular pithouses in the Mogollon and Anasazi historical sequences is in a way "preadaptive" to the move into rectangular surface structures in later centuries.

Broadly surveying the Southwest from A.D. 300 to 600, it is apparent that the underlying historical foundation and broad technoenvironmental conditions provide a basis for similarities across the different regions. It is equally apparent that a variety of factors have set these regions off into distinct historical trajectories, and these different histories must be taken into consideration in trying to understand patterns of regional similarities and differences. The importance of the different histories that first emerge at this time period becomes increasingly evident as we move up to the next period from A.D. 600 to 900 and begin to see very different historical and evolutionary developments.

Without trying to assign an absolute date or offer a new stage concept, the A.D. 600 to 900 period witnessed the continued differentiation and consolidation of the three major cultural traditions, Anasazi, Mogollon, and Hohokam.[7] There were more localized traditions emerging in different parts of the Southwest—the Sinagua in central Arizona and the Trincheras in northern Mexico, for example—but these were not of the same extent or magnitude as the "big three."

Ceramics continue to be one of the most visible marker of the major traditions, and they are now quite different and distinct in each.[5] The regional decorated ceramic traditions are characterized by significant commonalties of designs and by synchronous change in these designs over time. These pan-regional ceramic similarities are indicative of both a broad network of communication and interaction across the three areas, and in a real sense stand as historical records of the three different traditions. Not "readable" historical records, but nevertheless material manifestations of a communication system and in a sense symbolic of the emergent multiple histories of the Southwest.

In other classes of material culture, the differences in the respective histories are much clearer in this period than earlier. In architecture, for example, the Anasazi move from pithouses into above-ground architecture, and establish much more formally organized villages. Hohokam and Mogollon stay in pithouses, with the Hohokam retaining the roundish form, and the Mogollon shifting from roundish structures to rectangular/squarish structures. These architectural patterns are indicative of the continuing differentiation of the three traditions, and represent the product of different decisions made by the people of the areas in response to particular demographic/environmental conditions. All three groups were facing similar kinds of problems in extracting sufficient resources from the environment to support their growing populations, but they pursued different options in their efforts to solve these problems.

In the Hohokam area, for example, the environment and topography provided ideal conditions and stimulus for the development of intensive irrigation agriculture and the people took the option of intensifying production through irrigation to provide more resources. In contrast, in the Anasazi area large-scale irrigation was much less of an option, and the people had to pursue other means to produce more resources for the growing population. Intensification in this area appears to have consisted of placing more and more land under cultivation and increased attention to fields (as manifested in increased sedentism). The move into above-ground architecture must also be seen in terms of the particular Anasazi adaptive response to resource pressures. They needed to find a means to average out temporal and geographical variability in the availability of both domesticated and wild resources, and surface structures increased both the capacity and efficiency of storage for the Anasazi people (see Whalen[58] for a discussion of this transition in the Mogollon area). Although the Hohokam did move into above-ground architecture several hundred years later, the historical circumstances surrounding this move were significantly different from those found in the Anasazi area in the eighth and ninth centuries. It may be possible to develop models to explain the move from pit houses to surface architecture cross-culturally and cross-chronologically. However, explanation of this general pattern at the local level must include the different historical roots of the various local traditions.[12]

[12]Linda Cordell: This is a good example of the kind of confusion generated by "historical explanations." To offer some perspective, I cite an example from paleontology. Stephen Jay Gould,

The differences in architecture and ceramics are but part of the cultural and historical differentiation of the three traditions in this period. We also see distinctive mortuary practices, communal/ceremonial structures, and community layout. Differences in trade and interaction patterns represent a particular point of historical divergence for the three traditions. The Hohokam continued to interact with the Mesoamerican sphere while, for the other areas, there was at most incidental contact.

The process of differentiation that we see separating the Hohokam, Mogollon, and Anasazi continues with each of these regions through the 600 to 900 period, and the grand traditions seen on the largest scale do not remain coherent, homogeneous historical entities. We see the rise of localized traditions, such as the Kayenta and Mesa Verde Anasazi, the Tucson Basin and Tonto Basin Hohokam, and Forestdale and Mimbres Mogollon, and each of these has its own history at the subregional level. It may be possible to address some issues, such as the emergence of unit pueblos, on a regional level, without reference to the specific historical paths of the local traditions. Other issues, such as settlement pattern or ceremonial practices, must be addressed within the context of local historical and environmental conditions.

Up until A.D. 900 in the Southwest, we see a continuing process of regional and subregional differentiation and the diversification of an increasing number of localized historical traditions.[13] In this sense, the common historical trunk in the Southwest continues to play a role in shaping the development of cultural systems at the regional and subregional levels. The variability seen in the multiple historical branches nevertheless has significant influence on the specific manifestations and adaptations of the supraregional patterns. In the following period, the differences in the historical branches lead to profound shifts in the trajectories of the different regional traditions because the similarities between these regional traditions obscure fundamentally different phenomena.

well known for his love affair with contingencies, is perhaps at his most ebullient in an essay on "George Canning's Left Buttock." In this ode to the accidents of history, he recounts in all its convoluted wonder, the relationship between a duel in which George Canning is shot in the ass and Charles Darwin's invitation to sail on the *Beagle*. The point being that Darwin's voyage and his observations of nature from the *Beagle*, which Gould considers the "sine qua non of Darwin's revolution in thought," were the result of "quirks of circumstance."[17] However, Gould then states: "A theory of evolution would have been formulated and accepted, almost surely in the mid-nineteenth century, if Charles Darwin had never been born, if only for the simple reason that *evolution is true*."[17] (italics mine).

Rarely can archaeology retrieve data at the level of George Canning's left buttock, but a cross-culturally, cross-chronologically valid explanation for the move from pithouses to surface architecture within a broader explanatory framework might also be *true*!

[13]Stephen Lekson: Oddly enough, Haury and a few others (including me) see a convergence, not a regionalization, in Mogollon and Anasazi at A.D. 800 or so. Haury might say Mogollon was swamped by the Anasazi (an argument that should not be cavalierly dismissed). I say they were both the same thing anyway. Mimbres, at A.D. 1000, looks like "Anasazi" with the addition of longer growing seasona and a nice little creek to irrigate from, and the absence of sandstone to build with. It's all the same thing—people making a living in the piñon-juniper zone—with a few local twists.

CHACO, MIMBRES, AND HOHOKAM: A.D. 900–1200

The years from A.D. 900 to 1200 saw the development, and demise, in each of the major traditions of elaborate regional centers that formed the core of highly centric, regional systems. These centers did not result from some dominant or common trend in each tradition, but rather emerged from the elaboration of local developments in one of the subregions of each tradition. The three regional systems, Hohokam, Chaco (Anasazi), and Mimbres (Mogollon), are all quite large by Southwestern standards. The Chaco Regional System covers an area of about 75,000 km^2,[14] the Hohokam regional system an area of about 100,000 km^2[15] and the Mimbres an area of about 56,000 km^2.[27] The boundaries of each of these systems are marked by the distribution of architectural features, ceramics, or site types. The boundaries so defined are seemingly more distinct than those of the earlier traditions and give the impression of a separation between societies outside the system and those within it. This situation differs from earlier times during which traditions faded one into the other and there were no clear boundaries. Each of these systems also has a clear core or center, with a peripheral area surrounding that core. The defining characteristics of the system seem to originate in this center and the peripheries appear to be linked to it. Most archaeologists think that these systems united multiple ethnic and cultural groups. Despite the size of these three systems most of the Southwest—definitely in terms of area and probably in terms of population—lies outside of these systems.[16]

The apparent commonalties between these systems, especially the Hohokam and Chaco, invites us to treat their appearance, growth, and demise as a manifestation of an underlying process that can be described in terms of a common set of abstract principles. We cannot demonstrate strong (or any!) patterns of interaction and exchange between the Hohokam and Chaco to account for commonalties between the systems. While some of these commonalties spring from common constraints and similarities in material condition, we need to recognize that each of these systems represents different histories. Much of their commonality is a superficial product of different historical but convergent evolutionary processes in each area.

[14]Stephen Lekson: Some of us think that the Chacoan pattern covered a region at least twice that large. See also Vivian.[55]

[15]See Crown,[8] p. 224.

[16]George Gumerman: At the same time the A.D. 1000 to 1150 period was one of extreme differentiation and localization outside of these areas.

HOHOKAM

The core or center of the Hohokam regional system lies in the Phoenix basin.[8,15,20,23] This area has large sites that contained up to 500 scattered houses and that were internally differentiated with domestic and ceremonial precincts. Settlements existed in a hierarchy: villages with multiple ball courts, villages with a single ball court, and settlements lacking ball courts. A distinctive mortuary assemblage, that includes censors, palettes, etched shell, and shell trumpets, occurs in Hohokam cremations. By the end of this time period, the irrigation network extends to its maximum size and includes canals of up to 32 kilometers in length and 23 meters in width, that carry water to fields up to 11 kilometers from the river. The Hohokam strung their settlements out at regular intervals along these canals; the larger villages include ball courts and platform mounds. There is little evidence for an overall political organization that united all of the basin. Ceramic styles, iconography, and aspects of material culture (such as ball courts, copper bells, and macaws) suggest indirect linkages to Mesoamerica, or at best to a chain of sites in Sonora.[39]

A social and ideological glue appears to bind the system together, with ball courts being the physical focus of these relations. By the 1100s the over 206 ball courts contiguously distributed in over 165 sites define the limits of the system.[17] In addition, a few ball courts occur at possible trade outposts, just beyond the edges of the system. The Phoenix Basin Hohokam obtained marine shell from the Gulf of California, manufactured it into shell jewelry, and exchanged it over an area slightly larger than the range of the ball courts. Most shell production took place in the Papaguerian periphery with some production in the Phoenix and Tucson Basins. Despite the ideological and exchange relations that connected the peripheral areas, considerable variability exists between these areas and the core. The material culture of peripheral areas becomes increasingly different from the Phoenix basin over this period so that this variability increases over time.[33] The formation of new social forms in our subsequent period originates in this variability.

CHACO

The core of the Chaco Regional System lies in Chaco Canyon at the center of the Chaco Basin on the Colorado Plateau.[55] Here the Chacoans built 12 "Great Houses," massive buildings ranging in size from around 200 to 800 rooms.[28] These Great Houses look like pueblos with square or rectangular rooms and round kivas. The majority of the rooms are much bigger than any others that we see in the prehistory of the upper Southwest and most appear to have been storage rooms. The lack of firepits in most of these rooms has led many researchers to conclude that these structures housed far fewer people than the number of rooms would

[17] See Crown,[8] p. 233.

suggest.[18] A variety of artifact types, including cylinder vessels, shell, cloisonné, parrots, and copper bells occurs exclusively or predominantly in the Great Houses in the canyon. Most of the "Great Houses" contained one great kiva, and two of them included two of these large communal structures. Scattered among these "Great Houses" are a variety of special architectural features including platform mounds and tri-walled structures.[19] In addition, three isolated great kivas occur along the canyon wash. Smaller village sites also occur in the canyon, primarily along the wash's southern bank.

In the canyon, agricultural practices were intensified. Small-scale but sophisticated water control systems collected water from the mesa tops and delivered it to fields in the canyon below.[54,55] Seven roads that link the canyon to outlying settlements radiate out from the canyon. These roads may be several meters wide and up to 80 kilometers in length. They run in straight lines across mesas and washes and are built to maintain level surfaces.[20] The settlements at the ends of these roads (and beyond) show a consistent pattern. They contain Great House architecture and are built in the midst of small villages like those in the canyon.

The canyon seemingly acted as a magnet for goods from within and beyond the system. Large quantities of painted ceramics entered the canyon from the outliers. The Great Houses used thousands of fir and pine beams, the vast majority of which had to be brought in from mountains at the ends of the road network. Turquoise came from the Cerrillos mine near Santa Fe. Shell from the Gulf of California most likely came via the Mimbres area. The most exotic objects—copper bells, macaws, and cloisonné—originate in Mesoamerica. Mesoamerican objects and architectural features (such as colonnades) appear as rare elements in the Chaco culture in contrast to their integral role in the Hohokam pattern.

MIMBRES

The Mimbres regional system was considerable smaller and less developed than the Hohokam and Chaco systems but still bigger than other local traditions and possessing similar characteristics of centrality.[27][21] The core of the Mimbres system

[18]Stephen Lekson: There's a nice stone-lined hearth in every "kiva" at Chaco Canyon, but apparently those don't count.

[19]Linda Cordell: As I understand Great House architecture, the Chacoan Great Houses and the tri-walled structures were not built at the same time—about 100 years or more may separate them. Also, not all Outliers were built in the midst of small villages. Finally, a lot of ceramics were moved into Chaco from some place but not, apparently, from any of the Outliers.

[20]Linda Cordell: The roads may maintain level surfaces. That they were built for this purpose is unknown.

[21]Ben Nelson: Chaco and Hohokam seem to constitute a different order of social relations than Mimbres. I'm not all that comfortable with the notion of packaging them all together. Mimbres is an early instance of aggregation, but other than its remarkable art style, it has no corporate character or suggestion of supra-village sociopolitical integration like the other two.

lay in southwestern New Mexico where pueblos of up to 300 rooms were built along side fields irrigated by canals drawing water from the rivers. The most distinctive marker of the Mimbres system is a finely made black-on-white pottery painted with fanciful zoomorphic and anthropomorphic designs.[4] The Mimbres people trekked to the Gulf of California to gather marine shell that they made into jewelry in the Sonoran style. They may have traded this jewelry into the Chaco system.

INTERPRETATIONS

Of these three systems the Hohokam and Chaco have received the most attention and explicit comparison in the literature[9] and we focus the remainder of our discussion on them. We can see a number of commonalties in cultigens and technology between the Hohokam and Chaco. Both areas depended upon the triad of corn, beans, and squash and therefore had to address the biological requirements for growing these plants and the nutritional consequences of dependence upon them. The two regions were much the same in terms of similar limitations on transportation and technologies of exploitation.

The differences between the regions can be attributed their specific histories. Aspects of Chacoan culture, such as black-on-white pottery, kivas, ceramic styles, and pueblo construction, are directly related to the Anasazi past that was Chacoan history. Similarity, red-on-buff pottery, cremation burial and ceremonialism, and clay figurines were inherited from the Hohokam past. Also, from that past came several hundred years of Hohokam links to Mesoamerica, whereas the first good evidence of Mesoamerican items in the northern Southwest is at Chaco Canyon in this period. Thus it is clear that Mesoamerican features appear integral in the Hohokam while they are overlays in Chaco Canyon. Without reference to these factors, we cannot account for the fundamental differences in architecture, style, and material culture between the two systems.

Whereas commonalties between Chaco and the Hohokam are clear and important, it is also apparent that distinct historical processes shaped the lives of the inhabitants of the two regions. Indeed, we propose, admittedly in a speculative vein, that these differences arose in particular ecological circumstances and gave rise to certain kinds of social choices. As Crown and Judge[9] point out, while the Chaco system developed to manage scarcity, the Hohokam flourished by managing abundance.

In the ninth century in the San Juan Basin, the vast majority of the population was organized in small social groups, perhaps lineages or clans. Since the environment of the San Juan Basin is marginal to corn agriculture, its suitability for cultivation fluctuates widely from year to year. Any agricultural society that lives in the region must deal with episodic scarcity. Increases in population, long-term trends of environmental deterioration, or both will increase the extent of scarcity. One effective adaptation to this scarcity is a high degree of mobility, which we see in the Anasazi period as these lineage or clan groups move in and out of areas in response to increasing or decreasing moisture conditions.

By the end of the ninth century, such mobility may have been constrained due to increasing population density. Groups that moved into marginal environments in a wet period might not have been easily able to move back in a succeeding dry period because of crowding in the better watered areas. Most archaeologists agree that the Anasazi tradition included a number of linguistic, cultural, and ethnic groups.[6] Such social boundaries as may have existed would also have inhibited the movement solution to scarcity. Groups in the more marginal areas of the San Juan Basin, like the Chaco Basin, would have faced the dilemma of coming up with a social means of managing scarcity, as opposed to movement, before those groups in more well-watered regions like the San Juan River Valley faced this dilemma. Indeed, if, as several authors have suggested[16] Chacoans were Keresen speakers, they would have faced linguistic boundaries to the north and south, in which regions probably lived Tanoan and Zuni speakers.

Most theories for the origins of the "Chaco Phenomenon" include some mix of population pressure and environmental fluctuation.[9,55] They are concerned with how the Chacoans came to manage scarcity. In the historic western pueblos, the management of scarcity created a social contradiction. Although cooperation was necessary for the economy and society to work, resources could not be distributed evenly because, in times of extreme scarcity, starvation and destruction of the social order would be the result of such egalitarian distributions. The resolution of this contradiction lead to the Hopi type of social organization which is neither egalitarian nor stratified.[29] Powerful clans controlled both the best agricultural lands and the ceremonial cycle of the villages, while poor clans held inferior lands and had only minor roles in the ceremonial cycle. In good times, social relations and ideology stressed egalitarianism, cooperation, and peaceful relations between all members of the community that urged all to work for the common good. When insufficient rains fell or the frost came too early, the powerful and ceremonially more important clans had food and stayed in the village, while poorer clans were forced out to hunt or gather or depend upon the charity of the Navajo or other pueblos.

The people of Chaco faced the same problem as did the Hopi, since the need to cooperate, as so often discussed in reference to the water control systems, came into contradiction with the need to exclude people from the society in the event of extreme scarcity. Their resolution of the problem was different from the Hopi and seemingly involved a centralization of ritual and resource storage combined with a system of outliers. Thus, resources could be moved into the core and people could be removed as fluctuation in scarcity required. It might be erroneous, therefore, to interpret the "complex" social organization of Chaco Canyon by such a simple notion as ranking when it is clear that the Hopi developed a very complex structure of cooperation, inequality, power, and ideology to maintain their simplicity.

The resolution of the contradiction might have also sowed the seeds of the collapse of the system. By linking the canyon to better watered areas, such as the San Juan River, the Chacoans could have, in part, managed scarcity by drawing on the abundance in those regions. Growth also could have occurred for reasons other than those responsible for the rise of Chaco. Movement of population out of the canyon and increased production for the canyon would have encouraged population growth and pressure on resources leading to scarcity in outlying populations which probably could have been managed if it hadn't been for the pressures and obligations imposed by the canyon center. Drought would have served to hasten this process while wet years would have retarded it.

It must be recognized that if the origins of the Chacoan system are to be found in the management of resource scarcity, this was not the only community struggling with resource scarcity in the San Juan Basin. There were other such "supra-communities" struggling with similar problems, and they picked different solutions. In the case of Chaco, the solution they selected initially in response to environmental conditions had tremendous ramifications that rippled far beyond the "goals" of the initial socioeconomic experiments.

When the Hohokam began canal irrigation, they also created the possibility of abundance in the Phoenix Basin. The major limiting factor for agricultural production was water. Using canals, however, two crops a year could be produced.[1] Investment of labor in extending canals or improving the water delivery through them would bring substantial rewards from production. This stands in contrast to the marginal returns that Anasazi water control features might yield. As long as there was land that could be reached with gravity-fed canals and sufficient water was present in the rivers, abundance could be produced and increased through greater labor investment. The Hohokam in the Phoenix basin, therefore, faced the problem of managing labor to increase abundance. The basic principle of organization may have therefore been competition for labor as opposed to cooperation to manage scarcity.

In this context, in the ninth century, courtyard groups showed a consistent pattern of growth and decline. They began as a single "founder" house and then additional houses are added to form the courtyard over a span of 20 to 50 years.[42] Some founder houses failed and no courtyard group formed. Some courtyard groups were larger than others and some survived for longer periods of time, up to two generations, than others. These patterns suggest a process by which a courtyard leader tried to increase the labor force of the courtyard through fertility and by attracting new members. Courtyards with more labor could invest that labor in canal-based agriculture and gain abundance. The variable success of courtyards, as measured by their continued growth, suggests that competition existed between the groups. Such competition would have fueled growth as canals were extended and populations enlarged.

In the ninth century, these courtyard groups clustered in villages with a communal crematorium and cremation mounds. Wilcox[62] has suggested that this pattern reflects an ideology of ancestor worship with communal cremation ritual. During this century the pattern of religion, ideology, and ritual changes. Large ritual structures, ball courts, were constructed and they may have involved more people in public rituals providing an integrative mechanism to link competing courtyard groups. This ritual probably involved competition since the ball game throughout Mesoamerica involved a competition that resolved conflicts between social groups and extensive betting between social groups. The ball court game and ritual then became a mechanism for extending ritual, trade, and competitive relations over what would become the Hohokam regional system.

A contradiction existed in the social relations that fueled growth and the ecological relations in the region.[35] The growth engendered by the competition between courtyard groups reached a point sometime in the twelfth century where more water could not be obtained. In this context the critical variable for growth becomes control of land and water. New courtyard groups could not be founded while existing courtyard groups could solidify any advantages gained by labor in land or water into permanent advantages. A transformation of the social system and ideology, which involved the collapse of the ball court ritual and the freeing of peripheral areas of these ties to the core.

OUTSIDE OF THE SYSTEMS

Most of the Southwest lay outside of the three regional systems. In the rest of the Southwest local traditions continued and they may have become more bounded and distinct through this time period. The distinctions of Mogollon, Anasazi, and Hohokam become blurred beyond recognition in this variability. These local traditions and the peripheries of the regional systems survived the collapse of those systems and in some cases thrived from that collapse. The general pattern that we see in the next period develops out of the local pattern and variability that characterizes these traditions and peripheries.

These three highly centralized systems all breakdown in the middle of the twelfth century A.D. There is a major reorganization of sites in the Hohokam at about A.D. 1150 with a shift to above-ground architecture, a decline in the ball court ritual complex, changes in mortuary practices, and stylistic changes in artifacts.[8,15,20] The archaeological pattern of the peripheries in the Hohokam sphere had been diverging from the Phoenix basin core and by A.D. 1150 we see a variety of regional traditions rather than a highly centric system in the region.[33] The core of the Chaco system, Chaco Canyon itself, loses population in the late twelfth century and everyone has left by the early fourteenth century.[55] The major sites at the core of the Mimbres system are abandoned by the middle of the twelfth century.[38] As was the case with the Hohokam, the peripheries of both the Chaco and Mimbres systems become increasingly distinct, new regional centers arose, and these peripheries entered into new sets of social and cultural relations remaking the social structure of the Southwest.

POLYCHROME TRADITIONS: A.D. 1200–1450

In the centuries following the demise of the Chacoan, Mimbres, and Hohokam systems, several areally extensive regional networks appear in the Southwest, each associated with a distinctive ceramic style. The distribution of these ceramic styles freely crosscuts earlier regional and subregional boundaries, and is a further manifestation of the role of the earlier regional systems in breaking down those boundaries. Also, large areas of the Southwest are abandoned and sweeping gaps begin to open up between sedentary farming populations.

On the surface these networks have certain similarities: they all make relatively sophisticated decorated ceramics; they are all distinct from each other and from their antecedents; they all entail wide-spread population aggregation; their ceramics are widely traded outside their production centers; and a common set of design motifs cross cut the ceramic types. These similarities imply that these are structurally or functionally similar kinds of cultural systems, e.g., ethnic groups, alliances, or religious societies. However, when each network is examined separately, it appears that they represent two different kinds of phenomena: ethnogenesis (the formation of distinct ethnic groups) and multiethnic sociopolitical interaction spheres. The Jeddito, Zuni, and O'otam networks appear to fit the former pattern of emergent discrete ethnic groups, while the Salado, White Mountain, and Rio Grande networks appear to reflect interaction spheres cross-cutting (but never uniting) multiple ethnic boundaries.

The Mesoamerican societies in Sinaloa and Durango continued to prosper after A.D. 1200 but go into decline and disappear by A.D. 1350. In the Southwest the strongest evidence of Mesoamerican contacts both in terms of goods and styles shifts to the major regional center of Casas Grandes in northwest Chihuahua. This center develops after A.D. 1200 in what had been a peripheral area of the Mimbres system, and is abandoned by the late fifteenth century (for dating see Dean and Ravesloot[11]). It is the largest site in the prehistoric Southwest with massive, multistory, adobe, apartment blocks numbering over 2,000 rooms, ball courts, a ceremonial precinct, and an estimated population of almost 3,000.[12] The people of Casas Grandes also produced goods that had previously only come from Mesoamerica (e.g., copper bells and macaws).

NETWORKS IN THE LOWER SOUTHWEST

The largest of these networks was the Salado network which seems to come together in the mid- to late thirteenth century. This network was anchored on the west by the Classic Period society of the Phoenix Basin and on the east by Casas Grandes. The reorganized Phoenix Basin continued to grow with a consolidation of the canal system, the concentration of population in over 40 villages, the largest of which contained 35 compounds spread out over 2.5 hectares, elite residences on platform mounds, and special administrative sites.[8] A great arch of multistoried pueblos and large compound villages of up to several hundred rooms stretched between the two regional centers. The network appears to have been based on a weakly linked system of exchange and elite intermarriage between smaller independent regional polities.[62] The material culture of the system included a shared polychrome pottery style, and a set of elite goods, turquoise on shell mosaic, copper bells, macaws, Strombus shell trumpets, asbestos, and some type of shell beads. Utilitarian items including pottery, however, tended to vary between regions depending upon cultural and environmental factors.

This network did not develop in a vacuum but in relation to other networks of relations adjacent to it. To the north of the Salado network, a Little Colorado network stretched from western New Mexico into Arizona along the Mogollon uplands.[53] This network appears to be similar to the Salado, a weakly linked system of exchange and perhaps elite inter-marriage that cross cut different local polities and cultural boundaries. The Little Colorado network consisted of a string of large pueblo sites with up to hundreds of rooms. In what had been the southern peripheries of the Hohokam system and the Trincheras tradition, an O'otam network appears, defined by the distribution of a brownware pottery, fortified hill sites (Cerros de Trincheras), and a relatively uniform material culture inventory over the entire area.[13,33] It has been suggested that this network represents the efforts of people who share a common ethnic identity or language (in this case O'otam) intensifying these bonds to organize in opposition to the power center of the Salado network.[33] A similar ethnic-based Hakataya network may also have formed to the northwest of the Salado system.[[22]]

We have only a limited knowledge of what is happening in contemporary societies further south. On either side of the Sierra Madre Occidental, there were Mogollon-like local traditions, the Rio Sonora on the west slope and the Loma San Gabriel on the east. Each of these traditions consisted of scattered small settlements with unpainted brownware potter. In both Sonora and Chihuahua, a gap of two to three hundred kilometers with no major regional centers separated the regional networks of the Southwest from the northernmost Mesoamerican regional centers.[33]

NETWORKS IN THE UPPER SOUTHWEST

The upper Southwest witnessed a dramatic series of population movements and reorganization at the end of the thirteenth and the beginning of the fourteenth century with the abandonment of the four corners area. Sedentary agricultural populations moved off the Colorado Plateau and into the Rio Grande Valley, leaving farming societies only in northeastern Arizona (Hopi) and far western New Mexico (Zuni). Multiple distinct local traditions (presumably representing different ethnic and/or

[[22]] Linda Cordell: In a paper well known to one of the chapter's authors, Michael R. Waters[57] argued persuasively that the Hakataya, as defined in the literature, did not exist. Now, they are apparently an ethnic group.

linguistic groups[23]) moved out of the Four Corners region and merge into a relatively few already established traditions to the south and east. Interestingly, with few exceptions (such as the Maverick Mountain occupation at Point of Pines), there is very little direct evidence of the actual movement of Four Corners people into these established cultural and population centers. What we see is a sharp decline in the population of the Four Corners region and a concomitant increase in the population in these outside areas, such as the Rio Grande Valley. Concurrently a new religion, the Katsina religion, appears cross-cutting the various ethnic groups of the upper Southwest.[2] This religion included a distinct set of symbols and beliefs, and rituals, most notably masked dancers. Some of the icons and beliefs of the religion ultimately originate in Mesoamerica, but they are reworked in the Southwest into a religion that is distinctively Pueblo.[2,5] Along with this new religion came new forms of social organization and pueblos became quite large and remarkably similar in layout. We see at this time the coalescence of a Pueblo culture and the establishment of the historic division between eastern and western pueblos.

If we take the Jeddito yellow ware complex as a manifestation of the process of ethnogenesis in the Hopi, it must be seen as beginning with the massive population movements at the end of the thirteenth century. Along the Little Colorado River and on the Hopi Mesas, populations grew as people left Black Mesa, the Grand Canyon, and the Flagstaff region. With new immigrant populations, improving environmental conditions and relatively empty borderlands on two or three sides, the Hopi ethnic and social personae emerges from already existing ethnic and linguistic distinctions. Following Adams,[2] the Hopi persona arises specifically in the context of the developing Katsina cult which was integrated with the older tribal initiation and Soyal ceremonies. The Katsina cult is an integrative mechanism that serves to unite diverse displaced groups into coherent local ethnic/tribal groups. Thus in northern Arizona, with the surrounding blanket of the Katsina Cult, we see the coalescence of what are known today as the Hopi, and Jeddito yellow ware is the material representative of the emergent Hopi ethnic group. In this case, the distinctive black-on-yellow pottery is largely made only in the Hopi heartland and traded out. Its widespread distribution can be seen at least partially as a response to the culture vacuum created by the widespread abandonments, and may also reflect the formation of some kind of alliance relationships between the Hopi and neighboring peoples. We are not prepared at this point to argue the question of a far-reaching Jeddito alliance,[53] but wish to point out that it represents the concatenation of a discrete set of historical circumstances and cannot be understood independent of these circumstances with only a model of alliance formation based on economic regularities and political domination.

The Hopi are not the only people who were reorganizing and not the only people using the Katsina cult to achieve their reorganization. This set of processes was going on at Zuni, in the Acoma region, and in parts of the Rio Grande Valley, for example, as these areas also develop local ceramic, cultural, and historical

[23] George Gumerman: Maybe.

traditions, but they never achieve the widespread distribution of the Jeddito yellow ware. Although in each case the people were being subjected to similar social, demographic, and environmental conditions, their respective adaptations were individually distinct and historically contingent.

Along the Rio Grande a regional network developed that linked different ethnic and linguistic groups. All along the Rio Grande in New Mexico, large pueblos were built, a few with up to 1,000 rooms.[6] These pueblos consisted of multiple multistoried apartment blocks organized around central plazas with one or two large kivas. A suite of pottery types, the Rio Grande glazewares, were manufactured in some villages and then traded throughout the region. Each major village appears to have been politically independent from others, but alliances between villages may have existed. There is much shifting of population within the region and some evidence of warfare. The western Pueblos, Hopi and Zuni, clearly were interacting with the Rio Grande but do not appear to have been part of the Rio Grande network. Each of these areas had their own ceramic style and internal sets of relations.

THE TRANSITION TO HISTORIC PERIODS (A.D. 1450–1600)

During the fifteenth century the face of the Southwest was once again transformed. The Salado network and the Little Colorado network both broke down and sedentary farmers no longer lived in those regions. The Classic Period societies of the Phoenix basin suffered a severe population loss and a remnant population lived on until the middle of the century. These people may have gone west where they joined Yuman immigrants from the desiccated Lake Cahuilla on the lower Colorado River. In the mid- to late fifteenth century, intruders razed Casas Grandes and the survivors most likely crossed over the Sierra Madre Occidental where they became the Opata wedge splitting the upper and lower Pima.[43] After the mid-fifteenth century, people no longer built farming villages in the Mogollon uplands and these populations probably moved north to join the Hopi and Zuni. The O'otam in southern Arizona and northern Sonora continued to live in the region, but there appears to be a drop in the number and size of settlements.[13,33] Further south the Rio Lerma–San Gabriel populations probably became the Tarahumara. To the northwest of the Phoenix basin, in the Hakataya tradition, settled villages were abandoned and nomadic upland Yuman populations remained.

This transition was not as dramatic in the upper Southwest. The Pueblo world continued to grow and develop along the lines laid out in the previous period. The collapse of regional centers in the lower Southwest and the retraction of Mesoamerican societies in northern Mexico did, however, leave the upper Southwest more isolated from Mesoamerica than it had been.

ORAL TRADITIONS

Southwestern archaeologists are not the only ones writing histories of Southwestern prehistory and, if we are to admit multiple histories of the Southwest, we must recognize and consider how the Indian peoples of the Southwest tell their own history. The various native peoples of the Southwest account for their histories in a way very different from the archaeological account that we have presented here. These traditions primarily address the spiritual well-being and wholeness of the people. They establish the identity of the people and specify the proper relationship of the people to each other, to other people, to nature, and to the supernatural. They integrate this spiritual concern with many of the factors that we have used in our stories—droughts, introduction of new beliefs, the fissioning of social groups, contingency, and conflict. They are of interest to us here not only because they are alternative histories but also because they sometimes make direct and specific reference to prehistoric sites of this period. In general, they make general references to locations of ancestral sites to the north, east, west, or south. In many cases[14,56] these references haven been given overly specific locations by investigators. The fallacy comes from treating these traditions as less precise or mythicized versions of the accounts that western scholars wish to write rather than treating them as different ways of knowing the past.[24] Although we cannot consider all such traditions in this short article, we present here the Zuni origin story as related by Edmund Ladd.

In the Beginning

In the beginning
there were no humans on the surface of the earth.
Every day Sun Father came up in the east traveled high over Mother Earth
pausing overhead at high noon
and then descended into the western ocean and it became night.
All night long
Sun Father travels under Mother Earth to arrive in the east
in time to bring a new day.
But the days were empty.
There was no dancing,
no laughter,
no singing.
There were no prayers,
no offerings.
Every day as Sun Father traveled high above Mother Earth

[24] George Gumerman: I do not believe that the Native American stories are alternative histories. They are belief systems that are "known," just as Noah and the ark is "known."

he could hear the cries of his children deep in the womb of Mother Earth.
One day as Sun Father was passing overhead he paused at high noon.
He created the Twin Gods, and said to them,
"Go, Go into Mother Earth and bring my children into my light."
The Twin Gods obeyed Sun Father.
After many trials and tribulations
they brought Sun Father's children up from the Four Worlds below,
to his light.
That was The Beginning— Chimegann/Kya.

The exact place of origin is not known,
but it was somewhere to the West (to the West of modern day Shiwinn/a).
People say, based on the word of the beginning, that
it was in the Grand Canyon.
Names of places associated with the travel
in the Search for the Middle Place are known—
Once the search began, from the place of origin,
the people traveled north and eastward moving they said "every four days."
A calendar had not yet been developed,
therefore, what the ancients, The eno:te:que, were really saying was
"We moved every four years."
At first all the people traveled together living in many different places,
then the time for the people to divide and separate came.
The Gods presented the people with two eggs.
One, a very beautiful, shiny, turquoise blue egg.
The other brown, white, gray, yellow speckled egg.
Not very attractive.
The Gods gave the A:shiwi, first choice.
The other people (who were related to the A:shiwi) took the other.
The Gods said to the A:shiwi,
"If you had chosen the gray egg, you would have traveled to to the South,
to the lands of eternal summer where the Mu/la, the Macaw, lives.
But because you chose the Blue egg
you will travel to north the cold country
where the K/walashi, The Crow, lives."
When the A:shiwi continued their travels to the north and east,
moving every four years, in search of the Middle Place.
It came to pass that during these travels various curing societies originated,
the clans,
the rain priesthood,
the bow priesthood,
the kiva societies,
and the summer and winter ceremonial calendar,

based on the Winter and Summer sun,
the movement of Sun Father and Moon Mother,
were created.
When the Center Place was found, "Halona:etiwann/a,"
all the people became settled.
Some to the north,
Some to the east,
Some to the south,
And some to the west.

Today, from the place of origin
we can see many places where the ancients, the Eno:te:que,
stopped over on their journey in their search for "The Middle Place."
These places are still held sacred in the collective memory of the A:shiwi.

CONCLUSION

In conclusion, this panel wishes to stress that by focusing on "historical processes in the prehistoric Southwest," it advocates no radical reshuffling of the sequence of prehistoric events and does not claim to have made any methodological advance in basic interpretations of Southwestern data. Indeed, our overview on Southwestern prehistory finds abundant support for several major trends in analysis. In the first place, no history of the Southwest can ignore the important geographic and environmental conditions that shaped the behavior of prehistoric peoples. We do insist, however, that if long-term trends in social organization and culture change must be viewed in the perspective of the Braudelian "longue durée," Southwestern peoples also chose among various ways in which they constructed their lives. Such options were both constrained and opened through external contacts with distant regions and through the negotiations of social intercourse with nearer neighbors.

None of these "findings" will surprise either professional archaeological observers or informed on-lookers. That having been said, we hold firmly to a commitment to "historical process" as offering an important direction in Southwest studies. For too long, archaeologists in the Southwest have written prehistories of the region (or major subregions) on an admirable level of scientific abstraction. In these prehistories, the topics of periodization, subsistence, and settlement patterns are discussed and maps of the human landscape through time are established.[25]

[25] Kieth Kintigh: My fundimental problem is that I still don't understand what the authors mean by history and historical process as distinct from well-established aspects of traditional or "New" archaeology. Having read their synthesis, which I found interesting and generally reasonable, it is not at all clear how their focus on historical process really informed on their presentation. If what

We celebrate the recent advances that have been made on this level. At the same time, however, we note that these worthwhile prehistories, while mentioning sites and referring to subregional exceptions and anomalies have (perhaps necessarily) tended to overly categorize and so dehumanize the past. In promoting a historical approach then, we do not propose that our recent and valuable regional syntheses be set aside. And we surely do not suggest a return to the level of site-reporting and unfettering the imagination of pseudo-scholars to invent the past. We conclude, rather, by affirming our respect for the archaeological record and the legion of professionals who have painstakingly reconstructed it. The history of the Southwest must and can be told on multiple levels—interregionally, regionally, and the village level. The topics of our chapters must unpack institutional interconnections, and especially stress the ideological and even the spiritual components of social life. And our histories must, in a more comprehensive manner than we have done here, include the rich traditions in which Indian people of the Southwest tell and interpret their own past.

Our advocacy of "historical processes," in sum, is a commitment to the scale of what everyone finds interesting in the Southwest: the farmers of the desert, the dwellers in the cliffs, the builders of pueblos.[26] These prehistoric peoples need histories that are both human and scientific. These histories, furthermore, ought to be relevant, not least to peoples living in the Southwest today. For it is their history that is our prehistory.

they are arguing is that cultural trajectories are contingent (which I think is obvious, though some probably disagree) and that it is useful to look at things on a regional scale, I, of course, agree. Much of what they are doing here does not strike me as controversial.

[26]Stephen Lekson: There is a real danger here (already evident in Southwestern museums and parks) of slipping into a narrow-sighted region-worship. The Southwest is neat and the Pueblos are neat, but are they "neater" or in any meaningful way "unique" compared to the rest of humanity? This should be anthropology's strength: the broad vision, the global sample, the human context. But that vision, in the Southwest, has been narrowed by our fascination with the Southwest—a national Indian theme park. Tunnel vision affects anthropologists who should know better, and it is responsible for most of the theoretical and methodological silliness that makes Southwestern archaeology so discouraging. Do we have larger anthropological interests or are we only culture voyeurs, tourists in someone else's past?

REFERENCES

1. Ackerly, Neal W., Jerry B. Howard, and Randall H. McGuire. *La Ciudad Canals: A Study of Hohokam Irrigation Systems at the Community Level.* Anthropological Field Studies 17. Tempe: Arizona State University, 1987.
2. Adams, E. Charles. *The Origin and Development of the Pueblo Katsina Cult.* Tucson: University of Arizona Press, 1991.
3. Anyon, Roger. "Protecting the Past, Protecting the Present: Cultural Resources and American Indians." In *Protecting the Past: Readings in Archaeological Resource Protection*, edited by G. S. Smith and J. E. Ehrenhard. Caldwell: The Telford Press, 1991.
4. Brody, J. J. *Mimbres Painted Pottery.* Albuquerque, NM: University of New Mexico Press, 1977.
5. Brody, J. J. *Anasazi and Pueblo Painting.* Albuquerque, NM: University of New Mexico Press, 1991.
6. Cordell, Linda. *Prehistory of the Southwest.* New York: Academic Press, 1984.
7. Cordell, Linda, and George Gumerman, eds. *Dynamics of Southwest Prehistory.* Washington: Smithsonian Institution Press, 1989.
8. Crown, Patricia L. "The Hohokam of the American Southwest." *J. World Prehistory* **4** (1990): 223–255.
9. Crown, Patricia, and W. James Judge. "Introduction." In *Chaco and Hohokam: Prehistoric Regional Systems in the American Southwest*, edited by Patricia Crown and W. James Judge, 1–10. Santa Fe, NM: School of American Research Press, 1991.
10. Dean, Jeffrey. "Aspects of Tsegi Phase Social Organization: A Trial Reconstruction." In *Reconstructing Prehistoric Pueblo Societies*, edited by William Longacre, 140–174. Albuquerque, NM: University of New Mexico Press, 1970.
11. Dean, Jeffery, and John Ravesloot, eds. "The Chronology of Cultural Interaction in the Gran Chichimeca." In *Culture and Contact: Charles C. Di Peso's Gran Chichimeca*, edited by Anne Woosley and John Ravesloot. Albuquerque, NM: University of New Mexico Press, 1991.
12. Di Peso, Charles C. *Casas Grandes: Preceramic—Viejo Periods.* Flagstaff, AZ: Northland Press, 1974.
13. Doelle, William H., and Henry D. Wallace. "The Changing Role of the Tucson Basin in the Hohokam Regional System." In *Exploring the Hohokam: Prehistoric Desert Dwellers of the Southwest*, edited by George J. Gumerman, 162–196. Albuquerque, NM: University of New Mexico Press, 1991.
14. Fewkes, Jesse Walter. "Tusayan Migration Traditions." In *Nineteenth Annual Report of the Bureau of American Ethnology*, part 2, edited by J. W. Powell. Washington, DC: U.S. Government Printing Office, 1900
15. Fish, Paul R. "The Hohokam: 1,000 Years of Prehistory in the Sonoran Desert." In *Dynamics of Southwest Prehistory*, edited by George Gumerman and Linda Cordell, 19–63. Washington: Smithsonian Institution Press, 1989.

16. Ford, Richard I., Albert H. Schroeder, and Stewart L. Peckham. "Three Perspectives on Pueblos Prehistory." In *New Perspectives on the Pueblos*, edited by A. A. Ortiz, 22–40. Albuquerque, NM: University of New Mexico Press, 1972.
17. Gould, S. J. *Bully for Brontosaurus, Reflections in Natural History.* New York: W.W. Norton, 1991.
18. Gregory, David. "Form and Variation in Hohokam Settlement Patterns." In *Chaco and Hohokam: Prehistoric Regional Systems in the American Southwest*, edited by Patricia Crown and W. James Judge, 159–194. Santa Fe, NM: School of American Research Press, 1991.
19. Gumerman, George J., ed. *The Anasazi in a Changing Environment.* Cambridge: University of Cambridge Press, 1988.
20. Gumerman, George J., ed. *Exploring the Hohokam: Prehistoric Desert Dwellers of the Southwest.* Albuquerque: University of New Mexico Press, 1991.
21. Haas, Jonathan. "The Evolution of the Kayenta Regional System." In *The Sociopolitical Structure of Prehistoric Southwestern Societies*, edited by Steadman Upham, Kent Lightfoot, and Roberta Jewett, 491–508, Boulder: Westview Press, 1989.
22. Haas, Jonathan. "Warfare and the Evolution of Tribal Polities in the Prehistoric Southwest." In *The Anthropology of War*, edited by Jonathan Haas, 171–189. Cambridge: Cambridge University Press, 1990.
23. Haury, Emil. *The Hohokam.* Tucson: University of Arizona Press, 1976.
24. Irwin-Williams, Cynthia, ed. "Picosa: The Elementary Southwestern Culture." *Amer. Antiquity* **32(4)** (1967): 441–457.
25. Irwin-Williams, Cynthia, ed. *The Structure of Chacoan Society in the Northern Southwest: Investigations at the Salmon Site—1972.* Eastern New Mexico University Contributions in Anthropology 4(3). Portales: ENMU, 1972.
26. Kelley, J. Charles. "The Mobile Merchants of Molino." In *Ripples in the Chichimec Sea: New Considerations of Southwestern-Mesoamerican Interactions*, edited by F. J. Mathian and R. H. McGuire, 97–124. Carbondale: Southern Illinois University Press, 1986.
27. Lekson, Stephen. *Analysis and Statement of Significance on the Mimbres Culture in Southwestern Prehistory.* Mesilla Park, NM: Human Systems Research Inc., 1989.
28. Lekson, Stephen. "Settlement Patterns and the Chaco Region." In *Chaco and Hohokam: Prehistoric Regional Systems in the American Southwest*, edited by Patricia Crown and W. James Judge, 31–56. Santa Fe, NM: School of American Research Press, 1991.
29. Levy, Jerrold. "The Demographic Consequences of Social Stratification in an 'Egalitarian' Society." Paper presented at the annual meeting of the American Anthropological Association, New Orleans, 1990.

30. Martin, Paul S., and Elizabeth Willis. *Anasazi Painted Pottery in Field Museum of Natural History.* Anthropology, Memoirs, Field Museum of Natural History, Vol. 5. Chicago, 1940.
31. Martin, Paul S.. and Fred Plog. *The Archaeology of Arizona.* Garden City, New York: Natural History Press, 1973.
32. Mathien, Francis Joan, and Randall H. McGuire, eds. *Ripples in the Chichimec Sea: New Considerations of Southwestern-Mesoamerican Interactions.* Carbondale: Southern Illinois University Press, 1986.
33. McGuire, Randall H. "From the Outside Looking In: The Concept of Periphery in Hohokam Archaeology." In *Exploring the Hohokam: Prehistoric Desert Dwellers of the Southwest*, edited by G. J. Gumerman. Albuquerque, NM: University of New Mexico Press, 1991.
34. McGuire, Randall H. *A Marxist Archaeology.* Orlando: Academic Press, 1992.
35. McGuire, Randall H. *Death, Society and Ideology in the Hohokam Community of La Ciudad,* A.D. *800 to 1100.* Boulder: Westview Press, in press .
36. McGuire, Randall H., E. Charles Adams, Ben A. Nelson, and Katherine Spielmann. "Drawing the Southwest to Scale: Perspectives on Macroregional Relations." This volume.
37. McGuire, Randall H., and Michael B. Schiffer. "A Theory of Architectural Design." *J. Anthropological Archaeology* **2(3)** (1983): 277–303.
38. Minnis, Paul E. *Social Adaptation to Food Stress.* Chicago: University of Chicago Press, 1985.
39. Nelson, Richard S. "Pochtecas and Prestige: Mesoamerican Artifacts in Hohokam Sites." In *Ripples in the Chichimec Sea: New Considerations of Southwestern-Mesoamerican Interactions*, edited by F. J. Mathien and R. H. McGuire, 154–182. Carbondale: Southern Illinois University Press, 1986.
40. Plog, Fred. "Exchange, Tribes, and Alliances: The Northern Southwest." *American Archaeology* **4** (1984): 217–223.
41. Reff, Daniel T. "The Demographic and Cultural Consequences of Old World Diseases in the Greater Southwest, 1519-1660." Ph.D. disssertation, University of Oklahoma, 1986.
42. Rice, Glen E. *A Spatial Analysis of the hohokam Community of La Ciudad.* OCRM Anthropological Field Studies 16. Tempe: Department of Anthropology, Arizona State University, 1987.
43. Riley, Carroll L. *The Frontier People.* Albuquerque, NM: University of New Mexico Press, 1987.
44. Roseberry, William. *Anthropologies and Histories.* New Brunswick: Rutgers University Press, 1989.
45. Sahlins, Marshall D. *Islands of History.* Chicago: University of Chicago Press, 1985.
46. Sahlins, Marshall David, and Elman R. Service, eds. *Evolution and Culture.* Ann Arbor: University of Michigan Press, 1960.
47. Schiffer, Michael. "Archaeology as Behavioral Science." *Amer. Anthropologist* **77(4)** (1975): 836–848.

48. Schiffer, Michael. *Behaviroal Archaeology.* New York: Academic Press, 1976.
49. Service, Elman R. *Cultural Evolutionism Theory in Practice.* New York: Holt Rinehart & Winston, 1971.
50. Speth, John D. "Do We Need Concepts Like 'Mogollon,' 'Anasazi,' and 'Hohokam' Today? A Cultural Anthropological Perspective." *The Kiva* **53(2)** (1988): 201–204.
51. Steward, Julian. *Theory of Culture Change.* Urbana: University of Illinois Press, 1955.
52. Trigger, Bruce. *A History of Archaeological Thought.* Cambridge: Cambridge University Press, 1989.
53. Upham, Steadman. *Politics and Power: An Economic and Political History of the Western Pueblo.* New York: Academic Press, 1982.
54. Vivian, R. Gwinn. "Conservation and Diversion: Water-Control Systems in the Anasazi Southwest." In *Irrigation's Impact on Society*, edited by Theodore Downing and McGuire Gibson, 95–112. Anthropological Papers of the University of Arizona 25. Tucson, 1974.
55. Vivian, R. Gwinn. *The Chacoan Prehistory of the San Juan Basin.* Orlando: Academic Press, 1990.
56. Waters, Frank. *Book of the Hopi.* New York: Ballantine Books, 1963.
57. Waters, Michael. "The Lowland Patayan Ceramic Tradition." In *Hohokam an dPatayan: The Prehistory of Southwestern Arizona*, edited by R. H. McGuire and M. B. Schiffer, 275–297. New York: Academic Press, 1982.
58. Whalen, Michael. "Cultural-Ecological Aspects of the Pithouse-to-Pueblo Transition in a Portion of the Southwest." *Amer. Antiquity* **46(1)** (1981): 75–92.
59. Wheat, Joe Ben. "Mogollon Culture Prior to A.D. 1000." *Memoirs of the Society for American Archaeology.* No. 10.
60. White, Leslie. *The Science of Culture: A Study of Man and Civilization.* New York: Farrar, Straus, and Giroux, 1949.
61. White, Leslie. *The Evolution of Culture.* New York: McGraw-Hill, 1959.
62. Wilcox, David R. "The Evolution of Hohokam Ceremonial Systems. In *Astronomy and Ceremony in the Prehistoric Southwest*, edited by J. B. Carlson and W. J. Judge, 149–167. Papers of the Maxwell Museum of Anthropology 2. Albuquerque, 1987.
63. Wills, Wirt. *Early Prehistoric Agriculture in the American Southwest.* Santa Fe: School of American Research Press, 1988.
64. Wolf, Eric R. *Europe and the People Without History.* Berkeley, CA: University of California Press, 1982.

Jerrold E. Levy
Department of Anthropology, University of Arizona, Tucson, Arizona 85721

Ethnographic Analogs: Strategies for Reconstructing Archaeological Cultures

Southwestern archaeologists have, in the course of a century of research, developed a remarkably ambivalent attitude toward the use of ethnographic analogs. At the outset this was anything but the case. There were fewer data to assimilate and investigators tended to be equally concerned with ethnography, linguistics, and archaeology. Interpretations of archaeological data were made by inferring directly from the ethnographic present. Unfortunately, many of these inferences were based on faulty assumptions. Despite the fact that ethnologists were aware of the diversity of Pueblo languages, Fewkes and others assumed that Pueblo culture was homogeneous and stable, and myths were assumed to have historic validity.[1] Then, from around 1915 to 1950, archaeologists reacted to this simplistic use of ethnographic analogs by rejecting them entirely, turning their attention instead to technical problems of dating.[19] According to Longacre, even the reasonable hypotheses advanced by Steward[16] and Strong[2] were largely ignored. After 1950, however, there was a renewed interest in the reconstruction of culture history with emphasis on nonmaterial organizational developments. Eggan[2] provided a refinement of Steward's reconstruction of the evolution of Pueblo society and also provided some generalizations about the nature of social organization adapting to its total environment.

[1] Longacre,[12] p. 2.
[2] Strong,[17] p. 54.

Archaeologists soon used Eggan's formulation as the framework for interpreting their data without testing it against the archaeological data.

Longacre saw two major trends during the 1960s. There was a continued interest in the culture history of Puebloan societies, and some investigators continued to feel that the most important data for getting at the nature of culture and social organization of the distant past would come from a careful analysis of social systems in the ethnographic present. According to this view, the basic structure of modern Pueblo society would provide indications of the more stable and, therefore, more ancient aspects of social institutions. Ethnographic, archaeological, and linguistic data in combination would provide the key to unravelling the culture history and development of social organization. At the same time, there was also a growing emphasis on the discovery of generalizations regarding the evolution of cultural systems. The basic goal was to contribute to the understanding of the nature of culture and of cultural processes. According to this view, culture history is not the ultimate end of research but its by-product, and archaeology would contribute substantially to general culture theory.[3]

Recently, this volatile love affair with ethnographic analogs, at least those derived from within the Southwest, has cooled considerably. Upham[23] has criticized attempts to reconstruct the culture history of the Southwest by making interpretations based on the "ethnographic present." The assumption of persistent cultural forms, he argues, is doubtful at best, considering the three centuries of Spanish rule and the possibility of demographic collapse due to early small pox epidemics. In addition, Hopi data cannot be used to make inferences about other Pueblo societies given their great linguistic diversity as well as differences in social organization.

The most serious question raised by this criticism is what sort of ethnographic analog is better suited to the task, for it is a given that there is no interpretation of archaeological data without the use of an ethnographic analogy of some sort. Upham and others, using a "systems" approach and making cultural generalizations, seem to me to advocate a random selection of ethnographic cases which appear to suit their purposes. In part, I believe this is due to a penchant for classifying archaeological cultures according to cultural evolutionists' taxonomies. Once having decided, for example, that the archaeological sites in question represent the chiefdom level of society, the assumption may then be made that the nonmaterial aspects of the archaeological cultural have most of the characteristics of those societies classed as typical chiefdoms. But this approach has its own pitfalls, perhaps even more dangerous than those encountered in the simple application of contemporary Pueblo data to prehistoric sites. The recently published collection of papers from a School of American Research Advanced Seminar reflect the trend of the past fifteen years to concentrate on regional research. What Plog and Upham call regional alliance systems raise questions about the social and political mechanisms integrating interactions among large numbers of settlements. The interconnected sites of the Chaco and the complex irrigation networks of the Hohokam suggest

[3] Longacre,[12] p. 10.

systems of social, political, and economic integration not found among the ethnographically known societies of the Southwest. Chiefdoms, nascent states, and elites have all been suggested as having been present in the prehistoric Southwest. To the best of my knowledge, however, no serious effort has been made to develop a coherent strategy to guide the selection of ethnographic cases from which reasonable analogies may be derived.

Consider first the reasoning involved in reconstructing culture history as contrasted with that of generating cultural generalities. Every known society is the product of a series of historical events. Independent development, borrowing, and migration are all involved. Historical reconstruction recognizes this fact and does not emphasize one process at the expense of the others. In effect, the development of a given geographic area may be unique and not representative of general processes of cultural evolution or ecological adaptation found throughout the world. The search for cultural generalities, on the other hand, is concerned with independent development, that is, cause and effect relationships that may be inferred once a means to control for the affects of diffusion have been effected. Methods for each of these efforts have been devised by ethnologists.

A procedure for reconstructing past forms of social organization was developed by Murdock[4] that derived general rules of change from a world sample of cultures. Applied to several societies speaking related languages, the proposed reconstructions were considered reasonable only if they converged. The reasoning was analogous to that of reconstructing proto-languages. Lacking historical examples, Murdock's sequences cannot elucidate the time it takes to make specific transitions; where data are available we find examples of very rapid change as well as of great persistence. The reasons for these differences are not easily discovered. Driver's (see Jorgensen[8]) method of continuous area comparison helps distinguish traits that have developed independently from those diffused throughout an area. Finally, of course, historical linguistics may offer some idea about the length of time taken by societies within a language group to separate from each other.

Each of these methods may be helpful in bringing the gap between the ethnographic present and the archaeological site under consideration. Murdock's method is especially appropriate in those cases where archaeology can give reasonable assurance that the ethnographic culture is directly descended from the archaeological. Examining ever widening geographic areas, while not providing direct analogies to the past, often reveals the general characteristics of a continent or subcontinent that have developed over long periods of time due to diffusion as well as environmental constraints.

Selecting analogies from other areas of the world to interpret a specific site immediately presents problems and should only be attempted in the event methods of historical reconstruction have been exhausted and have cast little light on sites that appear to have almost nothing in common with the ethnographic societies in

[4] Murdock,[14] Chapter 8, Appendix A.

the area. An example of the dangers inherent in this approach might be appropriate here. Let us imagine that a Southwestern archaeologist has excavated a site with unequivocal evidence of intensive agriculture, i.e., irrigation and permanent fields, and that he wishes to speculate about the social organization of the people who inhabited it. From Textor's[20] "A Cross-Cultural Summary," for example, he learns that 90% of societies, in the world sample that practice intensive agriculture have patrilineal descent. The temptation to infer that the archaeological society in question most probably had patrilineal descent would be strong. Such an inference, however, would have little likelihood of being correct if the site was identified as western Anasazi.

The world sample attempts to control for Galton's problem, that is, the effects of common history and diffusion. Cultures rather than societies are the units of analysis. Thus, all Tewa villages would be counted as one unit, and if Hano is selected to be in the sample, Hopi is not because the connections are so recent. Moreover, an effort was made to achieve some balance in the number of cultures from each continent. In effect, this is not a random sample which would not control for Galton's problem. Rather it is a selected sample attempting to minimize historical connections and to be representative of the cultures of the world so that "functional relationships" between technology and social or political organization can be formulated and the causal relationships inferred.

If the claim is made that a western Anasazi site had patrilineal descent, evidence must be presented to show either that the matrilineal Hopis, Zunis, and Keresans are not the direct descendents of this archaeological culture or that a patrilineal system has become matrilineal. But a direct transition to the matrilineate is completely impossible for any typical patrilineal society[5] Of course, it is possible for the transition to occur with an intermediate period of bilaterality. The evidence for the persistent presence of factors promoting matrilineality must then be found. In short, ignoring the ethnographic evidence and historical connections creates difficulties regardless of the aspect of nonmaterial culture one is seeking to reconstruct.

Even more problematic is the selection of a single ethnographic example from some other area of the world with no consideration of its representativeness. In sum, ethnographic analogs from areas other than within or adjacent to the area in which the site is located is the method of last resort. If, finally, descendents of the Hohokam or the Chacoan Anasazi cannot be identified and the characteristics which demand explanation appear so unique as not to be found anywhere in the ethnographic Southwest or even on the North American continent, then it might be reasonable to look to the world sample for help. The results of such an inference, however, would be more reasonable if the possibility of transitions from them to the present of the culture area could be demonstrated.

With these caveats in mind, let us look first at the proposition that social stratification and elites existed in some prehistoric periods. If, for example, the

[5] Murdock,[14] pp. 216–219.

network of Chacoan sites was created for economic purposes, it is reasonable to posit some sort of managerial class. Two questions arise given the environment of the area: how likely is it that elites functioned as managers and controllers, and how can we know that the network functioned for economic purposes? Johnson[7] has commented on the marginal nature of agriculture in the area and its inability to accumulate food surpluses, a general prerequisite for the emergence of elites. Nevertheless, ethnographic Pueblo societies were stratified: can a reconstruction of the development of the Hopi system cast some light on the issue?

Murdock derives the Hopi matrilineal system from a bilateral system much like that of the Great Basin Shoshoneans.[6] His reconstruction agrees with that of Eggan.[4] The Hopi had matrilineal clans ranked by the number of ceremonies and ceremonial offices they controlled. There is an almost perfect correlation between the ceremonial rank of a clan and the quality of the agricultural land it controlled. Hopi tradition itself recounts how the clans that arrived first at the site of Oraibi obtained the best lands on the alluvial floodplain of the Oraibi Wash, and how later arrivals were allotted increasingly poorer quality land. The population of clans varied over time so that if a clan grew to press on its field holdings, there was no guarantee that all its members had equal access to the land. The prime, or senior, lineage of each clan controlled both the religious ceremony, the clan house where ceremonial paraphernalia were stored, and the authority to assign field plots. The preeminent clan was Bear clan whose senior male was the village chief. Bear clan lands were at the upper reach of the flood plain and were the least prone to drought. Prime lineages of several high- and middle-ranking clans were given plots of Bear clan land in recognition of the ceremonial services they performed in the Soyal ceremony controlled by Bear clan. The loyalty of the prime lineages of these clans was obtained by means of economic as well as ceremonial incentives.

The society, based on an unequal distribution of access to the key agricultural resource, was kept together by an egalitarian ideology which stressed cooperation and, of course, the importance of the ceremonial system for continued survival. Eggan[7] noted that "without a political superstructure, the clan and phratry groups tend to assert their position at the expense of the village," and that Hopi society "has been held together by kinship ties, marriage bonds, and associational structures which cut across clan lines." Moreover, within the clan, the ranking of prime, alternate, and marginal lineages tended to erode the unity of clans during periods of stress.[8] This system, in all likelihood, developed during the period of site aggregation during the thirteenth-century, when the Black Mesa area was abandoned and the Hopis developed large sites along the southern edge of Black Mesa and along the Little Colorado River.

The question then arises whether the prime and alternate lineages of high-ranked clans formed an elite. There is no ethnographic evidence for formal systems

[6] Murdock,[14] p. 336.

[7] Eggan,[2] pp. 116–120.

[8] Eggan,[3] pp. 122–128.

of food sharing beyond the household. Moreover, Hegmon[5] has modeled survival strategies for Hopi households and found that a system of restricted sharing among a small number of households (generally the size of a lineage) was consistently the best strategy, involving the lowest risk of failing to fulfill a household's needs for corn. Finally, an analysis of the 1900 U.S. census of the Hopi,[25] calibrated with Titiev's[21] household census of Oraibi and Kennard's[11] genealogies of Third Mesa, reveals that women of the prime and alternate lineages of the high-ranking clans had higher fertility rates, averaging 10 births per woman 33 years of age and over in 1900, than all other women who averaged only 7.5. At the same time, however, the survival of the children born to these high-ranking women was significantly less (28.6%) than the survival rate of 47.3% for children of all other Third Mesa women. This is the opposite of what one expects from elites. I suspect that the highest status women were trying to produce female heirs and tended to shorten birth spacing (perhaps after a male child), displacing the older sibling from the breast and exposing it to a variety of infectious diseases. By comparison with the Navajos on the Hopi reservation in 1900, the Hopis had higher fertility rates but very much poorer survival rates of children among women 33 years of age and above.[6] These comparisons are consistent with data presented by Martin[13] and suggest to me that despite the addition of sheep to the Hopi diet after the Spanish period, we cannot infer the presence of an elite managing trade in economic surpluses and integrating whole regions in an alliance network for purposes of withstanding local crop shortfalls and the like.

The question of elites may be approached further by asking whether observed regional uniformities, including regional distributions of such pottery types as Salado and Jeddito, can be accounted for by less complex social forms than those proposed for alliance systems. Leaving the Chaco and Hohokam aside for the moment, we can look at ethnographic culture areas to see what mechanisms were at work to create cultural similarities across ethnic and linguistic lines. Jorgensen[9] derived the culture areas of western North America by comparing 172 tribes on 292 variables covering eight major domains of culture. The Southwest was the area of maximal cultural variation. The many tribes there averaged only about 43% similarity. The Pueblos shared 65%, the eastern Pueblos about 74%, and the western Pueblos about 72%. Similarly, the tribes of the Northwest Coast, which stretched from Alaska to Cape Mendocino, California, showed only about 55% similarity. The northern tribes (Tsimshian, Gitskan, Tlingit, and Haida) shared about 74% and the north central tribes (from the Makah north to the Haisla) shared about 72%.

The similarities that define these regions as "culture areas" are well known and consist of such material items in the Northwest Coast as art styles, house types, and tools. Elements of social organization differed widely although social stratification of nobles, commoners, and slaves was found throughout. In the north there were matrilineal exogamous clans and moieties. Among the Haisla and various Kwakiutl groups of the central portion, there were bilateral kindreds. Nowhere was there a form of political organization that incorporated groups larger than the village although, in the nineteenth century, some trading alliance were formed. Linguistic divisions were also deep despite the cultural homogeneity.

In respect of cultural uniformity, the Southwest looks very similar. Informal trade networks and sharing of religious traits are well documented, but each Pueblo village was politically autonomous. The diffusion of traits throughout each area was accomplished despite intervillage competition and conflict, a trait for which both Martin[13] and Wilcox find some evidence in the archaeological record. In fact, Jorgensen[10] concludes that throughout western North America, sodalities functioned to make accommodations among sovereign residential kinship groups rather than through the development of complex political organizations. Village polities, even where centrally organized and tightly controlled as among the Eastern Pueblos, did not extend their control over neighboring villages of the same ethnic group. In sum, considerable cultural uniformity may be attained without the development of complex political structures that transcend the sovereignty of the village.

In his discussion of the deep linguistic cleavages found among the Tanoan-speaking Pueblos of the Rio Grande, Jorgensen[9] observes that such diversity was "commonplace wherever resource bases and subsistence technologies allowed people to settle for long periods, regardless of the proximity of neighbors." The need to protect the agricultural resource did not encourage unrestricted movement or intercourse, and trading was most often done through a few intermediaries. That Tewa and Jemez, for example, share only 47% of their basic vocabularies argues against Tanoans once having participated in a more intensive and organized form of interaction than is known ethnographically.

One of the characteristics of alliances mentioned by Upham is the exchange of mates. A system of political control by elites across village lines would have been strengthened by marriage alliances among elite families from different villages. Indeed, in the Northwest Coast, village agamy or exogamy was the rule. Among the Pueblos, however, village endogamy was strictly enforced. If intervillage alliances were cemented by marriage among elite families, the decline of such a system would have resulted in a agamous communities rather than the strict endogamy that we find even among the bilateral Tanoans. Jorgensen[10] argues that "concentrated and abundant resources, in combination with other factors, appear to have operated *against* the development of political organizations larger than 'inflexible' residential kin groups or 'inflexible' small tribelets."

There are, however, examples of political units that embraced a number of settlements although none of them gave their leaders as much control over community and personal affairs as did the Keresan and Tanoan villages. The River Yumans of the Southwest and the Valley Nisenan of California had political organizations that made decisions affecting several villages for some purposes. The "districts" of the Valley Nisenan united villages for the performance of collective ceremonials and formed alliances for defense. River Yuman districts farmed and protected their

[9] Jorgensen,[9] p. 70.

[10] Jorgensen,[10] p. 197.

common boundaries, "but centralized control was almost completely limited to the conduct of intertribal warfare."[11] The leaders of River Yuman districts were the war leaders whose positions were attained rather than ascribed. Although they were the most respected of all types of leaders, they could not allocate civic duties or confiscate or assign farmland or gathering sites. The hereditary chiefs of the local clan segments would sometimes redistribute scarce food among fellow clansmen, but mostly they exerted their influence by suasion. In no case did a "district" embrace settlements of differing ethnicity or language although Mohave war leaders, for example, could enter into alliances with the neighboring Yumas.[12]

In the eastern Woodlands the Iroquois League and the Creek Confederacy provided political unity for groups of linguistically related tribes. These political structures, however, achieved their purposes by extending kinship and residence principles to embrace villages and tribes. Among the Iroquois, the extended family residential unit was the longhouse and the five tribes of the original confederacy were designated by their position in the "Longhouse." The Senecas, the westernmost tribe, were the "Keepers of the Western Door" and negotiated trade agreements with other tribes. The easternmost Mohawks were the Keepers of the Eastern Door" and were the military authority. The remaining three tribes occupied places between them with the centrally positioned Onandagas as the Fire Keepers and, later, "Wampum Keepers." The arrangement resembled that of families in the ordinary longhouse.[13] The five tribes held a united council each year at Onandaga. The clans of each tribe were represented on the great council and were chosen by the senior women of each clan. The representatives from each tribe were grouped by moiety and relationships among them were governed by the kinship relationships determined by moiety membership. In the councils a series of discussions was held among the chiefs of each tribe and their decisions passed on to those of the other tribes for discussion until consensus was reached. The system of checks and balances rested in part on *not* making all the ablest men chiefs of the confederacy, and war chiefs could not be League chiefs.[14] In sum, great care was taken that power did not concentrate in the hands of a few powerful men.

According to Iroquois tradition, the League was formed to mediate conflict among the member tribes. History suggests that it was successful in this and that it was also able to formulate coherent policies toward the outside world. It is significant for our purposes to note that this political structure was created without the abandonment of egalitarian ideas, the creation of elites, or the creation of a centralized structure that overrode the sovereignty of participating tribes and clans. Even the principle chief was known as "first among equals" and, according to Morgan,[15] "the founders of the Iroquois Confederacy did not seek to suspend the [clan]

[11] Jorgensen,[10] p. 196.

[12] Jorgensen,[9] pp. 222–223.

[13] Tooker,[22] p. 418.

[14] Tooker,[22] p. 429.

[15] Quoted in Tooker,[22] p. 426.

divisions of the people, to introduce a different social organization; but, on the contrary, they rested the League itself upon the [clans], and through them, sought to interweave the race into one political family." Also of interest is the fact that this elaborate political structure was created to coordinate the affairs of five independent tribes with a combined population, in 1660, of only 2,200.

The Creek Confederacy also united a number of independent "towns" under the leadership of a general council.[18] Clan and moiety principles were extended to allocate leadership positions in the general council and moiety symbolism was also utilized to characterize whole towns. There is some evidenced that, at one time, there were preeminent towns and that the highest ranking chiefs had more power than those of the Iroquois League. There is also some indication of a rotating leadership of some of the towns similar to that of the Tewa sodalities which alternate between the moieties. In other respects, however, the political structure does not appear to have been more authoritarian than that of the Iroquois although the data should be reexamined with some care in light of the questions posed by archaeologists. So, for the same reason, should the Natchez system of social classes and satellite towns be reexamined in detail.

In my opinion, much of what archaeologists have interpreted as evidence for regional systems in the Southwest might have been accomplished by relatively simple political structures although the existence of confederacies similar to those of the eastern Woodlands might also be speculated upon. In this context it would be important to take a new look at the suprakinship functions of moieties in North America. Unilineal descent systems and moieties go hand in hand in North America. Where agamous moieties are found, they are either on the fringe of an area characterized by unilineal descent and exogamous moieties or are patently the result of population dislocation and decline as among the patrilineal Prairie tribes.[16] The Tewa have the remnants of patrilineal clans and still have patrilineal moieties which were most probably exogamous at one time. There are vestiges of moieties among the Pimas and Papagos which were most probably patrilineal at one time. Murdock derives Pima social organization ultimately from "Normal Dakota" which is characterized by patrilineal descent, and Russell[17] notes the vestiges of patrilineal clans. The moieties of the Eastern Keresans, on the other hand, appear to me to have been borrowed from the Tewa because they are not equally balanced in terms of prestige and function and do not appear to ever have functioned as descent groups agamous or otherwise.

If the Pimas and Papagos are the direct descendents of the Hohokam, one could speculate on the existence of a network of villages utilizing extensions of moiety and clan to achieve cooperation in the management of the extensive irrigation networks. The ethnographic Pima rotate community responsibilities for hosting major ceremonies, and communities worked communally and reciprocally to open the major

[16] Driver and Massey,[1] pp. 408–415.

[17] Russell,[15] p. 197.

canal while individual communities opened their tributary ditches.[18] Intersettlement cooperation was achieved without centralized control and the headman could direct the men of his community only through suasion and respect, he had no powers of control or coercion. Of course, Pima irrigation was neither as extensive nor as sophisticated as that of the Hohokam, and one cannot rule out the possibility that more complex systems existed in the past. Similar speculation concerning the Chacoan sites would also depend on identifying their ethnographic descendents.

The advances made by Southwestern archaeology in recent years are impressive, both in respect of technique and the accumulation of data sufficient for the analysis of whole regions. It is, I suggest, time to develop more sophisticated use of ethnographic analogs. This will not be an easy task. Archaeologists and cultural anthropologists have become increasingly specialized, and "historical" ethnology no longer occupies a central place in the curriculum of most departments. Not only has the "four field" approach been abandoned for the most part, but archaeology students desirous of learning the ethnography of their area of interested are not likely to find courses suitable for their purposes. The burden of rectifying this ought not to rest entirely on the archaeologists' shoulders. Rather, focii of interest in culture history should foster team efforts and encourage archaeologists and ethnologists to develop integrated course offerings and revivify old research agendas.

REFERENCES

1. Driver, Harold E., and William C. Massey. "Comparative Studies of North American Indians." *Transactions of the American Philosophical Society* **47(2)** (1957): 165–456.
2. Eggan, Fred. *Social Organization of the Western Pueblos.* Chicago: University of Chicago Press, 1950.
3. Eggan, Fred. *The American Indian: Perspectives for the Study of Social Change.* Chicago: Aldine, 1966.
4. Eggan, Fred. "Shoshoni Kinship Structures and Their Significance for Anthropological Theory." *J. Steward Anthropological Soc.* **11** (1980): 165–193.
5. Hegmon, Michelle. "Risk Reduction and Variation in Agricultursal Economies: A Computer Simulation of Hopi Agriculture." *Resh. Econ. Anthro.* **11** (1989): 89–121.
6. Johansson, S. Ryan, and S. H. Preston. "Tribal Demography: The Hopi and Navajo Populations as Seen Through Manuscripts from the 1900 U.S. Census." *Social Science History* **3** (1978): 1–33.
7. Johnson, Gregory A. "Dynamics of Southwestern Prehistory: Far Outside-Looking In." In *Dynamics of Southwest Prehistory*, edited by Linda S. Cordell

[18] Jorgensen,[9] p. 155.

and George J. Gumerman, 371–389. [A School of American Research Advanced Seminar Book]. Washington, DC: Smithsonain Institution Press, 1989.

8. Jorgensen, Joseph G. *Comparative Studies by Harold Driver and Essays in His Honor.* New Haven: HRAF Press, 1974.
9. Jorgensen, Joseph G. *Western Indians: Comparative Environments Languages and Cultures of 172 Western American Indian Tribes.* San Francisco: W. H. Freeman, 1980.
10. Jorgensen, Joseph G. "Political Society in Aboriginal Western North America." In *Themes in Ethnology and Culture History: Essays in Honor of David F. Aberle*, edited by Leland Donald, 175–226. The Folklore Institute. Meerut, India: Archana Publications, 1987.
11. Kennard, Edward A. *Genealogies of Third Mesa.* Manuscript, Santa Fe Laboratory of Anthropology, Santa Fe, New Mexico (nd).
12. Longacre, William A. "A Historical Review." In *Reconstructing Prehistoric Pueblo Societies*, edited by William A. Longacre, 1–10. [A School of American Research Book]. Albuquerque: University of New Mexico, 1970.
13. Martin, Debra L. "Patterns of Health and Disease: Stress Profiles for the Prehistoric Southwest." In *The Organization and Evolution of Prehistoric Southwestern Societies*, edited by George Gumernam. Santa Fe, NM: School of American Research, 1992.
14. Murdock, George Peter. *Social Structure.* New York: The Free Press, 1949.
15. Russell, Frank. "The Pima Indians." *Annual Report of the Bureau of American Ethnology* **26** (1908): 3–389.
16. Steward, Julian H. "Ecological Aspects of Southwestern Society." *Anthropos* **32** (1937): 87–104.
17. Strong, William Duncan. "An Analysis of Southwestern Society." *Amer. Anthropologist* **29** (1927): 1–61.
18. Swanton, John R. "Social Organization and Social Usages of the Indians of the Creek Confederacy." In *Forty-Second Annual Report of the Bureau of American Ethnology*, 25–472. Washington, DC, 1928.
19. Taylor, Walter W. "Southwestern Archaeology, Its History and Theory." *Amer. Anthropologist* **56** (1954): 561–575.
20. Textor, Robert B. *A Cross-Cultural Summary.* New Haven: HRAF Press, 1967.
21. Titiev, Mischa. "Census Notes from Old Oraibi." Manuscript, Peabody Museum of Archaeology and Ethnology, Cambridge, MA (nd).
22. Tooker, Elisabeth. "The League of the Iroquois: Its History, Politics, and Ritual." In *Northeast Handbook of North American Indians*, vol. 15, edited by Bruce E. Trigger, 418–441. Washington, DC: Smithsonian Institution, 1978.
23. Upham, Steadman. 1987 "The Tyranny of Ethnographic Analogy." In *Coasts Deserts and Plains: Papers in Honor of Reynold J. Ruppe*, edited by Sylvia Oaines and O. A. Clark. Anthropological Research Paper 38, Arizona State University.

24. U.S. Bureau of the Census *1900 Enumerators' Schedules.* Prepared by Elisabeth Tooker. Washington, DC, Microfilm Series T623 (Arizona reels).

Steadman Upham
The Graduate School, University of Oregon, Eugene, OR 97403

Systems Modeling and Political Evolution: A Position Paper

The title of the initial Advanced Seminar that was organized by George Gumerman and Murray Gell-Mann reveals as much about the beliefs and accomplishments of the participants as it does about the optimism of the organizers. The working title of that seminar was "The Organization and Evolution of Southwestern Society." It is certain that the participants at that session represent some of the best scholars currently working on problems related to the cultural evolution and social organization of different Southwestern groups. But the title of that initial seminar only reflected the singular number; that is, the title and organizing theme of the seminar was "The Organization and Evolution of Southwestern *Society*," a title that clearly implies some kind of underlying organic unity to the diversity of Southwestern groups and the course of Southwestern prehistory. It is thus no coincidence that my original contribution to that seminar[46] addressed the implied conclusion of this title. I and my co-authors rejected the notion of societal and/or cultural singularity and used our disaffection as a point of departure for discussing alliance formation in Southwestern prehistory.

As a member of the original group that attended that session at the School of American Research, however, I now find a number of the conclusions of the seminar unusually attached to that *singular* notion of Southwestern society. It is the case that at particular times during Southwestern prehistory, A.D. 750, A.D. 1025, A.D. 1140, and A.D. 1285, there was remarkable convergence of demographic, economic, social, and political processes across very wide regions. Although I still reject the notion of a "Southwestern *society*," I believe (and, in fact, have argued) there is evidence for the formation of regional and pan-regional organizations—systems, if you will—that can be found in the convergence of processes like population aggregation, the intensification of agricultural production, developing productive specialization, and regional exchange.

These introductory remarks are intended to direct attention to two issues. First, there is embedded in the notion of "Southwest *society*" the idea of social connectivity that transcends viewpoints emphasizing the local level. I find this embeddedness appropriate as it obviates the notion of village autonomy that has pervaded much writing on Southwestern prehistory (see Plog[36] for a concise critique of the concept of village autonomy). Second, there were hinge points[12] in Southwestern prehistory that resulted in the formation of larger and more inclusive social, economic, and political entities. The size and inclusiveness of these formations changed through time, but by larger, I mean developments that were regional or pan-regional in scale. If we are to understand how these hinge points came about and how these past social, political, and economic relationships were structured, we must be able to identify, describe, and explain "systems"...which brings me to the theme of this paper. When I think about systems modeling in archaeology, the topic that George Gumerman and Murray Gell-Mann asked me to cover, I automatically think about the organization and evolution of *political* systems. This is my bias and it is clearly reflected in the following statements and discussions. I focus on political systems because they encompass social and economic relations and provide at a minimum indications of a society's integrative mechanisms, systems of resource acquisition and distribution, and concepts of labor and labor management. In short, political systems reflect the nature of access to all kinds of natural, human, and social resources and the symmetry or asymmetry of social and economic relationships. As a result, political systems fundamentally reflect the systemic organization of a society.

Given my stated bias, I offer in this paper a framework for studying political evolution that provides a generic basis for examining change. Much of this discussion is drawn from one of my recent papers[47] that deals with these issues. Before describing the various dimensions of my "systems model," however, I offer a statement of my own biases regarding anthropological efforts to explain the evolution of social and political systems:

1. All studies of social and political evolution ultimately have as their underlying goal, the description and explanation of changes in the complexity of cultural systems. Such studies, if they are to contribute to general evolutionary theory, must also be explicitly concerned with measuring the rates at which changes in complexity take place. I believe the former statement defines the primary anthropological goal of "explaining change" while the latter statement defines the term "process." Both concepts are essential if anthropology is to be considered a social science.

2. The term "process" as defined above is simply another term for history. The goal of studies concerned with social and political evolution is to identify and explain changes in the complexity of cultural systems that are the product of shared developmental histories. Like cladism, the purpose of studying social and political evolution is to identify homologous similarities of development. History defined in this manner is thus scientific and of great relevance to anthropology. By pursuing issues related to social and political evolution, anthropologists are actually contributing to a general theory of history.
3. Entire cultural systems are not units of analysis in studies of social and political evolution. Although the goal is to generalize about the evolutionary history of such systems, they are too inclusive to be used as analytical units. In much the same way that biologists focus on change in a few key characters to define different species, anthropologists should focus on change in a few key societal dimensions in their efforts to develop a theory of history and explain the evolution of culture and human behavior.
4. Because cultural systems are too inclusive to be used as analytical units, taxonomic approaches that seek to identify "stages" of cultural evolution are scientifically inadequate to explain changes in the complexity of cultural systems. Consequently, the use of terms like band, tribe, chiefdom, and state in studies of social and political evolution are counterproductive since they define constellations of variables that may or may not covary and certainly cannot be measured.
5. Social and political evolution is a mosaic process; changes in the complexity of one variable may or may not be independent of changes along other axes of cultural variability. Because of this fact, different processes of development (e.g., demographic, economic, social, political, etc.) change differentially and exhibit different rates of change. The use of multivariate analytical techniques to model change processes presupposes that the rates of change in all variables are equivalent. Consequently, the use of such models is inappropriate *unless* it can be independently demonstrated that rates of change among variables are autocorrelated.

These five statements suggest a clear direction for studying questions related to political evolution. In the following section of this paper, I outline what I call a generic framework for explaining the development of political systems (see also Upham[47] for an extended discussion of these ideas). I use the term "generic" because of the fact that all of the issues I am concerned with are common to or characteristic of all cultural systems, regardless of the system's size and complexity.

THE GENERIC FRAMEWORK

Most anthropological studies of political evolution have used the emergence of state-level polities as a baseline from which all other such developments are measured and evaluated. I have argued, however, the goal of evolutionary studies is to identify and explain changes in the complexity of cultural systems that are the product of shared developmental histories. Accordingly, the state *per se* cannot be an analytical unit since state-level polities, as a class of phenomena, do not have shared developmental histories. Although such an idea is theoretically plausible, it is statistically unlikely that the constellation of traits and institutions comprising the state at different times and places would evolve in an homologous fashion. Rather, the developmental regularities sought by anthropologists must be identified at a more particularistic level. The specific axes of variability that relate to people and production within such systems thus become the foci of analysis. I define four such axes that are relevant to the complexity of any cultural system: demographic, economic, social, and political.

DEMOGRAPHIC VARIABILITY

Demographers use a variety of measures to describe the structure of human populations and the way that structure has changed through time. Age/sex ratios, fertility and mortality rates, migration, and the like are all viewed as essential measures in comparative demographic studies of modern populations. Because of demographic research, anthropologists are keenly aware that modern states have very different population structures, fertility and mortality rates depending on the position of the state in the developed or developing world. Such populations also have very different developmental histories owing to the divergent historical forces of colonialism and modernization. Consequently, the identification and explanation of changes in the complexity of cultural systems resulting from demographic processes that are the product of shared developmental histories would appear too complex to be of value in the search for structural regularities.

It is the case, however, that two axes of demographic variability are properties of all human populations: population size and population density. Although these measures are rarely used analytically by demographers in the study of modern populations, they are of key interest in the study of social and political evolution. This interest results from the fact that measures of changing population size and density, when viewed diachronically, reflect rates of population growth or decline, aggregation or dispersal. Because particular political and social systems are really nothing more than solutions to specific problems of management and production, the size and density of human populations on the landscape are directly linked to the complexity of the decision making system. In other words, the complexity of social and political systems, and especially decision making, are regulated in a fundamental manner by the size and density of populations. Thus, I suggest that a beginning point for all studies of social and political evolution is to compile longitudinal data on changes in population size and density.

Elsewhere,[45] I have sought to clarify the relationship between increases in social and political complexity and the two demographic variables of size and density by reevaluating a detailed comparative study of social and political organization undertaken by Feinman and Neitzel.[14] Based on the data they compiled for 106 ethnographically known New World groups, I constructed a series of statistical analyses to evaluate how various indicators of political control (number of administrative levels, craft production, form of political control, etc.) were correlated with the demographic variables.[1] Although my results remain tentative, overall population size was strongly correlated (correlation coefficient = 0.929779, explaining more than 86% of the variance) with seven variables related to centralized (and sometimes coercive) political control. More significantly, however, was that fact that population threshold values were identified for each of the variables. The mean threshold value for all of the variables was approximately 10,500 people, meaning that many of the attributes of political complexity identified by Feinman and Neitzel have a statistically higher probability of occurrence in total regional populations larger than this figure. From this analysis, it became clear that *population density* and not merely total population size was also an extremely important variable. Unfortunately, the ethnographic data were not sufficiently fine grained to examine this axis of variability.

There is little question that population density is an important variable in explanations of change in social and political systems. It is also the case that to measure population density, especially during periods in prehistory when emergent social and political hierarchies formed, archaeological data must be of superior quality. At the present time, archaeologists use a variety of different proxy measures to calculate total population size (number of rooms, floor area within rooms, site size, number of vessels, vessel capacity, etc.), each of which has both strengths and weaknesses. Extrapolating total population counts based on these proxies to measures of population density, however, presupposes a knowledge of the past that would be difficult to validate even under the best of conditions. Consequently, another kind of indicator is required to examine population density prior to the advent of detailed census data. I suggest that along with total population, the total population size of individual communities can also prove useful.

Naroll[31] has shown, for example, that communities larger than about 500 people generally have more centralized political leadership. This finding is not remarkable in that as community size increases, face-to-face interaction also increases and the potential for disputes and conflict is greater.[15,21] Community populations larger than 500 must also work out strategies for resource allocation and mate selection.

[1]The seven variables correlated with total population size in Feinman and Neitzel's data are a leader's control of storage, special burials for leaders, treatment of the leader with obeisance, provisioning the leader with special food, special residences for leaders, increased size of the administrative organization, and increased size of the largest community.

I find Naroll's result particularly intriguing since a community of 500 also is the approximate size of the minimum local population in which endogamy is possible on a routine basis.[53] Community endogamy can facilitate marriage strategies used by small local groups to gain access to or maintain control of strategic resources. Consequently, when communities larger than approximately 500 people exist in conjunction with total regional populations of 10,500, the potential for emergent social and political hierarchies to form is greatly increased.

I must point out that by advocating a primacy for demographic data in the search for structural regularities, I am not suggesting that demographic prime-mover arguments or Boserupian solutions are required to explain the emergence and subsequent evolution of specific political or social systems. On the contrary, one must also explain why the size and density of populations change. Recourse to pat prime-mover arguments or Boserupian models is antithetical to the general anthropological goal of explaining change. I also point out that there may be an illusion of "magic numbers" in the above discussion. It should be recognized that these population figures are simply estimates. Such numbers are not a panacea for explanation, nor are they substitutes for empirical data. It is the case, however, that threshold values in state variables exist and when such values are met (within certain prescribed ranges), quantum leaps in the form and structure of social and political systems occur.

ECONOMIC VARIABILITY

Since the time of Karl Marx, social scientists have actively sought to demonstrate how economy and the mode and relations of production either determine or are determined by various political and social configurations. It is again the case, however, that particular economies and production strategies, as aspects of social and political systems, are merely solutions to problems of management, and that management needs pertaining to production are intimately related to demographic variability. There is little question that, for example, strong connections exist between the relations of production in any economy and the size and quality of the productive base.[16] These relationships, when filtered through populations of varying size and density, give rise to distinctive economic forms and institutions. When such connections are considered on a worldwide basis, the incredible variability of relationships found among just these few variables would seem to preclude anthropologists' ability to offer more general, evolutionary formulations. Yet there is one change in the relations of production that divides all economies into two distinctive, wholly separate groups: the change from subsistence production to production for exchange. This change thus identifies a boundary, and that *boundary* can be viewed as a general property of economic systems.

The boundary between economic systems based on subsistence production and those based upon exchange is not arbitrarily placed. Because the relations of production are reflections of management considerations more broadly construed, fundamental changes in such relations mirror like changes in the form and structure of management and decision making. Frederick Engels[13] may have been the first to recognize how strategies of production for exchange stimulated specialization in labor, the emergence of social classes, and the development of private property. But his ideas have become unfortunately politicized and because of this fact, foster polemic in scientific arguments. Leacock,[24,25,27] in her work on gender hierarchy and hierarchy in general, has taken a less doctrinaire (but no less political!), more social scientific approach to this issue and is perhaps most recently responsible for recognizing the importance of this boundary. Based on her work, one can define two mutually exclusive organizational modes that coincide with subsistence production and production for exchange, respectively. They are found in systems of group dependence (usually hunting and gathering familial groups or patribands) on the one hand, and systems of hierarchy on the other (suprahousehold systems). In the vast middle ground between these two organizational modes are found dyadic (male-female) structures of dependence and decision making that exist as horticulturally or pastorally based households (although not necessarily sedentary ones). As Netting[32,33,34] and Wilk and Netting[52] have shown, the household as a primary unit of production, distribution, transmission (inheritance), reproduction, and consumption is enormously variable across cultures. Yet it is within the context of these dyadic structures of dependence that the change from subsistence production to production for exchange occurs. To understand the emergence of more complex vertically and horizontally specialized decision-making structures (bureaucracies) that characterize all suprahousehold systems, investigation must then center on the relation of production found in the household.

It is clear that by focusing on the important boundary between subsistence production and production for exchange, a host of other economic processes become relevant. The intensity of agricultural production, storage, productive specialization, as well as the local and regional procurement, and supply and distribution of commodities are all variables that must be measured. Moreover, all of these variables must be considered in conjunction with demographic processes.

SOCIAL VARIABILITY

A third major axis of variability relevant to the complexity of any cultural system pertains to social organization and to the social arrangements found in human societies. It is clear from the past work of ethnographers that the variety of social institutions found cross-culturally is immense. Systems of kinship, cross-cutting associations and sodalities, patterns of post-marital residence, and so on are so diverse in the literature as to appear nearly unique for each human group. Although the previous sentiment is no doubt an overstatement, such uniqueness has been compounded by the facile manner that anthropologists describe such systems as "localistic" developments, and by anthropologists' willingness to identify unique traits and institutions over more general feature of social organization. The resulting welter of organizational forms found in the ethnographic literature would thus seem to preclude more general formulations.

Again, however, there is a dimension of social variability that stands apart from the details of any specific organizational form, and mirrors specific changes in the complexity of social organization. That dimension is social symboling, embodied in the concept of "style," and it can be used to distinguish significant evolutionary changes in the complexity of human social systems. These changes in complexity can be identified and measured based on the way various cultural and ethnic groups employ stylistic criteria to signal individual or group identity. The exchange between Sackett[39] and Wiessner[51] on the interpretation of stylistic data has called needed attention to this important area of anthropological inquiry. More recently, S. Plog[38] has sought to clarify and elaborate on the ideas of Sackett and Wiessner and show how stylistic data are relevant to the evolution of social and political systems. Based on his work and on the work of Sackett and Wiessner, I distinguish between three different kinds of "stylistic" behavior: isochrestic, symbolic or iconographic, and heterochrestic. At the most basic behavioral level, isochrestic variation refers to stylistic criteria that are both idiomatic and diagnostic of ethnicity.[2] Yet isochrestic variation (meaning literally "equivalent in use") specifically refers

> ...to the fact that there normally exists an appreciable range of equivalent alternatives, of equally viable options, for attaining any given end in manufacturing craft products. Style enters the equation when it is recognized that the choices artisans make along the range of options potentially available to them tend to be quite specific and consistent, and that these are directed largely by the craft traditions within which the artisans have been enculturated as members of social groups (ibid.).

In other words, stylistic variability that is isochrestic derives from the unique properties of each cultural group's historical development and reflects patterns learned by rote by all participating members of that social system. Conkey[10,11] has argued that the kind of standardization in material culture resulting from isochrestic behavior is not found prior to the Mousterian (ca. 100,000 B.P.). Today, isochrestic variability is a measurable property of all classes of material objects, but is not a *primary* means of signaling cultural or ethnic identity. Because of that fact, isochrestic style similarities and differences are of little interest in the search for structural regularities.

The second stylistic category, stylistic variability that is symbolic or iconographic in content, corresponds to Wiessner's "emblemic" designation and refers to:

[2]See Sackett,[39] p. 157.

> Styles (that have) a behavioral basis in the fundamental human cognitive process of personal and social identification through stylistic and social comparison. In this process, people compare their ways of making artifacts with those of others and then imitate, differentiate, ignore, or in some way comment on how aspects of the maker or bearer relate to their own social and personal identities. Style is thus not acquired and developed through routine duplication of certain standard types, but through dynamic comparison of artifacts and corresponding social attributes of their makers. Stylistic outcomes project positive images of identity to others in order to obtain social recognition.[3]

No doubt Leslie White[48] would have viewed the kind of stylistic behavior identified by Wiessner and S. Plog beginning with the expression of parietal cave art during the Upper Paleolithic about 40,000 years ago. Symbolic content as a primary form of stylistic behavior is enduring. Wiessner, for example, has shown how modern !Kung San projectile points symbolically communicate individual and language group identity.[50] Many other examples of symbolic or emblemic styles can be found today as a primary means of expressing individual or cultural identity. Often symbolic variability takes the form *of* iconographic representation. I define iconography as the use of styles that are fixed by convention to convey the identity of a specific subject (an individual or group) and the set of conventions or principles that govern the use of such imagery. Wiessner[4] and S. Plog[5] have both suggested that iconographic variability is a "special case of stylistic behavior" that is symbolic in content, information rich, and aimed at a specific target population. The most important characteristic of both symbolic and iconographic behavior is that they communicate identity and membership homogeneously, at the level of the entire cultural or ethnic group; they are boundary setting, but their use is inclusive at the level of the entire group.

The final category of stylistic behavior refers to variability that is *heterochrestic*, a neologism literally meaning "unequal in use." As the opposite of isochrestic, I have reluctantly chosen this term to represent stylistic identities that are differentially shared within and between cultural and ethnic groups. Such differential sharing of heterochrestic styles is conscious and directed and, like iconographic variability, is governed by rules and conventions. Heterochrestic behavior, however, demarcates subgroups that possess and maintain distinct stylistic identities. Most commonly these identities are status related and access to heterochrestic styles is determined by membership criteria that are ascribed.

As a primary form of stylistic behavior, heterochrestic variability is found in societies that publicly identify the rank or status of individuals or groups. The corollaries to this kind of public identification may vary, but the possession and

[3]See Wiessner,[51] 161.

[4]See Wiessner,[51] p. 161.

[5]See Plog,[38] p. 3.

use of heterochrestic styles always defines a distinct status for an individual or subgroup within a larger interacting population (the most studied are styles that define "elites," but heterochrestic styles also demarcate slaves, serfs, peasants, or the "poor" from other segments of a population). Like symbolic or iconographic styles, the identification of heterochrestic behavior is as much a study in the distribution of specific styles as it is an exercise in defining their information content, and the rules or conventions that govern their use. Many different examples of heterochrestic variability have been identified in both the archaeological and ethnographic literature. Unlike broadly available symbolic or iconographic stylistic elements, differential access and use of heterochrestic styles always results in inhomogeneous distributions of key style elements.

The distinction between symbolic and heterochrestic styles defines a boundary between two distinct social and political forms. In any society, style in material culture (or speech) is a social resource. Because public style displays are always for public consumption, they reflect directly on the nature of access to resources in a society. The boundary that separates symbolic from heterochrestic styles also distinguishes the boundary between unrestricted and restricted access to social resources. In all societies, when access to social resources is restricted there are corresponding restrictions, albeit of varying intensities, to natural resources that divide a given population into subgroups of varying rank or status. The emergence of these subdivisions, from the standpoint of social and political evolution, is of paramount interest.

Many anthropologists have identified these fundamental changes in styles and access as a meaningful sociopolitical boundary. A few anthropologists have also sought to describe the corresponding organizational features of the two broad groups of societies that result from this type of classifying. Leacock and Lee,[6] for example, discuss the essential forms of "group dependence," and I believe their definition conforms to the kind of social and political systems that rely solely on the use of symbolic styles. The attributes, found most typically in hunting and gathering societies, include "egalitarian patterns of sharing, strong anti-authoritarianism, an emphasis on the importance of cooperation in decision-making, conflict resolution, sharing, exchange, the allocation of rights to lands and resources, and the socialization of children, and the ritualization of potential conflict between the sexes" (ibid.). It is significant to note that the virtual opposite of many of these attributes defines what have been described as state-level polities.[17] These latter systems are dominated by heterochrestic style behaviors. Occupying the vast middle range between these two forms of human organization are most of the societies that have been studied or recorded by anthropologists. It is among these groups that the use of symbolic and heterochrestic styles is of greatest interest, since the public identification of rank or status that is correlated with a restriction of access to social and natural resources represents a quantum leap in the complexity of social and

[6] See Leacock and Lee,[27] pp. 7–8.

political systems. Styles and their context, then, become a focus for inquiry into the evolution of social and political organization.

POLITICAL VARIABILITY

The final axis of variability in the generic framework I have constructed pertains to political decision making and to the way decision-making systems articulate with the three axes of variability described above. In a previous study,[7] I defined the political processes in the following manner:

> I equate the decision-making process with the political process, defining the latter as the ability of an individual or group to instigate and link together the behavior and acts of others by the power of persuasion or coercion so as to create a necessity for action within the perceived common good. The concern...then, is with the managerial structure, the decisions of which affect the behavior and economy of the general population.

In a more recent study, F. Plog and I have sought to elaborate on this definition to clarify how political decision making might demarcate societies into broad classes that are managerially distinct. We also sought to develop a model of political decision making that would have material correlates and, consequently, would be useful to archaeologists.

> Although status and role are pertinent concepts for the study of political systems, the archaeological record, to the extent that it reflects political organization, is the product of managerial decisions broadly construed. In any society, the most basic of those decisions involves access to space, access to human and natural resources, and access to social resources including statuses, organization, and symbols. Across societies access to space, natural resources, and social resources varies *in the extent of its restriction to particular individuals or groups and whether the restrictions are consensual or cooptative.*[8] *emphasis added.*

In much the same way that symbolic and heterochrestic variability are manifest in content, use, and ultimately, in homogeneous or inhomogeneous distributions of particular style elements, decision-making by consensus or cooptation produces different kinds of distributions of material culture that can be used to distinguish major differences in the complexity of decision-making systems. Because political decision making often pertains to problems of access to various kinds of resources,

[7]See Upham,[44] p. 4.
[8]See F. Plog and Upham,[35] p. 201.

distinguishing between consensual and cooptative systems also distinguishes between systems of free and restricted access. As I argued for styles and social resources, I would again argue that this fundamental difference in access identifies two separate trajectories of social and political evolution.

There is general consensus among anthropologists, a few dissenters notwithstanding,[7] that most hunting and gathering societies now and in the past can be characterized by a sociopolitical ethos that emphasizes egalitarian patterns of sharing and free access within the group to basic natural, human, and social resources. On the other hand, most anthropologists would also agree that the large, vertically and horizontally specialized bureaucracies found in some preindustrial and industrial societies are characterized by a sociopolitical ethos that emphasizes disparities in access to natural, human, and social resources. Between these two extremes the vast majority of societies are found. At just what point political decision-making changes from a system of consensus to one of cooptation is an empirical question that can be determined by studying patterns of access to key resources.

OTHER CONSIDERATIONS

The four axes of variability and the pathways of evolutionary development might be discussed in terms of mutually exclusive outcomes. For example, one outcome, undifferentiated political systems, refers to the basic, and possibly the only abiding, form of human organization: *systems of group dependence.* Systems of group dependence are historically linked to small hunting and gathering populations (although they occur under other subsistence modes as well) and, as a class of phenomena, have existed since the Lower Paleolithic. It should be pointed out that a few anthropologists have argued that other kinds of social configurations, especially those formed during what Braun and Plog[37] identify as the process of tribalization, would conform to the definition of an undifferentiated political system. The second outcome, emergent social and political hierarchies, represents the only alternative to systems of group dependence. These emergent hierarchies are a class of phenomena and are best described as *systems of inequality.* Obviously the intensity of both group dependence and inequality varies between systems and through time so that there are more and less intense egalitarian patterns of sharing or inequality in different social and political systems of the same class.

The general patterns of world history suggest that a prerequisite for the development of emergent social and political hierarchies is a sedentary or semi-sedentary lifestyle, in which substantial portions of each year are spent in permanent settlements. Social and political systems characterized by emergent political hierarchies thus probably existed in very small numbers before the origins of agriculture and other strategies of food production. Based on archaeological and ethnographic data, their occurrence would be predicted in resource-rich environments, like those seen in temperate zones along the Pacific coast of North America at the close of the Pleistocene. Following the origins and spread of agriculture, the number of emergent social and political hierarchies increased rapidly.

Traditionally, sedentarization is linked to groups involved in food production, groups residing in those regions where sufficient natural resources permit the establishment of permanent settlements, or, more recently, groups in the throes of modernization or development. As such, sedentarization is normally associated with (a) certain kinds of environments (those either conducive to agriculture or naturally resource rich), (b) surplus production (at least production sufficient to carry a group through the four seasons of the year), or (c) the availability of resources or capital from more developed groups, a slightly different kind of "natural resource."[9] I term this process "sedentarization through abundance." Many discussions of sedentarization through abundance, especially in the archaeological literature, are thus directed to the study of purely environmental considerations (availability of water, amount and quality of arable land, abundance of natural resources, climate, etc.). If one were to survey the anthropological literature on sedentarization, this pathway to sedentary life would appear virtually exclusive.

A focus on purely environmental issues and on sedentarization through abundance, however, obscures a very important issue. It is the case that another alternative, "sedentarization through impoverishment," is just as common, if not more common, than the traditionally accepted pathway. Sedentarization through impoverishment has been occasionally described in the anthropological literature.[1] Sedentarization through impoverishment can occur in a variety of ways but most often begins when population increases and a given landscape becomes "packed." Among pastoralists, a packed landscape decreases the amount of available pasturage, stimulates herd reduction strategies, and can result in the eventual loss of animals. If unchecked, entire herds can be lost and households can be forced to join existing settlements, almost always in dependency relationships with other households or suprahouseholds and in circumstances of greatly reduced status. Among hunter-gatherers, a packed landscape decreases foraging range, increases competition for increasingly scarce resources, and may stimulate intensified procurement strategies. In prehistory, many such hunting and gathering populations may have responded to increased population densities by joining agricultural or pastoral communities. Such communities existed as "magnets" on the landscape in much the same way that trading posts or mission stations do today in remote regions.[28,29] Exactly which communities were selected by hunter-gatherers may relate to long-term social relationships between foragers and farmers and to the kinds of "alliances" described by Bender.[10] Hunter-gatherers, like pastoralists joining sedentary communities, also existed in positions of greatly reduced status.

Arguments seeking too strong a link between environmental variables and the process of sedentarization often mistake the environmental setting as causal. As I

[9] See Kenyon,[22] p. 35; MacNeish,[30] p. 531; and Braidwood and Braidwood[5] p. 278.

[10] See Bender,[4] pp. 210–213.

intended to show above, sedentarization through impoverishment is fundamentally a social and demographic process.[11] As Hitchcock[12] points out,

> Simple availability of resources (is) insufficient to bring about residential stability for an extended period of time....Long-term residential stability comes about when a group's mobility options are restricted due to the fact that there are too many other groups occupying the habitat.

CONCLUSION

Significant changes in the complexity of social and political systems, resulting in the fundamental transformations discussed in this paper, occur in only a few of the many societal dimensions studied by anthropologists. I have sought to identify the four basic axes of variability that are both preconditions and indicators of cultural evolutionary change. I would argue that change in these four axes, in effect, "drives the system," and that other more particularistic societal dimensions respond to these axes of variability and change in kind. In this sense, elements of any cultural system that fall outside the four basic axes of variability I have described can be considered as epiphenomena.

I have argued that demographic variability as it relates to total population size and to population density, viewed by its proxy indicator of community size, is the most significant axis of variation because it provides a set of preconditions that must be met before significant evolutionary change can occur. When these preconditions are met, a threshold is crossed and economic, social, and political systems also begin to change. Clearly, the rates at which change occurs in these latter three societal dimensions is an empirical question of great interest. In the natural sciences, systematists have devoted substantial energy to classifying phenomena based on homologous rates of change.[41,42,43] Given the perspective of the present paper, I argue a major goal for anthropology is the development of what might be considered "cultural taxa" based on homologous rates of change in economic, social, and political systems that derive from different demographic inputs. It is at this level that the idea of shared developmental histories becomes truly meaningful.

I have also suggested that the major social indicator of change from systems of group dependence to emergent political hierarchies is stylistic and that the change from symbolic to heterochrestic style behaviors clearly identifies this important boundary. Identifying the heterochrestic or symbolic "signature," especially for archaeologists who do not work with living systems, is explicitly a distributional

[11]See Cohen,[9] p. 83.

[12]See Hitchcock,[18] p. 231.

exercise whereby the context of use and discard for items of material culture are critical dimensions of any reconstructions of past social and political systems. Cultural anthropologists, on the other hand, can investigate the material patterns along with the jural systems governing the rules and conventions related to access and use of styles marking social identity.

In the generic framework, the remaining two axes of variability, economic and political, are located structurally between the demographic preconditions and the stylistic manifestations of social and political complexity. I have suggested that the change from subsistence production to production for exchange, for example, occurs in the context of the household. Wilk and Netting[13] have shown that the most important kind of variation in production within the household occurs in scheduling, which they define as "the timing (in the yearly cycle) of productive tasks and the sequencing (the order) of the tasks themselves" (ibid.). They go on to identify simple and diverse production systems. In simple productive systems, the sequencing of tasks is relatively uncomplicated and productive activities can be ordered without conflict. When household production systems are diverse, however, the performance of major productive activities must be simultaneous. Consequently, scheduling conflicts can arise. Theoretically, a variety of solutions to scheduling conflicts are possible, but Wilk and Netting show convincingly that one solution, to increase the size of the household, is far more common than any other.

Because scheduling conflicts most often arise when populations increase, there is a general relationship between labor demands within the household and local and regional demographic conditions. Moreover, there are indications that specialization and differentiation in task performance is a strategy used by large households with diverse productive strategies to linearize scheduling conflicts.[2,3,19,40] Specialization in task performance, a basic condition of production for exchange, may thus result from scheduling conflicts in diverse production systems and be related to local and regional demographic conditions. I have argued in this paper that cross-cultural data and other features of the generic framework suggest the change from subsistence production to production for exchange would be far more likely to occur when village and total regional population threshold values were met or exceeded. These demographic preconditions are linked to the household through labor demands and scheduling. Once specialization in task performance begins to occur, individuals are separated from the full range of subsistence activities and a related process, the separation of food producers from consumers, can also be seen to have its roots in the household. This kind of separation presents a unique challenge to any decision-making system, and especially to those predicated on egalitarian patterns of sharing. How does one evaluate (and valuate) human labor performed at different tasks? Are all tasks of equal value or are some more valuable than others? While a complex issue, the basis for the change in decision making from consensus to cooptation and the concomitant dimension of access to all human, natural, and

[13]See Netting,[54] pp. 7–8.

social resources resides in the process of valuating human labor performed at different tasks. The way such decisions are ultimately made is manifest in the social styles of any society.

In this paper I have attempted to bring together a series of tangible and measurable attributes of human societies that are sensitive to one of the most important changes in the human cultural past: the change from systems of group dependence to systems of inequality. From a systems modeling perspective, I believe the four axes of variability I have identified must be considered if our goal is to generalize about the course of cultural evolution. It is important that in our attempts to model social and political systems and in the broader search for structural regularities, we are not deflected by particularistic details of social form and adaptation. Yet it is equally important that those details inform the broader picture we seek to paint.

REFERENCES

1. Barth, F. *Nomads of South Persia: The Basseri Tribe of the Khamseh Confederacy.* New York: Humanities Press, 1961.
2. Befu, H. "Ecology, Residence and Authority: The Corporate Household in Central Japan." *Ethnology* **7** (1968): 25–42.
3. Befu, H. "Origin of Large Households and Duolocal Residence in Central Japan." *Amer. Anthropologist* **70** (1968): 309–320.
4. Bender, Barbara. "Gatherer-Hunter to Farmer: A Social Perspective." *World Archaeology* **10(2)** (1978): 204–222.
5. Braidwood, R. J., and L. J. Braidwood. "The Earliest Village Communities of Southwestern Asia." *J. World History* **1** (1953): 278–310.
6. Braun, David P., and Stephen Plog. "Evolution of 'Tribal' Social Networks: Theory and Prehistoric North American Evidence." *Amer. Antiquity* **47** (1982): 504–525.
7. Cashdan, Elizabeth A. "Coping With Risk: Recipiocity Among the Basanva of Northern Bostwana." *Man* **20** (1978): 454–474.
8. Cashdan, Elizabeth A. "Egalitarianism Among Hunters and Gatherers." *Amer. Antiquity* **82** (1980): 116–120.
9. Cohen, M. N. *The Food Crisis in Prehistory: Overpopulation and the Origins of Agriculture.* New Haven: Yale University Press, 1977.
10. Conkey, M. W. "Style and Information in Cultural Evolution: Toward a Predictive Model for the Paleolithic." In *Social Archaeology: Beyond Subsistence and Dating*, edited by C. Redman et al., 61–85. New York: Academic Press, 1978.
11. Conkey, M. W. "Context, Structure, and Efficacy in Paleotithic Art and Design." In *Symbols as Sense*, edited by M. L. Foster and S. H. Branndes, 225–248. New York: Academic Press, 1980.
12. Cordell, Linda S., eds. *Dynamics of Southwest Prehistory.* Washington, DC: Smithsonian Institution Press, 1989.
13. Engels, F. *The Origin of the Family, Private Property, and the State*, edited by E. Leacock. New York: International Publishing, 1972 [1884].
14. Feinman, G., and J. Neitzel. "Too Many Types: An Overview of Sedentary Prestate Societies in the Americas." In *Advances in Archaeological Method and Theory*, vol. 7, edited by M. B. Schiffer, 39–102. New York: Academic Press, 1984.
15. Flannery, Kent V. "Culture History v. Culture Process: A Debate in American Archaeology." In *Contemporary Archaeology*, edited by M. Leone, 102–107. Carbondale and Edwardsville: Southern Illinois University Press, 1972.
16. Gall, P., and A. H. Saxe. "The Ecological Evolution of Culture: The State as Predator in Succession Theory." In *Exchange Systems in Prehistory*, edited by T. K. Earle and J. E. Erickson, 255–268. New York: Academic Press, 1977.

17. Haas, J. *The Evolution of the Prehistoric State.* New York: Columbia University Press, 1982.
18. Hitchcock, R. K. "Patterns of Sedentism Among the Basarwa of Eastern Botswana." In *Politics and History in Band Societies*, edited by E. Leacock and R. B. Lee, 223–268. Cambridge: Cambridge University Press, 1982.
19. Hughes, D. "Urban Growth and Family Structure in Medieval Genoa." *Past and Present* **66** (1975): 13–17.
20. Johnson, Gregory A. "Local Exchange and Early State Development in Southwestern Iran." Anthropological Papers 51. Ann Arbor: Museum of Anthropology, University of Michigan, 1973.
21. Johnson, Gregory A. "Decision-Making Organization and Pastoral Nomad Camp Size." *Human Ecology* **11** (1983): 175–199.
22. Kenyon, K. M. "Some Observations on the Beginnings of Settlement in the Near East." *J. Royal Anthropological Institute* **89(1)** (1959): 35–43.
23. Leacock, Eleanor B. "The Montagnais 'Hunting Territor" and the Fur Trade." American Anthropological Memoir 78. Menasha, WI, 1954.
24. Leacock, Eleanor B. "Class, Commodity, and Status of Women." In *Women Cross-Culturally: Change and Challenge*, edited by R. Rohrilich-Leavitt, 601–618. The Hague: Mouton, 1975.
25. Leacock, Eleanor B. "Women's Status in Egalitarian Society: Implication for Social Evolution." *Current Anthropology* **19(2)** (1978): 247–276.
26. Leacock, Eleanor B. "Relations of Production in Band Society." In *Politics and History in Band Societies.* edited by E. Leacock and R. B. Lee, 159–170. Cambridge: Cambridge University Press, 1982.
27. Leacock, Eleanor B., and Richard B. Lee, eds. *Politics and History in Band Societies.* Cambridge: Cambridge University Press, 1982.
28. Lee, Richard B. "!Kung Spatial Organization: An Ecological and Historical Perspective." *Human Ecology* **1(2)** (1972):125–147.
29. Lee, Richard B. "Population Growth and the Beginnings of Sedentary Life Among !Kung Bushmen." In *Population Growth: Anthropological Implications*, edited by B. Spooner, 343–350. Cambridge, MA: MIT Press, 1972.
30. MacNeish, R. S. "Ancient Mesoamerican Civilization." *Science* **143** (1964): 531–537.
31. Naroll, R. "A Preliminary Index of Social Development." *Amer. Anthropologist* **56** (1956):687–715.
32. Netting, Robert McC. "A Trial Model of Cultural Ecology." *Anthropology Quarterly* **38** (1965):81–96.
33. Netting, Robert McC. "Household Organization and Intensive Agriculture: The Kofyar Case." *Africa* **35** (1965):422–429.
34. Netting, Robert McC. "Household Dynamics in a Nineteenth-Century Swiss Village." *J. Family History* **4** (1979):39–58.
35. Plog, Fred, and Steadman Upham. "The Analysis of Prehistoric Political Organization." In *The Development of Political Organization in Native North*

America, edited by E. Tooker and M. Fried, 199–213. Washington, DC: American Ethnological Society, 1983.

36. Plog, Stephen. "Village Autonomy in the American Southwest: An Evaluation of the Evidence." *Society for American Archaeology Papers* **1** (1980): 135–146.
37. Plog, Stephen. *Stylistic Variation in Prehistoric Ceremics.* London: Cambridge University Press, 1980.
38. Plog, Stephen. "Sociopolitical Implications of Southwestern Stylistic Variation." In *The Use of Style in Archaeology*, edited by M. W. Conkey and C. A. Hastorf, 56–73. Cambridge: Cambridge University Press, 1987.
39. Sackett, J. R. "Style and Ethnicity in the Kalahari: A Reply to Wiessner." *Amer. Antiquity* **50** (1985): 154–159.
40. Sahlins, Marshall D. "Land Use and the Extended Family in Moala, Fiji." *Amer. Antiquity* **59** (1957): 449–462.
41. Simpson, G. G. *Tempo and Mode in Evolution.* New York: Columbia University Press, 1944.
42. Simpson, G. G. "Periodicity in Vertebrate Evolution." *J. Paleontology* **26** (1952): 359–370.
43. Simpson, G. G. *The Major Features of Evolution.* New York: Columbia University Press, 1953.
44. Upham, Steadman. *Polities and Power: An Economic and Political History of the Western Pueblo.* New York: Academic Press, 1982.
45. Upham, Steadman. "A Theoretical Consideration of Middle Range Societies." In *Archaeological Reconstructions and Chiefdoms in the Americas*, edited by R. Drennan and C. Uribe, 345–368. New York: University Press of America, 1987.
46. Upham, Steadman., Paatricia Crown, and Stephen Plog. "Alliance Formation and Cultural Identity in the American Southwest." Paper presented at the School of American Research, Santa Fe, NM, 1989.
47. Upham, Steadman, ed. "Analog or Digital?: Toward a Genetic Framework for Explaining the Development of Emergent Political Systems." In *The Evolution of Political Systems*, 87–118. London: Cambridge University Press, 1990.
48. White, L. *The Evolution of Culture.* New York: McGraw-Hill, 1959.
49. Wiessner, Pauline W. *Hxaro: A Regional System of Reciprocity for Reducing Risk among the !Kung San.* Ph.D. dissertation, University of Michigan. Ann Arbor: University Microfilms, 1977.
50. Wiessner, Pauline W. "Style and Social Information in Kalahari San Projectile Points." *Amer. Antiquity* **48** (1983): 253–276.
51. Wiessner, Pauline W. "Styles or Isochrestic Variation? A Reply to Sackett." *Amer. Antiquity* **50** (1985): 160–166.
52. Wilk, Richard R., and Robert McC. Netting. "Households: Changing Forms and Functions." In *Households: Comparative and Historical Studies of the Domestic Group*, edited by R. McC. Netting, R. R. Wilk, and E. J. Arnould, 1–28. Berkeley: University of California Press, 1984.

53. Wobst, H. Martin. "Boundary Conditions for Paleolithic Social Systems: A Simulation Approach." *Amer. Antiquity* **39** (1974): 147–178.

Steadman Upham,[*] Fred Plog,[**] Robert Reynolds,[†] and Joseph Tainter[‡]
[*]The Graduate School, University of Oregon, Eugene, OR 97403
[**]Department of Sociology and Anthropology, New Mexico State University, Las Cruces, NM 88003
[†]Department of Computer Science, Wayne State University, Detroit, MI 48202
[‡]U.S.D.A. Forest Service, Albuquerque, NM 87131

A Perspective on Systems Modeling in Southwestern Archaeology

Systems modeling has a long history in the natural sciences that predates by several decades the appearance and use of such modeling in archaeology. Since the 1960s, however, archaeologists have attempted to model cultural systems (or aspects thereof) in a variety of ways. The many efforts to model cultural systems reflect the need of archaeologists to construct parameters and boundaries for the events and processes that unfolded in the past and to simulate the operation of human cultural systems through time. Like most scientists, archaeologists employ simulation models as heuristic devices that can be used to examine the ways systems behave under specific conditions. Recent efforts to model cultural systems in archaeology have been dramatically affected by improvements in computer technology and simulation techniques, including the refinement of several different simulation programming languages.

In the brief statement that follows, our concern is not with the technical aspects of computer simulation or systems modeling, although these are important to any coherent modeling effort. Rather, our concern is to provide a concrete theoretical foundation and methodological structure upon which archaeologists may identify, describe, and explain increases or decreases in the complexity of cultural systems. Unfortunately, this daunting task may only be completed in the most rudimentary and partial way given our limited understanding of the operation of the many

Understanding Complexity in the Prehistoric Southwest,
Eds. G. Gumerman and M. Gell-Mann, SFI Studies in the Sciences of
Complexity, Proc. Vol. XVI, Addison-Wesley, 1994

variables involved. Fortunately, the position statement[46] that precedes this paper provides clear direction for such an effort, and we intend to build (albeit selectively) on the dimensions of the generic framework outlined by Upham. For this paper, we choose to focus specifically upon how one models increases or decreases in the complexity of *regional systems* and *alliances*, a topic that has been of considerable interest to archaeologists during the past decade.[5,30,44,49]

A variety of issues inform the discourse about regional systems and alliances that are especially pertinent to the present consideration of systems modeling. Many of these issues were spawned as a result of disagreements among archaeologists about appropriate theories and approaches for studying the past. Cordell[7] has provided the uninitiated with a compendium of the divergent theoretical positions current in archaeology today. Unfortunately, the creative chaos identified by Cordell underlies a variety of other more profound methodological differences among archaeologists. We believe, however, that the real differences that separate archaeologists today are not necessarily easily explained by reference to subjective labels—ecological, empirical, functionalist, post-processualist—that are given to different kinds of arguments and different analytical points-of-entry to explanations of change. Instead, we see the issues as being more directly related to what is possible for archaeologists to accomplish given the limitations of the archaeological record. Some archaeologists may claim that this is an epistemological issue (but, then, isn't everything?). For purposes of clarity, however, we prefer to treat the issues more programmatically in this brief statement on systems modeling.

EQUIFINALITY

Several papers in this volume underscore the problem of equifinality as imposing severe, if not insurmountable limitations on the explanatory potential of archaeology. The fact that two or more plausible explanations of the same observed data can be proposed but not refuted, however, should not be a cause for despair among archaeologists. Instead, we believe that exactly the opposite reaction is appropriate: that the perceived problem of equifinality actually underscores the robustness of archaeological data *within certain quite prescribed limitations.* These limitations include, but are not limited to the fact that (1) archaeologists deal with extinct cultural systems that can no longer be directly observed, (2) the archaeological record is a partial record of the past and does not contain the remains of entire cultural systems (as once asserted), and (3) archaeologists employ many indirect measures to describe phenomena that cannot be directly measured using archaeological data. An example of this latter limitation is the way archaeologists estimate past demographic regimes, and especially population size, using room counts, the floor area of residential architecture, or refuse volume at different kinds of sites.

We argue that the key to working within these limitations is to first recognize that they exist and, then, to design research strategies that are sensitive to these boundary conditions. We suggest further that the emphasis on equifinality in this context is misplaced and is used by archaeologists as a convenient club to beat down specific explanations of change that may not conform to received wisdom or textbook orthodoxy.

In its place, we believe that a renewed emphasis should be placed on devising indirect or "proxy" measures for phenomena that cannot be directly measured. By a proxy measure, we mean nothing more than an indirect measure or index that is easily calculated from data collected in the course of routine archaeological fieldwork or analysis. For example, archaeologists routinely map architectural features and in the process obtain relatively precise measurements on the shape and dimensions of a variety of different kinds of structures. These numeric data are often reported incidentally in journal articles and monographs, sometimes only appearing as secondary information on the legends of maps. Instead, these data (and a host of other similar measures) could be used to establish an index of space that in turn could function as a generalized proxy measure for construction intensity, population, or other similar variables. Such measures could facilitate comparative analyses between different geographic regions, between groups with different cultural histories, and for archaeological manifestations of different age.[35] Unfortunately, the use of such proxy measures does not allow direct comparison with the present (e.g., the population size of Awatovi at A.D. 1500 with the population size of modern Walpi), but allows more precise comparisons between archaeological cases. The creation and use of proxy measures will also permit archaeologists to assess (for the first time, in many cases) the intensity, magnitude, and duration of different kinds of cultural phenomena within and between regions and will provide the kind of data needed to evaluate competing explanations of change. Until archaeologists use and employ such measures routinely, we see little progress being made in resolving the very different scenarios suggested by various systems models.

Why do we see archaeological data as robust in the face of these seemingly profound limitations? In our estimation, the archaeological record is complex (that is, it is comprised of objects, relationships between objects, and contexts) and exists within a dynamic matrix that alters and transforms the record in ways that must be discovered. The archaeological record is thus measured only by using a very large number of variables, relationships between variables, *and* contingent conditions arising not only from the context of deposition, but also from the transformation of archaeological deposits during subsequent periods. Such complexity gives the archaeological record of specific regions a very robust character which is not unlike that of a complex ecosystem.

The creation and use of proxy measures is especially important for systems modeling because many of the key systems variables (like population size noted above) cannot be directly observed in the archaeological record. Similarly, other parameters of the system (for example, rates of change) are dependent on the precision and comparability of measurement techniques. While archaeologists will eventually realize the kind of precision seen in the other sciences, much can be made for an argument for standardizing measures and systematizing data-recording strategies. In the Southwest, this issue is especially critical to archaeological analyses that seek to address issues at scales larger than the artifact (in the extreme, one could argue that they even affect precision on this scale of analysis). As archaeologists become more interested in the larger questions dealing with increases or decreases in the complexity of entire cultural systems, the kind of rigor demanded by standardization of data recording and analysis is essential.

Nowhere is this problem more evident than in the recent discussion that has taken place among Southwestern archaeologists regarding the emergence of alliances and regional systems at different times in prehistory. In the remaining parts of this paper, we address the concepts of alliances and regional systems within the larger framework of complex systems and nonlinear dynamics. Our theoretical consideration is followed by a more detailed consideration of the pertinent variables and by a still more concrete exposition of an initial structure that can be used to model portions of extinct cultural systems.

REGIONAL SYSTEMS

As students, most of us were taught a simple means of conceptualizing prehistory. Time was the y axis and space the x axis. In each box created by the x–y intersects was the phase of a culture. We now know that this approach is flawed in relation to both data and theory. For example, there were times (e.g., the Archaic) when the Southwest was occupied by peoples with largely indistinguishable patterns of material culture and behavior. There were also times when the distribution of artifacts (e.g., Kana'a black-on-white) or features (e.g., Chacoan roads and public buildings) suggest a relatively unitary system across very large areas. Finally, there were times when regional differences in artifacts (Jeddito yellow ware, Zuni glaze ware, and the Gila polychromes) suggest a significantly regionalized pattern. Such data are not neatly placed in boxes; rather they suggest the swirling pattern of a storm.

A convenient means of conceptualizing spatial variation in a manner that recognizes these very different time/space distributions is that reflecting a synthesis of the efforts of Plog,[28,30] Wilcox,[47] Upham,[44] and Hantman.[15] This approach is sensitive to the size of social groups and the strength of interactive ties among members. We offer the following concepts from the above-cited works (modified slightly) that allow us to partition spatial variation and describe it in a social context.

SOCIAL NETWORK AREA. A social network[15] comprises an area varying from tens to hundreds of square kilometers in which a family or group of families interacted on a routine basis.

REGIONAL SYSTEM. A regional system is a neutral term (like "interaction sphere") that refers to geographical area(s) of hundreds to thousands of square kilometers in which populations interacted in unspecified ways. "Regional system" is a term used to denote an area and its population where we do not feel confident applying a more specific term like province.

PROVINCE. A province[28,34] is an area comprising thousands to tens of thousands of square kilometers in which the interaction between members of social networks was sufficiently intense so as to have produced observable homogeneity in material culture.

ALLIANCE. An alliance[44] may encompass portions or all of a single region or province, or it may incorporate several provinces. To postulate an alliance is to advance a hypothesis about the strength of social, political, and economic ties between the inhabitants of regions and provinces.

SCALE AND ATTRIBUTES

Social networks, provinces, and alliances are new concepts used to identify patterns that have long been recognized by archaeologists. We believe that these concepts are sensitive to differences in scale that have very important social, political, and economic implications. A primary difference between these new concepts and the kinds of distributional arguments advanced in the past is that they inject a social dynamic into arguments that were formerly completely distributional in nature. For example, the pottery handbooks developed by Harold S. Colton provide archaeologists the ability to identify and describe ceramic types over wide areas. Colton's systematic descriptions are difficult to improve upon even years after their initial formulation. It is because of this important work that contemporary archaeologists have been able to differentiate ceramic patterning at regional scales. Variability in ceramic distributions have been and continue to be used as ways to provide relative chronological controls in the absence of absolute dating. Moreover, the work of Colton has served as the baseline for virtually all culture historical studies completed since his detailed typologies were first published.

In the central and northern Southwest, one of the first important statements about the social content of regional ceramic distributions was made by Gumerman[14] in his study of the Hopi Buttes region. Gumerman not only identified a "strong pattern" in the ceramic distribution of Little Colorado white ware, but also evaluated settlement distributions and demographic data to posit the development of a localized interaction network at A.D. 1200. Looking back on Gumerman's work, it is clear that this research provided an outline for later formulations of social networks, provinces, and alliances. His research, along with that of Eddy,[9] Bluhm,[2] Longacre, and others, allowed later researchers to resolve more clearly both the extent and content of regional ceramic distributions. More importantly, it facilitated the development of models and hypotheses that were aimed at elucidating the social basis of interaction in prehistory.

As a result of the research noted above, archaeologists are now using a few classes of artifacts and attributes to identify the formation, development, and collapse of social networks, provinces, and alliances. The artifact classes most suited to this kind of research are the most durable, and hence enduring, of those found on Southwestern sites: ceramics and lithics. More specifically, ceramic and lithic attribute classes have been identified that are more sensitive to regional spatial and temporal variation than the artifact classes themselves. These include ceramic design styles (usually meaning metric measurement of design attributes, but also including more inclusive approaches that rely on surface color and motifs), ceramic technology (including the identification of clay, temper, surface treatment, and occasionally firing temperature), and lithic raw material (including proveniencing of source). These classes of artifacts and attributes have been used to suggest that the social configurations noted earlier were present in the Southwest at different times and places after A.D. 700.

In some cases, especially after A.D. 1100 when exchange relationships in the Southwest expanded, more specific kinds of artifacts (turquoise jewelry, copper bells, macaws, shell, minerals, and pigments) can be used to define these spatial units at different scales. In all cases, it is the homogeneity of different classes of artifacts and attributes in space and time that leads archaeologists to hypothesize the existence of these interactive patterns. The initial research of Upham,[44] Lightfoot,[19] Hantman,[15] and Neitzel,[23] as well as much of their later work has been devoted to identifying such patterns.

It is important to note that the theoretical framework we advocate does not link the data to a linear succession of forms, but rather to dynamics that are nonlinear. At some times and in some places, peoples in the prehistoric Southwest came together in a way that some have called "alliances,"[44] "strong patterns,"[30] or the "cookie cutter effect." Universally clear is the rarity of this pattern. Only once between A.D. 700–900 (the Kana'a-White Mound pattern described by Plog[28]) does it appear to have been as widely distributed as even over the entire Colorado plateaus. Following A.D. 900, more localized regional alliances emerged (for example, the Chacoan System). In only single epochs can the same claim be made for the southern deserts of the Southwest (A.D. 1150 to 1300 during the "Classic" Hohokam period) or for parts of that region and northern Mexico (A.D. 1300 to 1450 during the florescence of Casas Grandes). During later periods in the plateau Southwest (that is, after A.D. 1250), a few regional alliances appear to have emerged that encompassed very large areas.[45] More typically a regional system came to exist, as described in a later section of this paper.

A great deal of work also remains to be done to determine the precise nature of temporal variation in these patterns. Important to the effort are the realization that:

1. Most areas of the Southwest were characterized by one or several periods of time when people were not present or were involved in adaptive strategies leaving little impact on the archaeological record.
2. Archaeologists must improve understanding of the placement of sites in time as a probability, not a fact. The considerable potential for such an approach has been discussed by Plog[27] and Schact.[35]
3. While it is important to recognize that one cannot immediately infer temporal from spatial variation, there are important "phase space"[13] correlates needing study.

If one acknowledges variations in "strong patterns" or "strong boundaries" among groups, the issue of why such variation arises is immediate (attractors?). The possibilities are many. At one extreme is the argument for islands in seas of chaos—that we tend to make too much of apparent organization in random events. At the other extreme is the claim of strong causal factors—intensification of agriculture, productive specialization, or population growth, for example.

During the last decade, our understanding of settlement distributions and longevity in the Southwest has changed dramatically. Sites are no longer analyzed as if occupied for 200 years.[15] Very large sites are seen to be "rare events."[6] Major hiatuses in the archaeological record have been identified.[31] That sites appeared concentrated in a few well-watered drainages is now seen as more the product of the types of sites in which archaeologists had interest than the behavior of prehistoric peoples. Let us now consider some patterns warranting further exploration.

DISTRIBUTION

SARG, ARM, and AZSITE not withstanding, we have no overall notion of site distributions in the Southwest. We now know sites to be virtually everywhere, but our understanding of the spatio-temporal patterning of the distribution of those sites remains poor, especially if different types of sites are the subject of our study. Given the vast investment of time, energy, and money involved in efforts that range from SARG to the development of state-level site files, it is time to use the computer facilities routinely available to all of us to produce a refined distributional picture.

One barrier to such an effort is our remaining indecision regarding site types. Once again, the development of proxy measures based on the spatial dimensions of different kinds of sites may be appropriate here. Efforts to impose field-based classifications made by government agencies will not be useful in resolving this problem. Careful studies of the attributes of sites and analyses of the artifacts on those sites can result in important modifications of our understanding of the prehistory of the Southwest.

SIZE

That large sites are relatively rare occurrences in the archaeological record is now a given, but the spatio-temporal distribution of sites of different sizes remains unknown. Similarly, it is unclear when such sites were "aggregates" and when they were major political and economic centers. Previous studies[27] show that even small sites that have four standing walls can contain the majority of trade goods in a region. It is possible (but we think unlikely) that some very large sites were relatively unimportant politically and economically.

We must learn to know the difference. Large sites tend to be the best dated of Southwestern sites, although this is not universally the case. The potential for showing them, as Kidder[16] said, "twinkling like stars in the night sky" as they come and go is now a substantial potential. The analytical techniques now available to us for analyzing the patterning are of equal importance.

AN APPROACH TO SYSTEMS MODELING

In the systems modeling position paper, Upham provides a conceptual framework for studying cultural change.[44] In terms of this framework he identifies four basic dimensions within which one can describe the complexity of cultural systems. These dimensions are: demographic, economic, social, and political. The participation of an individual in a society can be described relative to the relationships allowed in each of these dimensions. The question addressed in this section is: how may one model the emergence of these structures as the result of the behaviors of the individuals who comprise the system?

The traditional frameworks for modeling cultural change have utilized various evolutionary paradigms as metaphors for cultural change at the individual level. For example, Reynolds has used Holland's Genetic Algorithm, an exploratory learning approach based upon Darwin's paradigm for evolution, as a framework for modeling the acquisition of techniques for incipient agriculture by individuals in the Valley of Oaxaca, Mexico.[30] The model was used to generate inductive generalizations about successful resource acquisition strategies based upon the experience of individual hunter-collectors. Many of these experiences were opportunistic in nature. Thus,

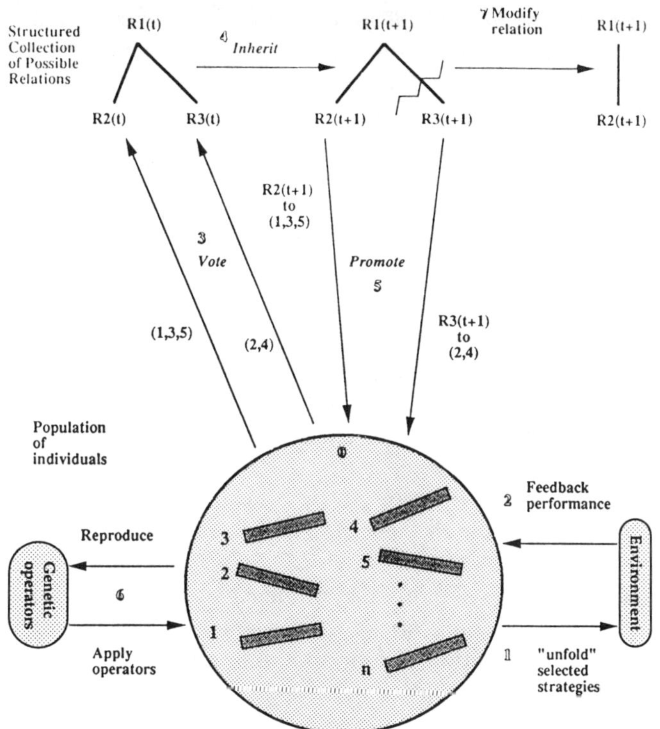

FIGURE 1 Schematic description of KBGA system.

the Genetic Algorithm was used to produce a structured base of knowledge from the unstructured or chaotic experience of individuals. The behavior of the modeled group compared well with the associated archaeological record.

Holland's approach is a generic one that can be used to model aspects of cultural evolution, but is capable of supporting an endless variety of other applications. Boyd and Richerson extended a model developed by Cavalli-Sforza and Feldman,[4] and based also upon the Darwinian paradigm, but designed specifically to deal with the acquisition of cultural traits by individuals in a population.[3] This approach was used to demonstrate a functional basis for the *waytakuy* llama marking ceremonies practiced by herding communities on the Punas of Ayacucho Peru.[10]

Both of the techniques described above were successfully used to model the acquisition of complex cultural traits by individuals. However, modeling the emergence of political, economic, social, and demographic structures based upon the experiences of individuals requires an additional component. This component is needed to channel the generalized experience of individuals into the formation or modification of shared cultural frameworks in which individuals participate. These frameworks describe relationships that pertain among individuals in a population. The relationships can pertain to kinship, political and economic decision-making, and social relations, as suggested by Upham. The experience of an individual should potentially be able to affect each of the four categories of cultural structures given by Upham.

Reynolds has extended Holland's Genetic Algorithm in order to support the generation of these shared relational frameworks based upon individual experiences.[31] The extension of the genetic algorithm integrates a technique for explicitly learning about conceptual structures, that is, relationships among individual units. This hybrid learning system uses the experience of the individuals modeled in the genetic algorithm component to direct the emergence of specific relationships among the behaviors of these individuals. These shared relational structures can, in turn, be used to influence the behavior of the individuals that are associated with them. This hybrid paradigm is called the Knowledge-Base controlled Genetic Algorithm, or KBGA for short.

In Figure 1 a schematic description of the KBGA is given. The basic sequence performed therein is now briefly discussed:

0. INITIALIZE. The population of individual problem-solvers is represented as a pool of structures in the diagram. Each structure is graphically represented in the diagram as a rectangle. The collection of supported relationships among individuals is represented as a collection of graphical structures.

1. TEST. In a given time step, a subset of individuals in the population is selected to interact with a given environment. Each of these selected individuals is "unfolded" in order to represent its interaction with the environment. Its interaction can be modulated by the extent to which it is associated with the set of shared conceptual structures.

2. UPDATE. The performance of an individual is then "fed back" to the population in order to update its performance.

3. VOTE. Next, the information about the performance of each individual during that time step is broadcast to those portions of the shared conceptual structures with which the individual is associated. This is called "voting" since the performance of an individual is used to update or "vote" for the utility of a given relationship that they have participated in.

4. INHERIT. The "inherit" process involves the actual updating of the performance information associated with the current set of relationships based upon the experiences of the individuals associated with each. In other words a portion of each relational structure inherits a performance or utility adjustment based upon the performance of those individuals who exhibit it.

5. PROMOTE/DEMOTE. The updated performance figures for each structure are then sent back to the population of individuals in order to "promote" those individuals associated with relationships, and "demote" those who are not.

6. MODIFYING THE INDIVIDUAL STRUCTURE. Individuals are then subject to modification using genetic operators based upon their individual performance in the environment as well as the performance of those shared structures with which they are associated.

7. MODIFYING THE RELATIONAL STRUCTURE. This phase involves the possible removal of unproductive relationships, and the addition of new, productive ones. A particular implementation of the KBGA, the VGA, has been used to solve a card inference problem posed by Nilsson,[24] a multiplexor inference problem posed by Wilson,[48] and a problem in the acquisition of software engineering knowledge posed by Reynolds.[31] In the VGA, the set of possible relationships is represented as a lattice of hypotheses. The learning activity involves systematically isolating the most general set of hypothesized relationships that are consistent with the observed data. The reader is referred to the paper by Reynolds given above for the details of the implementation. The generic version of the KBGA presented here is intended to serve as a basic framework in which the shared complex conceptual structures suggested by Upham might emerge. It might be used as a vehicle for assessing the complexity of cultural systems, a goal proposed by Gell-Mann. That is, what are the minimal algorithms required to produce certain types of shared structures in some standard environment?

MODELING A SIMPLE REGIONAL SYSTEM

While it may once have been common to view Southwestern communities as if isolated and independent, no archaeologist today can argue for such a position. To paraphrase John Donne, no Southwestern society was an island, entire unto itself. Each prehistoric Southwesterner was embedded directly in relations of kin group and community, and each community had ties to other communities and other regions. At the highest level, Southwesterners participated in trade networks that spanned much of North America.

In 1981 one of us (Tainter) began a project—the Elena Gallegos Project—that involved survey and excavation in several localities in the area of Albuquerque, New Mexico. These localities—Atrisco on the Rio Puerco, the lower Las Huertas Valley, and the Ball Ranch—were each, throughout prehistory, peripheral areas of land-use and settlement. The research design written in 1982 for the excavation phase of the project (reprinted in Tainter[38]), proposed that the intensity of use of these peripheral areas would be a response to population growth in the Albuquerque area. When the analysis of the results was finally completed, it was clear that the research design was utterly misguided. The patterns of settlement and use in these areas could not be understood in terms of local processes. To understand these peripheral areas required a context involving events and processes across large areas of northwestern New Mexico, specifically the formation of economic ties between high and low elevation territories; the extent of the Chacoan system; and the abandonment of the Four Corners area.[39] The lesson is that regional analysis is necessary to understand even areas of peripheral use.

To understand the operation and evolution of regional systems, we must be able to model characteristics of these systems in such a way as to yield quantifiable predictions. This involves identifying critical variables and system parameters, understanding the relations among variables, and estimating the rates of resource flow. Robert Reynolds has shown the power of such models in both ethnographic and archaeological simulations.[10,30] Earlier work along this line was done at the Massachusetts Institute of Technology by Jay Forrester's System Dynamics group.[11]

We will discuss the modeling procedure in terms of a case introduced on the first day of the workshop: the economic relations between higher elevation areas of western New Mexico, such as Acoma, and the lower elevation Rio Puerco of the east, in the area of Atrisco.[40] As noted in the opening-day address, the Atrisco area displays a strong, persistent pattern: nearly all of the pottery was imported and about 95% came from the west. This includes the area of Grants, the vicinity of Acoma, and the territory south to the Datil volcanics and the upper Rio Salado drainage basin. This pattern persisted from about A.D. 500 to 1300. The Atrisco sites represent both seasonal use and longer term settlement. Two factors probably account for the ceramic pattern. The first is seasonal foraging and planting by people otherwise resident in the areas from which the pottery originated. The second, which applies during times when Atrisco sustained long-term settlements, is that the people of the central Rio Puerco Valley had their major economic ties with people to the west and southwest. Atrisco, thus, appears to have been at the eastern end of an east-west network of transhumance and economic exchange.

The relationship between Atrisco and higher elevation areas to the west represents perhaps the most basic kind of regional system: the linking of areas with productivity cycles that may vary nonsynchronously. To include both kinds of areas in a seasonal round, or to plant crops in higher and lower elevation areas, or to form exchange ties between groups occupying such areas, are among the most basic things that a population can do to increase subsistence security. While this is not the only reason why regional systems might emerge, it is one of the most basic and ubiquitous reasons. It is a good place to begin the process of modeling a regional system. For convenience, we will refer to this very basic regional system as Acoma-Atrisco, even though at the western end an area larger than historic Acoma was involved, and at the eastern end we don't know how much land was included beyond the Elena Gallegos project area. In what follows, then, references to "Acoma" should be read as shorthand for "the higher-elevation terrain extending from Grants and Acoma south to the upper Rio Salado drainage basin."

In the period of interest, populations depended substantially on agriculture. Yet the contents of the sites at Atrisco make it clear that it was used for foraging as well as agriculture. For both types of use we must know about population densities, per capita consumption requirements, and fluctuations in environmental productivity.

As pointed out in another of the first-day addresses,[22] demographic reconstruction is an uncertain science (although the Southwest is among the best areas for such analysis). Many variables affect the range of possible estimates. Yet this uncertainty can be ameliorated somewhat by a systems modeling approach, provided that a model is developed that is robust enough to support simulation. It is often the case that general constraints imposed on the operation of each subsystem, when taken together, force the whole system to behave in a very constrained and predictable manner. By manipulating the variables that are used in a simulation model, we can see what demographic parameters produce credible results, and which don't. We can also test the extent to which the behavior of the modeled system is sensitive to the model's basic assumptions.

In the matter of consumption requirements, we are on firmer ground. There is much general knowledge about per capita requirements, in both calories and specific nutrients. Of equal importance, as a result of many years of studying faunal and botanical remains from Puebloan sites, we now know much about the Anasazi diet on the Colorado Plateau. Combining knowledge of diet and dietary requirements with maps of known and potential vegetation would allow us to predict an ideal seasonal round. The native vegetation of Atrisco has been greatly altered, and much of the area is, in fact, devoid of vegetation today. The archaeological sites, though, have given us much information about the resources sought there.[1,36,41]

Fluctuations in environmental productivity may be fairly easy to model, although the responses of natural resources to variations in rainfall are not known at the level of detail that would be desired. (Agricultural experiment stations may have some of these data.) The most important short-term factors would have been variations in the amount *and distribution* of rainfall, and in the length of the growing season. Over the long-term, erosion cycles were very important, particularly in the erosion-prone Rio Puerco Valley. The effect of human populations on the environment much also be a component of any realistic model.

A model incorporating these factors, and perhaps also including population growth rates, would allow us to predict year-to-year variations in how the Atrisco area was used, and also to understand in what periods long-term settlement was desirable. Among the things that simulating a detailed model would help us to understand are:

1. How often would higher elevation populations have needed to forage in the Atrisco area?
2. To what extent was this use conditioned by yearly variations in the quantity and distribution of rainfall? What were the effects of shifts in rainfall along an east-west gradient, or along a north-south gradient?
3. How much territory along the Rio Puerco would a higher elevation community need as part of its sustaining area?
4. During periods when Atrisco sustained sedentary occupations, was it possible to maintain long-term economic ties with the highlands on the basis of reciprocity? Or would there have been an imbalance in exchange that might lead one area to exchange services or other goods for foodstuffs? Exchange systems that are systematically unbalanced can lead to relations of status and obligation between participants.[37] Thus a simple consideration of land use and trade leads inevitably to questions of specialization and status relationships.

Unfortunately, even the simplest systems are never really simple, and this is particularly true in human societies. Any living system has support costs, and as elements are added and complexity increases, costs do also. In the Acoma-Atrisco land-use model, attention must be given to the costs of travel and transport. The costs of transporting food by human bearers are fairly well known and quite high. Kent Lightfoot has compiled data on this matter specifically for the study of Puebloan societies.[17] T. Patrick Culbert has estimated that in the central Maya Lowlands, food transported 100 km would have cost 33% of its value in consumption by bearers.[8] This is characteristic of transport in societies without mechanized vehicles. In the Roman Empire in A.D. 301, when Diocletian issued his Edict on Prices, it cost less to send grain by ship from one end of the Mediterranean to the other than to send it 120 km by cart.[15]

This matter leads to a further question that could be clarified by studying the behavior of a simulation model.

5. Was it more efficient to transport consumers to resources (i.e., Acoma to Atrisco), or resources to consumers?

This is an important question, for the answer to it will tell us whether to expect evidence of settlements, summer family camps, or task-group activity loci at Atrisco. The archaeological record, in fact, seems to show all three kinds of sites.[34,35,39] Perhaps this indicates that at times it was more efficient to move population segments to Atrisco, at times the opposite was true, and at times it was best to establish long-term settlements along the central Rio Puerco.

The final consideration is to assess the cost to Acoma-area people of maintaining their territorial claims to Atrisco. Atrisco is some distance from Acoma, it was often used only seasonally, and, depending on climatic factors, there may have been periods of up to several years when it was used little or not at all. Yet Atrisco was used by, or linked to, Acoma and other higher elevation people for perhaps 800 years. How would a territorial claim be maintained through all these centuries, and what were the costs of maintaining it? At this point the model becomes potentially much more complex, for it involves the relationship of Atrisco to surrounding areas. These include the upper Rio Puerco, which was densely settled, and the Rio Grande. To the south, the lower Rio Puerco seems to have been little used prior to A.D. 1300.[49] We discuss two ways that the territorial claim to Atrisco could have been maintained against other potential users:

1. Acoma-area people could have claimed the area for their exclusive use. This would be a costly strategy, for it would require constant monitoring against "poachers." Such monitoring could never have been 100% successful.
2. Acoma-area populations could have allowed *regulated* use of the area by other people. Such use would probably have been regulated by ties of kinship, gift-exchange, status obligation, or other vehicles of social alliance. The 5% of Atrisco pottery that came from the south, east, and north could reflect such use. On the face of it, this would be a less costly strategy than 1 (above), and one with positive benefits in establishing ties with other groups. The drawback would come in years of low productivity, when both Acoma groups and others might have wished to use the area simultaneously.

To add the factor of territoriality into our model would require building models of demography, land-use, and resource fluctuations for the areas abutting Atrisco on the south, east, and north. We cannot attempt to do so here, but the exercise has now taught us a lesson. It may be difficult to model any single southwestern region as a closed system existing in isolation. As in the building of any model, the question arises of how much of a system's environment needs to be included. Certainly, if our answer is "all of it," then simulation may not be a fruitful approach. If it seems that some factors can be removed, this can be done in two ways: either "top down," by adding factors in order of importance, or "bottom up," by removing factors one at a time, and each time rechecking the system's behavior.

Where previously we learned that Southwestern communities could not be studied as if isolated, so now we learn that even regions cannot be so studied, except by ignoring factors that may turn out be be quite important. This is a serendipitous outcome of the model-building process: by trying explicitly to recognize all factors that affect even a simple system, we learn that the system is not simple after all. In simulating a model, of course, boundaries must be drawn somewhere, and it may be that the factor of territorial maintenance is one factor that must be excluded. Yet we will exclude it in conscious recognition that we are doing so, and if the model proves inadequate to account for the archaeological record of Atrisco, we will know where to look to find the problem.

INITIAL STAGES IN MODEL BUILDING

The genetic algorithm described in a previous section was couched in abstract terms that are applicable to many kinds of systems. To model the Acoma-Atrisco land-use system, these terms will need to be translated into anthropological concepts. We will devote a few sentences to this.

Individuals are the fundamental units of analysis. In the present case they will be units by which information is collected. These are economic units, of which the most basic will be task groups. Based on *experiences* in the environment, individuals feed information back to more-inclusive elements, which form *shared structures*, or interaction among units. In anthropological terms, the shared structures will be the corpus of cultural data localized in kin groups, sodalities, or communities. Shared structures regulate or prescribe how task groups interact with each other and with the environment.

Individuals *vote* for the utility of a given structure by acting in ways that are or are not consistent with the relations prescribed by the structure. If the structure no longer prescribes appropriate behavior vis-à-vis the environment, it will be *updated* based on the current experiences of individuals. The shared structure is open to *alteration by an associated learning algorithm or structure generator* (i.e., learning).

Individuals associated with successful structures or portions of a structure will be *promoted*. That is, there is an increased likelihood that their behavior will be repeated. The converse holds as well.

In the previous section several problems were proposed concerning regional economic interaction among populations in parts of west-central and western New Mexico. The overall problem can be viewed as homomorphic to the problem of inferring the behavior of an arbitrary multiplexer circuit. The job of such a circuit is to facilitate the connections between certain input data lines and certain output data lines. If one assumes that each line corresponds to a part of a computer system, the question is what components/interactions are supported in the environment provided by the multiplexer? For N components the number of possible interactions is $N!$. Thus, the number of possible interactions increases exponentially as a function of the number of components.[48]

The multiplexer problem has been shown to be NP-hard. That is, it is one for which a tractable (polynomial time) solution is possible only on a nondeterministic machine. NP-hard problems are ones that are the most difficult to solve on computational systems. The problem of regional interaction given here is likewise NP-hard.

A specific implementation of the KBGA framework described earlier, the VGA, was used to solve efficiently several variations on the multiplexer problem. As a result the VGA shell was selected as the basic framework in which to express the simulation prototype to be described next.

In Figure 2, the problem of identifying regional interaction patterns is expressed in the KBGA framework. The relationships portrayed here pertain to the need for a population in a region to participate in particular activities with other economic units. The possible relationships correspond to the set of all subsets of populations in the region. This set can be organized hierarchically as a lattice, with the root corresponding to the hypothesis that all populations are related, and a leaf-node corresponding to the hypothesis of no interaction.

The lattice in Figure 2 can be termed a version space, in which all versions of the relationship(s) under consideration are displayed. The goal of the system is to isolate the most general set of relatively productive hypotheses (about relationships) from the set of all hypotheses in the version space. The model can then be used to determine the effects that different environmental configurations (e.g., fluctuations in environmental productivity) and patterns of economic behavior (e.g., seasonal rounds) have on relationships among populations. The relationships of best productivity (higher utility) can then be identified.

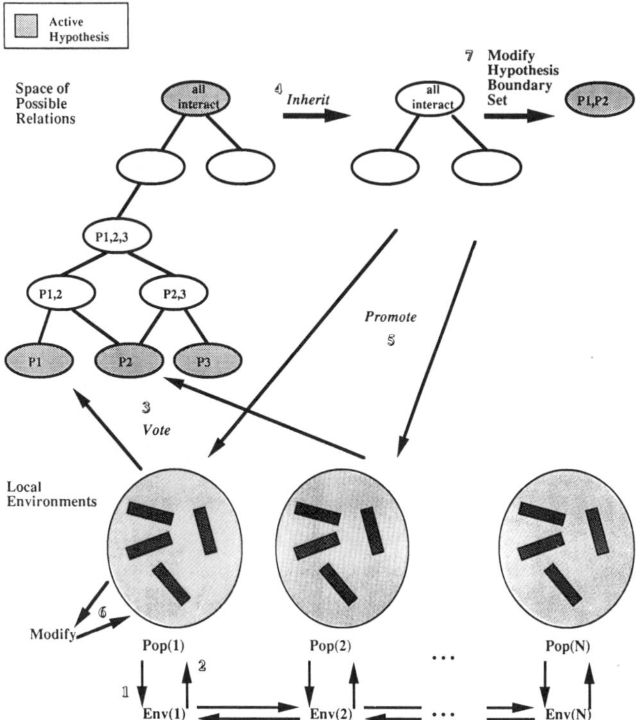

FIGURE 2 Expression of regional interaction model in terms of KBGA framework.

In order to do this, each population in the region can be viewed as a collection of resource acquisition strategies. Each population, POP(i), in the diagram is indexed from one to the number in the region, N. Each population has a pool of strategies, and each strategy corresponds to the behavior of at least one task group in the population. The number of strategies in each pool can be a function of the estimated density of that population, or can be estimated in other ways.

During the course of a year, a certain subset of strategies from each population can be used. Their relative probabilities of selection can be a function of the current or past state of the environment and the population. Each of the selected strategies is "unfolded" to act on the environment. This was labeled Phase 1 in the KBGA scenario (above), and is marked similarly in the diagram. Note that each phase in the diagram is indexed by the phase in the generic KBGA model to which it corresponds.

Each population has its own local environment that is indexed by the name of the population. Here, the local environment for Population 1, POP(1), is Environment 1, ENV(1), and so forth. The performance of each structure is assessed in terms of a basic payoff function that represents the results of interacting with its environment, and constitutes feedback to the population (see 2 in the diagram).

Note that the strategies in a population may refer to activities to be performed in another region. The arcs between the environments reflect the potential for individuals from one environment to use others. Although the current configuration allows all populations to exploit all environments, subsets of these areas may be removed for runs of a simulation to reflect the presence of physical or social boundaries to interaction.

The payoff associated with a group is best viewed as a vector. Each element of the vector reflects the number of units of a particular resource that are obtained. For example, performance can be viewed in terms of the number of calories, the amount of protein or other specific nutrients, or the amount of raw materials obtained. A strategy can, in fact, return some units of each.

Next, the behavior of individual task groups in the population is used to support or deny the hypotheses that certain relationships are associated with strategies of above-average productivity. This is Phase 3, or the voting phase. For example, if a strategy of POP(1) interacts with the environment from POP(2), then the performance of that strategy is used to update the performance associated with the (POP[1], POP[2]) node. The criteria for accepting a hypothesis of interaction corresponds to a given lower boundary in the performance value associated with a node.

The voting process identifies which nodes are to be updated (Arc 3). The updating process is the "inheritance" activity here, and is labeled 4 in the diagram. The updated performance information about each active hypothesis is then fed back to the strategies in each population associated with that hypothesis. This information is used to modify the performance of the most relevant strategies. That is, strategies leading to relatively productive relationships are promoted, and strategies leading to relatively unproductive relationships are demoted. This is identified by 5 in the diagram.

In Phase 6, individual strategies in each population can be reproduced and modified using the traditional genetic operator of mutation and recombination as a function of their relative performance.

In Phase 7, the current specification of the version space can be modified. The basic approach taken here is to consider only a subset of the version space to be active at any one time. This active set, or boundary region, is the set of hypotheses (nodes) exceeding the cutoff for acceptability. A maximally specific generalization (one stopping condition for the simulation) is generated when the boundary sets produced by the top-down exploration overlap with those generated from the bottom up.

A shell to support this basic framework has been written in *C*, and is running on a Sun workstation. What remains to be done to operationalize a simulation is to add in the application-specific knowledge that has been outlined here for the Acoma-Atrisco land-use system.

CONCLUSION

In order to operationalize the kinds of simulation models we envision with anything other than dummy data, we must first obtain precise information on the distribution, size, and age of sites in different regions. These data are minimum requirements and the Acoma-Atrisco regional system discussed in this paper in large part satisfy such requirements. More sophisticated modeling efforts, like the relatively simple model described above, also require information about different categories of material culture and their distribution, age, and duration of use. In the years ahead, archaeologists need to take advantage of the important advances in chronometrics and computerized databases (especially those with graphic interfaces) to develop accurate and reliable data that can be used in simulation models.

REFERENCES

1. Bertram, Jack B. "Faunal Remains from the Atrisco Sector." In *Draft Final Report of Investigations: The Albuquerque Sector of the Elena Gallegos Project*, edited by Joseph A. Tainter, 383–411. Albuquerque: USDA Forest Service, 1990.
2. Bluhm, E. A. "Mogollon Settlement Patterns in the Pine Lawn Valley, New Mexico." *Amer. Antiquity* **25** (1960): 538–546.
3. Boyd, R., and P. J. Richerson. *Culture and the Evolutionary Process.* Chicago: University of Chicago Press, 1985.
4. Cavalli-Sforza, L. L., and M. W. Feldman. *Cultural Transmission and Evolution: A Quantitative Approach.* Princeton: Princeton University Press, 1981.
5. Cordell, Linda S., and Fred Plog. "Escaping the Confines of Normative Thought: A Reevaluation of Puebloan Prehistory." *Amer. Antiquity* **44(3)** (1979): 405–429.
6. Cordell, Linda S., and G. J. Gumerman. *Dynamics of Southwestern Prehistory.* Washington, DC: Smithsonian Institution Press, 1989.
7. Cordell, Linda S. "The Nature of Explanation in Archaeology: A Position Paper." This volume.
8. Culbert, P. T. "The Collapse of Classic Maya Civilization." In *The Collapse of Ancient States and Civilizations*, edited by N. Yoffee and G. L. Cowgill, 69–101. Tucson: University of Arizona Press, 1988.
9. Eddy, F. W. *Prehistory in the Navajo Reservoir District, Northwestern New Mexico.* Museum of New Mexico Papers in Anthropology 15. Santa Fe: Museum of New Mexico, 1966.
10. Flannery, K. V., Marcus, J., and R. G. Reynolds. *The Flocks of the Wamani.* New York: Academic Press, 1989.
11. Forrester, J. "Counterintuitive Behavior of Social Systems." In *Simulation*, 1971.
12. Gleick, J. *Chaos: Making A New Science.* New York: Viking, 1987.
13. Gumerman, G. J. "The Archaeology of the Hopi Buttes District, Arizona." Ph.D. dissertation, Department of Anthropology, University of Arizona, 1969
14. Hantman, Jeffrey L. "Social Networks and Stylistic Distributions in the Prehistoric Plateau Southwest." Ph.D. dissertation, Department of Anthropology, Arizona State University, 1983.
15. Jones, A. H. M. *The Later Roman Empire, 284–602: A Social, Economic and Administrative Survey.* Norman: University of Oklahoma Press, 1964.
16. Kidder, A. V. "Southwestern Archaeological Conference." *Science* **66** (1927): 489–491.
17. Lightfoot, Kent G. "Food Redistribution Among Prehistoric Groups." *The Kiva* **44** (1979): 319–339.
18. Lightfoot, Kent G. *Prehistoric Political Dynamics: A Case Study from the American Southwest.* DeKalb: Northern Illinois University Press, 1984.

19. Longacre, W. A. "A Synthesis of Upper Little Colorado Prehistory, Eastern Arizona." In *Chapters in the Prehistory of Eastern Arizona II*, edited by P. S. Martin, Vol. 55 :201–15. Chicago: Fieldiana Anthropology.
20. Meadows, D. H., D. L. Meadows, J. Randers, and W. W. Behrens III. *The Limits to Growth.* New York: Universe Books, 1972.
21. Neitzel, Jill. "The Regional Organization of the Hohokam in the American Southwest: A Stylistic Analysis of Red-On-Buff Pottery." Unpublished Ph.D. Dissertation, Department of Anthropology, Arizona State Unviersity, 1984.
22. Nelson, B. A., D. L. Martin, A. C. Swedlund, P. R. Fish, and G. J. Armelagos. "Studies in Disruption: Demography and Health in the Prehistoric Southwest." This volume.
23. Nilsson, N., and M. Genesereth. *Logical Foundations of Artificial Intelligence.* Los Altos, CA: Morgan Kaufmann, 1987.
24. Nilsson, N. *Artificial Intelligence* Los Altos, CAL Morgan Kaufmann, 1988.
25. Plog, Fred. *The Study of Prehistoric Change.* New York: Academic Press, 1974.
26. Plog, Fred. "Prehistory: Western Anasazi." In *The Handbook of North American Indians, Volume 9*, edited by A. Ortiz, 108–130. Washington, DC: Smithsonian Institution Press, 1978.
27. Plog, Fred. *Cultural Resources Overview: Little Colorado Area, Arizona.* Albuquerque: USDA Forest Service, 1981
28. Plog, Fred. "Exchange, Tribes, and Alliances: The Northern Southwest." *Amer. Archaeologist* **4(3)** (1984): 217–223.
29. Plog, Fred. "Studying Complexity." In *The Sociopolitical Structure of Prehistoric Southwestern Societies*, edited by S. Upham [need names of other editors] et al., 103–128. Boulder: Westview Press, 1989.
30. Reynolds, R. G. "An Adaptive Computer Model for the Evolution of Plant Collecting and Early Agriculture in the Eastern Valley of Oaxaca." In *Guila Naquitz: Archaic Foraging and Early Agriculture in Oaxaca, Mexico*, edited by K. V. Flannery, 439–500. New York: Academic Press, 1986.
31. Reynolds, R. G. "The Control of Genetic Algorithms Using Version Spaces." In *Proceedings of IEEE Second Annual Conference on Tools for Artificial Intelligence*, Washington DC, 1990.
32. Ruppe, R. J. "The Acoma Culture Province: An Archaeological Concept." Unpublished doctoral dissertation, Department of Anthropology, Harvard University, 1953.
33. Schact, R. M. "The Contemporaneity Problem." *Amer. Antiquity* **49** (1984): 678–695.
34. Schmader, M. F. *Data Recovery Program and Resource Management Plan for Four Archaeological Sites in the Elena Gallegos Land Exchange, Rio Puerco Valley, NM.* Albuquerque: Rio Grande Consultants, 1988.
35. Schmader, M. F. *Archaeological Data Recovery at Five Sites Near Cerro Colorado in the Elena Gallegos Land Exchange, Rio Puerco Valley, NM.* Albuquerque: Rio Grande Consultants, 1990.

36. Scott, Linda C. "Pollen Analysis at AT 1A, 15C, 25C, 30C, and 12B." In *Draft Final Report of Investigations: The Albuquerque Sector of the Elena Gallegos Project*, edited by Joseph A. Tainter, 355–367. Albuquerque: USDA Forest Service, 1990.
37. Tainter, Joseph A. "Simulation Modeling of Inland Chumash Economic Interaction." *Archaeological Survey Annual Report*, Vol. 14, 79–106. Los Angeles: University of California Archaeological Survey, 1972.
38. Tainter, Joseph A. "Introduction and Research Design." In *Draft Final Report of Investigations: the Albuquerque Sector of the Elena Gallegos Project*, edited by Joseph A. Tainter, 1–18. Albuquerque: USDA Forest Service, 1990.
39. Tainter, Joseph A. "Synthesis of the Prehistoric Occupation and Evaluation of the Research Design." In *Draft Final Report of Investigations: the Albuquerque Sector of the Elena Gallegos Project*, edited by Joseph A. Tainter, 988–1051. Albuquerque: USDA Forest Service, 1990.
40. Tainter, Joseph A. "Southwestern Contributions on the Understanding of Core-Periphery Relations." This volume.
41. Toll, Mollie S. "Floral Remains from Late Basketmaker/Early Pueblo Sites in the Rio Puerco Valley." In *Draft Final Report of Investigations: the Albuquerque Sector of the Elena Gallegos Project*, edited by Joseph A. Tainter, 313–334. Albuquerque: USDA Forest Service, 1990.
42. Upham, Steadman. *Polities and Power: An Economic and Political History of the Western Pueblo*. New York: Academic Press, 1982.
43. Upham, Steadman, Patricia Crown, and Stephen Plog. "Alliance Formation and Cultural Identity in the American Southwest." Paper presented at The Organization and Evolution of Prehistoric Southwestern Society. School of American Research, Santa Fe, 1990.
44. Upham, Steadman. "Systems Modeling and Political Evolution: A Position Paper." This volume.
45. Wilcox, David R. "Changing Perspectives on the Protohistoric Pueblos, A.D. 1450–1700." In *The Protohistoric Period in the North American Southwest, A.D. 1450–1700*, edited by D. R. Wilcox and W. B. Masse, Vol. 24, 378–409. Tempe: Anthropological Papers of Arizona State University, 1981.
46. Wilcox, D. R., and L. O. Schenk. "The Architecture of the Casa Grande and Its Interpretation." *Arizona State Museum Archaeological Series, No. 115.* Tucson: University of Arizona, 1977.
47. Wilcox, D. R. "How the Pueblos Came to be as They Are." Manuscript on file, Arizona State Museum Library, Tucson, AZ, 1976.
48. Wilson, S. W. "Classifier Systems and the Animat Problem." *Machine Learning* **2(3)** (1987).
49. Wimberly, Mark, and Peter Eidenbach. "Reconnaissance Study of the Archaeological and Related Resources of the Lower Puerco and Salado Drainages, New Mexico." Tularosa, NM: Human Systems Research, 1980.

W. H. Wills
Department of Anthropology, University of New Mexico, Albuquerque, NM 87131

Evolutionary and Ecological Modeling in Southwestern Archaeology

ARCHAEOLOGY AND EVOLUTION

Archaeology is fundamentally about the way that people in the past modified their environment. The things that people deliberately made—we call them "artifacts" after the fact—are perhaps the most common subject of archaeological study, but incidental changes resulting from human activity, such as soil depletion in abandoned agricultural fields or vegetation changes from forest clearing are also the material remains of former societies. The general goal of all archaeologists, regardless of theoretical persuasion, is to understand human heritage through the physical indications of human activity. As in any historical science, archaeology is confronted with the difficult task of inferentially identifying cause and effect relationships when the cause cannot be observed.

However, unlike some historical disciplines to which archaeology is sometimes compared, such as geology, the most probable causes for the formation of the archaeological record do not correspond to general physical laws or principles. For example, geologic phenomenon in the past are interpreted from the perspective of uniformitarian principles that presume, to a large degree, that physical processes observed in the present are accurate analogs of prehistoric processes. Those

Understanding Complexity in the Prehistoric Southwest,
Eds. G. Gumerman and M. Gell-Mann, SFI Studies in the Sciences of
Complexity, Proc. Vol. XVI, Addison-Wesley, 1994

processes, in turn, are governed by relationships (gravity or friction, for example) which are predictable and regular (or thought to be). The origins of artifacts that constitute the record of human evolution have no such firm processual grounding. Artifacts are produced for perceived purposes rather than as the result of physical laws. Perception, deliberation, intentionality, choice—these are "causes" for the archaeological record that have no parallels in the physical sciences.

The human lineage exists in the fossil record for perhaps 3 million years, but anatomically modern skeletons appear only in the last 40,000 years. Stone tools, sometimes considered the first true indication of human culture, may have been made and used for nearly a million years but exhibit little sophistication or complexity until after the appearance of modern human skeletons. Tools—the things created by human from their environment to somehow make life more successful—do not really become complicated until the last 15,000 years. Consequently, a question that anthropologists frequently ask is what role did tools have in human evolution? Catalyst or by-product?

Tools are one aspect of the fascinating human capacity to alter, manipulate, and affect the future deliberately. Whether the result of these efforts is as intended is not nearly so intriguing as the evolution of the ability to envision a particular kind of future (even if it's only seconds away) and then set out to actively achieve it or, perhaps, to avoid it. Evolution has given humans a talent for anticipation and creation that is, for all practical purposes, utterly unique among living creatures.

During the last 10,000 years, extraordinarily rapid changes have taken place in the nonbiological aspects of humanity, those features that anthropologists refer to as "culture." Culture includes technology, economy, art, religion, philosophy, and just about anything else that is not obviously biological. These features exist as the product of organisms with particular abilities, but, so far as we know, they have no genetic bases. A problem for scientists attempting to explain the origin and evolution of biological and cultural changes in humans is that those very temporal patterns of change mean that uniformitarian assumptions about causality are probably valid only at extremely basic and simplistic levels. If human evolution (especially the last 40,000 years) is essentially about changing capacities for perception, what assurance do we have that models based on human behavior today can accurately account for prehistoric behavior?[1]

Do we think and respond to our environments as our ancestors did? If we do not, if our experiences have generationally cumulative effects (learning), or if our intellectual abilities have evolved, then uniformitarian assumptions about human behavior seem dubious. There is a clear, empirically prominent, progressive increase in the complexity of the material world found in the archaeological record of the human species that archaeologists reasonably assume reflects a changing cultural world. Does the record of changing cultural adaptations imply that assumptions about behavioral constancy are, at best, of limited value?

[1]See Trigger,[15] p. 19.

These questions suggest that a fundamental issue for archaeologists would be the discovery of the conditions that favored the evolution of the remarkable human ability to plan, forecast, and react to the future, especially the hypothetical future. It is an issue with interesting implications for evolution. For instance, is human evolution directed? Should we think of humans as being able to control their reproductive fitness? Or is reproductive fitness even a measure that is useful for understanding the material basis of human success?

Biologists assess the advantage of a trait by its persistence over time, or more precisely, its transmission from one generation to the next. But how should an archaeologist determine the advantage of traits—artifacts—that may be invented and discarded within a lifetime, or even within minutes? How does an archaeologist assess the differential persistence of tools that begin as one thing (a pottery vessel, for example) are broken and then transformed to serve some new function (a pottery vessel ground up for temper in another new pot)? What do we do with artifacts that change their meaning over time? An arrowhead may be simply a weapon in one generation, but a valued heirloom in the next, its meaning shifting from tool to symbol. How do we account for the possibility that identical artifacts may play vastly different roles in the hands of individuals having different statuses within a social system? What aspect of evolutionary theory accounts for the fact that a single object can convey advantages of different kinds to different people in different places at different times; that the use-life of a tool may exceed the life of the individual or society that produced it? Indeed, with whom or what should we associate an artifact in assessing its selective advantage? The individual who made it (or individuals), the range of individuals who might have used it (or benefited from its use), the society, the population, or the species?

The evolutionary significance of artifact association goes to the central issue in applying Darwinian concepts to the archaeological record. For the most part, archaeologists and anthropologists work with classifications that have no counterparts in biological taxonomies. Archaeologists are concerned with cultures, cultural traditions, style horizons, civilizations, states, hunter-gatherers, and a seemingly endless array of other classifications that incorporate social, economic, technological, and artistic variables. These anthropological concepts are used to describe variation in the complexity of human social systems as measured by the diversity of physical remains recovered from the archaeological record.

Some archaeologists[1,5,9] criticize the use of these anthropologically based concepts in studying human evolution. Their criticism centers on the argument that anthropologists should adopt more strictly Darwinian classifications if archaeology is actually going to address selective conditions in human development. The concept of "culture" is particularly onerous to these critics, since it really only gives a special label to the human behavioral repertoire when all living organisms exhibit some degree of behavior.

A counterargument to this criticism might be that all living organisms possess culture to some degree. That would bring the issue back to the complexity of human behavior, rather than the less profitable question of whether humans are somehow qualitatively different from other animals. Either way, the ultimate problem for archaeologists concerned with human evolution remains the basic concern with inferring behavior from artifacts. The history of biologically modern humans is clearly one of rapid, now certainly exponential technological and demographic change. How are we to understand the basis of that change?

EVOLUTION AND ECOLOGY IN SOUTHWESTERN ARCHAEOLOGY

People have lived in the American Southwest for at least 11,000 years and perhaps longer. From 11,000 to about 3000 B.P., Southwesterners were hunter-gatherers organized in small, probably extended family groups that ranged widely in search of food and other resources. The adoption of corn, beans, and squash from Mexico around 3200 B.P. provided important new resources but did not dramatically alter the hunting-gathering adaptation. Not until about 1500 B.P., shortly after the beginning of ceramic manufacture, is there good evidence for an increase in sedentism probably resulting from increased investment in farming. Rapid changes took place after this point, and by 1000 B.P. large population aggregates such as Chaco Canyon and Mesa Verde had formed. The appearance of large communities appears to have been associated with periodic episodes of widespread regional population abandonments and shifts to new, relatively unpopulated areas. Communities formed in new areas were often larger than those in abandoned regions. The arrival of Spanish explorers in the middle sixteenth century precipitated drastic population decline and community destruction among Native Americans from which they have not yet fully recovered.

The archaeological record of the prehistoric Southwest is one of the most extensively studied anywhere. From the standpoint of data quality, the American Southwest probably has more potential for examining the selective conditions of recent human evolution than any other region in the world. No other area enjoys the accuracy of our chronologies, the detail of our environmental reconstructions, the vast number of excavated sites, or the even vaster amount of material culture recovered from those sites. Few places in the world have the sheer density of professional archaeologists and anthropologists as the Southwest. And yet, for the most part, these data have rarely been sought or used to address the fundamental questions of evolutionary change that are raised by the human conceptual ability.

Which is not to say that Southwestern archaeology has been nonevolutionary in practice. Much to the contrary, archaeological research in the American Southwest is about evolution. Since the early days of this century, archaeologists here have explicitly adopted the goal of explaining change through time in human society as the proper research objective for their discipline. It would be extraordinary to find an archaeologist then or now whose research was not predicated on this simple notion.

However, many of the approaches to evolutionary change adopted by Southwestern archaeologists might not be easily recognizable to a biologist. Take, for example, the concept of a "culture area." Originally developed in the early part of the twentieth century by ethnologists interested in the geographical distribution of cultural traits among American Indian groups, the culture area was quickly appropriated by archaeologists for the same purpose. The culture area is based on the presumption of "natural areas" defined by vegetation patterns. Deserts are natural areas, the Eastern Woodlands another, the Arctic still another. The ecology of a natural area would dictate that only certain kinds of cultural patterns could exist there; you ought not to find corn farmers tilling away in the Boreal Forest. The Southwest was considered a natural arid area that included smaller natural areas, such as the Sonoran Desert, the Colorado Plateaus, or the Southern High Plains.[8]

Three major prehistoric culture areas have been defined in the Southwest: the "Hohokam" in the Arizona Deserts, the "Anasazi" in the Colorado Plateaus, and the "Mogollon" in the mountains of central Arizona and central and southern New Mexico. Initially, these culture areas were recognized as cultural "traditions," meaning distinctive patterns in material traits such as pottery and architecture. A rough association of traditions with broad geographic areas supported the notion of cultural areas as well. The idea of related natural and cultural areas was closely linked to the evolutionary approach proposed by ethnologist Julian Steward, who felt that cultural progress (evolution) could only be understood by reference to economic strategies, which were in turn conditioned by ecological factors. In order to explain the cause and effect relationships between economy and ecology, Steward[2] argued that one had to seek out cultural "laws" (generalizations really), which could be found in the comparison of multiple historical sequences of different cultures. In other words, one should control for ecology, then examine the histories of different areas for similar patterns of change. Similar sequences could be viewed, Steward suggested, as the equivalents of "laws" of cultural development.

Steward's belief that ecology determined economy (and hence social organization) was adopted by Southwestern archaeologists as an explanation for the geographical variability apparent in the archaeology of the Southwest. Different sequences of change in artifacts from one area to another could be attributed to differences in environmental "potential," typically described in terms of constraints on cultural development.[8]

For the most part, Southwestern researchers continue to seek causes of change in the archaeological record by reference to environmental factors. Other casual agents have been cited, especially contact (primarily trade and migration) between different cultural groups, but environment is without question the most compelling "motivation" found in the literature for temporal and spatial patterns of variation in prehistoric material remains.[7]

[2]Steward,[13] p. 88.

A significant attempt to understand the role of the environment in the prehistory of the Southwest took place in the 1970s with the formation of the Southwest Anthropological Research Group. SARG consisted of archaeologists concerned that archaeological information was accumulating at a tremendous rate but in an inconsistent manner that prohibited synthetic analyses. SARG researchers decided to attack this problem by devoting a portion of their separate research programs, regardless of problem orientation, to the specific question of understanding why people chose to locate their settlements in particular geographic locations. SARG research collected an astonishing amount of data on local environmental conditions around individual archaeological sites in order to determine what factors were responsible for the formation of the site. Nonenvironmental data were also collected and considered, such as the distance to other sites, but the primary focus in published results was how environmental conditions affected "decisions" about site location.[6]

The results compiled by SARG researchers fell short of initial expectations. A commonly encountered problem was that the research design adopted by SARG had difficulty recognizing "circumstances under which one would expect the relationship between site locations and environmental variables to be quite low."[3] SARG also ran head long into the issue of artifact association described earlier in this paper. As Dean[4] pointed out, SARG asked the question of why sites are found in certain places at certain times but often conflated "sites" with "people," sometimes described as population aggregates. Sites represent activities conducted at a place that left material remains. Activities that did not leave remains obviously are not represented, yet just as clearly might be equally or more important in the decisions made by people with regard to why that spot was occupied.

Researchers in the 1970s and 1980s increasingly sought causes for artifact patterning in the "social environment" as well as the "cultural environment." One example is the role of artifact style. Rather than view style differences, such as the designs on pots, as incidental by-products of cultural distance, some researchers took a more functional perspective wherein designs could be forms of social messages[11] and distinctive design patterns might result from efforts to create social boundaries.

In recent years most archaeological research in the Southwest has emphasized a combination of ecological and economic variables in explaining patterns in the archaeological record. While a majority of work still appears to have strongly environmentally deterministic tendencies, several researchers have attempted to document the actual systemic relationships among multiple variables. Spielmann,[12] for example, has developed models of economic mutualism to account for Pueblo-Plains trade systems based on resource complementarity. Other studies have presented optimal foraging models for preagricultural settlement systems and for differential patterns of prey selection by hunter-gatherers. In addition, computer simulation modeling has a growing role in suggesting possible processual relationships in the development of particular subsistence strategies.

[3]See Plog,[10] p. 70.

General evolutionary goals in Southwestern archaeology have, like ecological analyses, often drawn on Julian Steward's perspective about the goals of anthropology. For Steward, the ultimate anthropological objective is "to see through the differences of cultures to the similarities, to ascertain processes that are duplicated independently in cultural sequences, and to recognize cause and effect in both temporal and functional relationships."[4]

Steward's focus on historical sequences lends itself to an archaeologist's ability to delineate temporal changes in artifacts. By defining a prehistoric "culture" as a grouping of particular kinds of artifacts (or patterns) in a certain place, an archaeologist has only to demonstrate chronological relationships between such groupings to create a "cultural history." Steward would then seek regularities among different culture histories as the basis for generalizations about human progress (evolution), and thus archaeological reconstructions provide the first and fundamental step in evolutionary explanation. Syntheses of Southwest prehistory that adopt an explicit objective of documenting evolutionary change have almost always been based on compilations of culture histories.

Recently (1980s) the study of evolutionary change in the prehistoric Southwest has been dominated (albeit in discussion more than practice) by the issue of social "complexity." Social complexity means various things to different researchers, but it conventionally involves definitions of societal types or categories. Discussions have had basically two aspects: (1) the identification of relative degrees of social complexity among contemporaneous or sequential social systems, and (2) the identification of conditions that promote variation in social complexity. In both endeavors the outcome depends entirely on how complexity is defined.

One approach has been to adopt societal typologies developed by ethnologists, particularly ones with evolutionary implications. In the 1960s, anthropologists attempted to implement Steward's goal of defining cross-cultural regularities by establishing an evolutionary classification of social development, the essence of which was the assumption that social evolution exhibited a progressive increase in complexity. Less complex social forms, such as hunting and gathering bands, eventually gave rise to more complex farming societies, which in turn gave rise to states and civilizations. One of the typological distinctions in these evolutionary classifications is the degree of social inequality. In Southwestern archaeology, the emergence of social complexity as an evolutionary issue has been framed almost entirely in terms of differential access to power among members of a community.[16] Although some of the current literature concerns conditions under which social inequality might have some selective advantage, most appears to focus on the identification of status differences from archaeological data.

As a result, published research about social complexity in the Southwest has, so far, been largely a series of counterarguments supporting alternative interpretations of the same data. A prominent current example involves a single archaeological site in eastern Arizona whose prehistoric social organization is construed differently (and

[4]Steward,[14] p. 180.

usually bitterly) by researchers whose assumptions about the meaning of artifact patterning differ. And therein continues the most basic issue in archaeology—how do we assign significance to artifacts?

Unfortunately, the arguments about social complexity in the prehistoric Southwest often seem to miss or avoid the difficult point discovered by SARG researchers that artifacts are not people. When proponents of alternative interpretations of social complexity debate the concordance between artifact patterns and anthropologically established social typologies such a clans, lineages, elites, communities, villages, alliances, networks, or decision-making hierarchies, they equate physical objects with abstractions based on social interaction. Rarely do such studies explore in depth the ambiguity that exists in that equation.

Despite the difficulty that attends the application of social constructs to the archaeological record, there is probably little disagreement among archaeologists that a major trend in the prehistoric Southwest is toward increasing complexity. Over time there were clearly more people, more kinds of artifacts, and greater stylistic variation within artifacts. Archaeologists have yet to agree on precisely what such complexity means or how to explain it. Chaco Canyon is a prime example, where over 100 years of research and a truely stunning volume of literature have produced a mass of information about the social phenomenon that took place there in the relatively brief span between A.D. 900 and 1150, but not a consensus (or even majority) view for why.

An observation made during the School of American Research Advanced Seminar that preceded this conference was that Southwest archaeologists have been able to clearly identify periods in the past when dramatic and widespread changes took place in the material remains of prehistoric people. These periods of artifact change were followed by patterns that suggest a more complex social world, but we are uncertain how to interpret that complexity with the concepts traditionally used to describe cultural variation. In the following chapter, members of the Santa Fe Institute workshop on evolutionary modeling in the prehistoric Southwest adopt an economic perspective on complexity, one that concentrates on the specialization of roles and functions within adaptive systems rather than ethnographically defined societal types. We believe this economic view offers a systematic approach to understanding the distinctive periods of archaeological change that researchers agree upon.

REFERENCES AND SUGGESTED READINGS

1. Braun, David P. "Selection and Evolution in Nonhierarchical Organization." In *The Evolution of Political Systems: Sociopolitics in Small-Scale Sedentary Societies*, edited by S. Upham, 62–86. Cambridge: Cambridge University Press, 1990.
2. Cordell, L. S., and G. J. Gumerman. "Cultural Interaction in the Prehistoric Southwest." In *Dynamics of Southwest Prehistory*, edited by L. Cordell and G. Gumerman, 1–18. Washington, DC: Smithsonian Institution Press, 1989.
3. Cordell, L. S., and F. Plog. "Escaping the Confines of Normative Thought: A Reevaluation of Puebloan Prehistory." *Amer. Antiquity* **44** (1979):405–429.
4. Dean, J. "An Evaluation of the Initial SARG Research Design." In *Investigations of the Southwestern Anthropological Research Group: An Experiment in Archaeological Cooperation.* Flagstaff, AZ: Museum of Northern Arizona, 1978.
5. Dunnell, R. C. "Aspects of the Application of Evolutionary Theory in Archaeology." In *Archaeological Thought in America*, edited by C. C. Lamberg-Karlovsky, 35–49. Cambridge: Cambridge Press, 1989.
6. Euler, R. C., and G. J. Gumerman. *Investigations of the Southwestern Anthropological Research Group: An Experiment in Archaeological Cooperation.* Flagstaff, AZ: Museum of Northern Arizona, 1978.
7. Gumerman, G. J. *The Anasazi in a Changing Environment.* Santa Fe, NM: School of American Research Press, 1988.
8. Haury, E. "The Greater American Southwest." In *Courses Toward Urban Life*, edited by R. J. Braidwood and G. R. Willey, 106–131. Chicago, IL: Aldine, 1962.
9. Leonard, R. D. "Resource Specialization, Population Growth, and Agricultural Production in the American Southwest." *Amer. Antiquity* **54** (1989): 91–503.
10. Plog, F. "An Analysis of Variability in Site Locations in the Chevelon Drainage, Arizona." In *Investigations of the Southwestern Anthropological Research Group: An Experiment in Archaeological Cooperation.* Flagstaff, AZ: Museum of Northern Arizona, 1978.
11. Plog, S. *Stylistic Variation in Prehistoric Ceramics: Design Analysis in the American Southwest.* Cambridge: Cambridge University Press, 1980.
12. Spielmann, K. "Inter-Societal Food Production Among Egalitarian Societies: An Ecological Study of Plains/Pueblo Interaction in the American Southwest." Ph.D. dissertation, University of Michigan, 1982.
13. Steward, J. H. "Ecological Aspects of Southwestern Society." *Anthropos* **32** (1937): 86–104.
14. Steward, J. H. *Theory of Cultural Ecology.* 1955. Urbana, IL: University of Illinois Press, 1955.

15. Trigger, Bruce G. *A History of Archaeological Thought.* Cambridge: Cambridge University Press, 1989.
16. Upham, S., K. Lightfoot, and R. Jewett. *The Sociopolitical Structure of Southwestern Society.* Westview Press, 1989.

W. H. Wills,* Patricia L. Crown, Jeffrey S. Dean,† and Christopher G. Langton‡**
*University of New Mexico
**Arizona State University
†University of Arizona
‡Los Alamos National Laboratory and the Santa Fe Institute

Complex Adaptive Systems and Southwestern Prehistory

INTRODUCTION

This chapter addresses the general conference goal of exploring the relevance of complex adaptive systems to understanding the record of cultural evolution in the Southwest. As defined by associates of the Santa Fe Institute, complex adaptive systems are those that evolve or learn, including, for example, prebiotic chemical evolution, biological evolution, the evolution of ecosystems, languages, individual learning and thinking, the global economy, and computer programs.[1] The evolution of complex adaptive systems is not equivalent to Darwinian models of biological evolution; rather, organic evolution by natural selection is one kind of evolutionary process.

According to Farmer and Belin,[35] the evolution of both physical and biological entities reflects tendencies for matter to organize itself under certain conditions. The paradox of order arising from disorganized energy (heat from the sun) was a central theme that anthropologist Leslie White[145,146] invoked as the key to understanding human cultural evolution. Although rejected by some anthropologists as

[1]Gell-Mann,[47] pp. 3–4.

inconsistent with biological perspectives on evolution,[31] the organizational properties of matter under particular circumstances that interested White are the same phenomena that now concern scientists studying nonbiological evolution. The study of artificial life, defined as man-made systems that exhibit characteristics of natural living systems, holds forth the expectation that in the future the evolution of human-created systems will be based on the transmission of acquired characteristics from generation to generation.[2] This explicitly Lamarckian view of evolutionary change is not inconsistent with Darwinian biological change, since the source of variation on which selection operates is irrelevant. Researchers are suggesting that culture, as one case of evolutionary process that results in the organization of matter, is not adequately understood without a capacity for conscious design.

Human cultural evolution is therefore a prominent illustration of complex adaptive systems, since change through time is dependent upon the transmission of information between individuals and from one generation to the next.[9,14,77] In the discussion that follows, we assume that human culture is heavily invested in capacities for learning and that this characteristic provides much of the innovation responsible for cultural evolution. We are also concerned with the possibility that basic principles underlie the evolution of all complex adaptive systems and that, consequently, we may usefully consider analogies between cultural systems and other types of systems, particularly economies.

By cultural system we mean the functional integration of roles, institutions, technologies, behaviors, and beliefs shared by a group of people. Although anthropologists prefer to use the term "culture" to distinguish human behavioral and social characteristics from those of other animals, the concept of a cultural system is essentially the concept of an economic system as used by biologists. In biology, an economic system represents the transfer of matter, energy, and non-genetic information between an organism and its environment.[33] All organisms exist and survive and have differential reproductive success within the context of interaction with their environment and other organisms. Evolution is the outcome of these interactions. Our focus in this paper is understanding evolutionary change as the result of internal changes in the adaptive system and/or changes that alter the transmission of information.[3]

In the remainder of this chapter we address the problem of identifying transformations in the complexity of cultural systems through the introduction of novelty. We argue that complexity arises from specialization among members of a population and present examples from southwestern prehistory that seem consistent with our expectations. The introduction of technological innovations generally create conditions which favor greater specialization within populations.

We are not seeking functional explanations for change in the sense that having identified the way a system works, we then presume that the system evolved for that reason. Few, if any examples of adaptive systems are exceptionally efficient;

[2] Farmer and Belin,[35] p. 20.

[3] See Eldredge,[33] p. 352.

some configurations work well, others not so well, but all achieve some success.[6] Our interest is in specifying how changes occur and whether some patterns of change are attributable to the systems themselves (are autocatalytic in having positive feedback loops) or to external innovations. We begin by presenting a series of general principles that Santa Fe Institute associates propose are common among historical processes.

COMPLEX SYSTEMS MODELS

Each group of researchers (economists, biologists, chemists, linguists, physicists, geologists, archaeologists) that attends Santa Fe Institute workshops always arrives with detailed descriptions of some unique historical process. These researchers feel certain that universal principles are responsible for much of what occurred during the process, but, as it is a *unique* historical process, they are not able to distinguish between the effects of universal principles and accidental chance events that took place as the process evolved. What was due to chance and what to necessity?

To answer this kind of question, we must generalize from an *ensemble* of similar historical processes. Then features due to necessity would be common to all the processes in the ensemble, whereas those due to chance would be unique to each process. If these researchers could only "rewind" the tape of their unique historical process and start it over again from different initial conditions, or under different accidental perturbations, they would have an ensemble that would allow them to distinguish between the effects of chance and necessity.

Unfortunately, this is not possible. Anthropologists cannot rewind the history of cultural development in the Southwest and play it over again from slightly different initial conditions, any more than economists can replay a stock-market crash, or geologists can replay an earthquake. Researchers are stuck with a single, *actual*, historical sequence out of a presumed (but inaccessible) ensemble of *possible* historical sequences. They must attempt to derive general principles from single examples, a particularly unrewarding task. How can one go about unveiling these principles despite the problem of having single examples from which to generalize?

At the Santa Fe Institute, historical processes are studied in general, so that while each historical process is unique when viewed from within a particular academic discipline, it becomes just another sequence in the "Grand Canonical Ensemble" of historical processes when viewed across traditional academic disciplines. Thus we have an ensemble constructed from many unique examples, and we are able to generalize on the basis of this ensemble. It is becoming clear from the study of this Canonical Ensemble of historical processes that universal principles are indeed afoot in most historical processes, and we can begin to say what they are.

First, most historical processes involve populations of interacting individuals. The network of interactions supported within this population can, in most cases, be captured in the form of an interaction graph, which is often quite sparsely connected. The fact that we can often represent the dynamic system underlying these historical processes as an interaction graph provides us with a means for their numerical and mathematical analysis.

Second, it is clear that it is not just the strategies of individuals in the population that are important, but how these strategies behave in each other's presence. Most historical processes occur in populations whose individuals engage in *nonlinear* interactions with one another, so that the behavior of the whole population is, in a fundamental sense, truly more than the sum of the behaviors of its parts.

In many historical processes, the most important effects of individuals' behaviors are realized at the level of the group, rather than at the level of the individual. Therefore, it is important to understand how collective behavior can emerge and be selected for in populations of individuals, and it is even more important to understand how the diversity of behaviors that one observes at the level of individuals can be coherently integrated, and therefore understood, at the level of the collective behavior of the group.

Third, related to this behavior at the level of the group is the way in which new levels of complexity spontaneously emerge in historical processes. The complexity of historical processes often takes big jumps, not by making individuals themselves more *generally* capable, but rather by discovering new combinations of *specialized* individuals. Thus, increases in complexity tend to be accompanied, or fostered, by an increase in the diversity of specialized roles, rather than by an increase in generalized capability. Collections of cooperating specialists can often achieve things that are unattainable by collections of competing generalists. Many of the categories used by anthropologists to describe societies, such as clans, lineages, sodalities, organizations, etc., represent cooperative groups with specialized functions.

Fourth, such leaps in complexity via the emergence of new collections of specialized individuals are often accompanied by an explosion in diversity among the populations engaged in the historical process. Many variations on the basic theme of a new collection of specialists are quickly "discovered" or "invented," and most such variations survive and lead to further variations. This is because the basic innovation is often such an improvement on the "old way" of doing things that it is quickly incorporated throughout the population.

However, once the basic innovation has spread throughout the population in various forms, subtle differences between the variant realizations of the innovation begin to make a difference. Now the situation primarily involves competition between the innovation and the older forms that dominated before the innovation. This later stage, following upon the explosion in diversity, leads to a decline in diversity, as less successful variants of the innovation lose out in competition to more successful variants.

As Stuart Kauffman noted during the workshop: Once you invent the bicycle, almost any variation on it that retains its basic "bicycleness" is successful and will inspire yet other variants. However, once everyone has a bicycle, then it becomes apparent that some ways of building a bicycle are inherently better than others, and many initially successful but ultimately inferior designs will be dropped in favor of superior ones. This will eventually lead to a reduction in the diversity of bicycles.

If one attempts to draw a "genetic tree" of such a process, delineating the historical relationships among all of the variants and their descendents, one will see an initial explosion of branching, followed by a thinning-out process as various branches eventually die out. This process is commonly associated with the so-called "Cambrian Explosion" that accompanied the "innovation" of multicellular animals around 700 million years ago. Multicellular animals are clearly associated with an increase in the specialization of the cells that make them up. "Cambrian Explosion" style leaps in diversity, accompanied by increased specialization of the individuals involved, appear to be a generic property of historical processes.

Fifth, diffusion obviously plays a role in such processes. The rate of innovation dispersal is important. If innovations spread too rapidly through a population, fewer variants develop than if they spread more slowly and have a chance to be adapted to local circumstances before they are passed on. If innovations spread *too* slowly through a population, however, they may become so overly adapted to local contexts as to be inapplicable in other contexts. Thus, one can often hypothesize about the the dynamics behind the spread of innovations in particular historical processes from these generalizations about diffusion.

Sixth, as might be apparent, evolution in the most general sense seems to be an important operating principle in many historical processes. One might even go so far as to say that "evolution is what historical processes do." Although the specific mechanisms of "inheritance", and the specifing evolutionary forces operating on evolving populations of "schemas"[46] or "strategies," will be different in different historical processes, one is usually able to identify such mechanisms and forces without much trouble.

Seventh, such evolutionary processes are operating on what are clearly rugged or *coupled* fitness landscapes.[68] We often cannot get precise pictures of the surface topology of such landscapes (where the fitness peaks and valleys lie), primarily because these dynamic landscapes are changing all the time. What today is a hill for one population could very well be a valley tomorrow, as some other linked population climbs a different "hill" on the dynamic fitness landscape. Populations climb a very complex fitness landscape which is changed and distorted in complex ways by the very act of climbing.[106]

Finally, the construction of abstract models of specific historical processes often sheds much light on the processes themselves. Models partially alleviate the problem of not being able to "rewind" the tape to experiment with different initial conditions and perturbations. More important, good models force researchers to be explicit about the assumptions they make about the processes, often forcing them to ask new questions or actively "dig up" new data.

In fact, it may be that one never ends up with a working model of the process but nonetheless finds the very effort of constructing a model to have drastically altered one's perspective about that process. This has largely been true in the field of Artificial Intelligence, where no working model of intelligence has yet been constructed, but where the attempt to build such a model has radically altered researchers' perceptions of what "intelligence" is.

ARCHAEOLOGICAL APPLICATIONS

Many of the universal principles outlined above seem apparent in the archaeological record of the American Southwest. In particular, the explosion of diversity and subsequent "thinning out" of variants may characterize various developmental sequences in the economic organization of both hunter-gatherer and farming populations. Such patterns may also be evident in styles of artifacts and architecture. For instance, a new form of technology will lead to an "explosion" of experimentation as the potential of the novelty is "tested" in different niches. Improvements to the innovation are made rapidly at first and then decline. In some niches, the novelty provides increased viability for the population, while it proves unsuccessful in other niches. The less successful "experiments" fail, and diversity declines. Improvements are made more slowly. The number of variants depends in part on the rate of diffusion of the novelty.

Likewise, the aggregation of human populations into large communities may have been accompanied by increased diversity of specialized roles, as the adaptive success of the group depended less on individual capabilities and more on collective abilities of many individuals having different roles. It is interaction with other populations, particularly in the form of competition, that is likely to lead to such leaps in complexity.

Leonard[75] argues that prehistoric southwestern subsistence systems should be described as specialized relative to other systems if, in such a comparison, individuals utilizing that system derive proportionally more of their diet from fewer foods. Leonard's use of specialization is apparently meant to be analogous to the biological concept of species specialization, in which the behavioral and physiological adaptations of a species confine it to a narrower environmental niche than other species, which are termed "generalists" in contrast.[4] While we think dietary analysis of subsistence systems is especially valuable, we find an economic definition of specialization more appropriate for evaluating our model.

In economic theory, specialization is a product of the division of labor. Individuals and economic systems work to generate a limited set of products, whether food or goods. Division of labor contributes to increased efficiency and higher productivity[5] and leads to an important relationship between specialization and exchange. The more specialization occurs within an economic system, or between connected

[4] Pianka,[98] pp. 253–256.

[5] Baulmol and Binder,[7] p. 45.

systems, the greater the importance of exchange as a means for individuals to satisfy basic needs. Specialization implies a loss of household self-sufficiency.[6] One of the key functions of corporate groups of any scale or configuration is that members contribute to the success of the whole by the "exchange" of services and products. The complexity of an economic system, or a society, varies positively with the division of labor and the degree of economic exchange among constituent units.

Our goal in this chapter is to consider the role of innovations in promoting greater societal complexity in the prehistoric Southwest examining two important technological innovations: the introduction of domesticated plants and the initiation of pottery production. We begin that assessment with a consideration of the environmental context in which the evolution of Southwestern societies took place. Mediated by interactions with human behavioral systems, the natural environment is an important component of the fitness landscapes that provided the dynamic context of co-evolutionary processes. Adaptive options are limited by the harsh and often unpredictable environments of the Southwest, many of which are and have been marginal for the different prehistoric economies of the region. New behavioral adaptations were threatened by instabilities created by the changed systemic relationships that define the new fitness landscape and by concomitant alterations in that landscape caused by (1) other populations' responses to new configurations, and (2) changes wrought in the physical environment by the transformations in behavior-environment relationships. Furthermore, environmental fluctuations on scales ranging from years to centuries change fitness configurations independently of human action.

Most archaeological research in the Southwest deals with the remains of agricultural societies and therefore the marginal nature of the region for farming assumes great importance. The Southwest is located at the northern geographical limit for maize, beans, and squash cultivated with a Neolithic technology. Thus, Southwest agriculture is highly susceptible to environmental limiting factors that vary from place to place through time and which contribute materially to the ruggedness of human fitness landscapes. Technological and behavioral responses to the considerable variability of dynamic landscapes are important features of Southwestern sociocultural evolution. Any attempt to understand this evolutionary history must control, insofar as possible, all relevant spatial and temporal variability in the environmental factors that affect agricultural systems.

Although it is common to conceptualize environmental variability in terms of climatic variation, a wide range of nonclimatic factors is equally significant. Nonclimatic environmental variability is important in itself, and as it mitigates, accentuates or is affected by climatic variation. In any given adaptive situation, a number of environmental variables or combinations of variables can affect agricultural systems. Climatic interpretations of Southwestern cultural evolution are therefore inadequate as the sole source of understanding social and economic change.

[6] Rice,[107] p. 261.

Environmental variability can be conceptualized in terms of the spatial and temporal structure of different variables. While a virtually endless array of environmental variables exists, herein we focus on those which most directly affect agricultural production. These include the distribution and quality of arable land, the availability of water, and the length of growing seasons. Suitable farmland is widely distributed in the semiarid Southwest, and thus water and growing season are the principal factors in agricultural production.

The temporal dimension of these variables can be divided into three broad, nonexclusive classes.[25,23,24] First is stability, with stable variables being those which have not changed significantly over the last 3000 years. Second, are low-frequency processes (LFP) with periodicities, either regular or irregular, longer than 25 years (one human generation). Despite the long periodicities of low-frequency natural processes, however, their effects can be abrupt, such as the onset of arroyo cutting. Third are high-frequency processes (HFP) that occur at intervals of less than 25 years, and include daily, seasonal, annual, and longer variations. The spatial dimension of environmental variability ranges from regional (the entire Southwest), to subregional (the Sonoran Desert or Colorado Plateau), to areal (the Phoenix Basin or Rio Grande Valley), to local (Chaco Canyon or the Tucson Basin).

Stable environmental factors tend to have broad spatial relevance (subregional, areal). The most obvious stable variables are those which determine the general environmental potential and limitations of the subregional physiographic divisions of the Southwest region (desert, mountains, plateau). Other stable factors (e.g., bedrock geology) exhibit areal and locality patterning. An important characteristic of stable factors on any spatial scale is that their present stages accurately reflect conditions throughout the last 3000 years.

LFP variability tends to occur across smaller scales than the most extensive stable factors. Relevant LFP variability in the Southwest includes long-term climatic trends, the rise and fall of alluvial water tables, the deposition and erosion of alluvial sediments, and changes in the elevational limits and composition of plant communities. Some of these factors (e.g., climatic trends and fluvial processes) may vary on subregional or regional scales, while others (such as elevational boundaries of plant communities) mostly occur at areal or locality levels.

HFP variability tends to exhibit even more restricted spatial distributions than LFP variability. Limited primarily to rapid climatic fluctuations, HFP usually take place on areal and locality scale, with considerable between-subregion variability. As in LFP variability, current states of HFP variables are not accurate indicators of past conditions.

In the Southwest, accurate reconstructions for two types of LFP variability (fluvial processes and effective moisture) and one type of HFP variability (climate) are available from, respectively, alluvial chronostratigraphy, palynology, and dendroclimatology. These studies have been confined principally to the Colorado Plateau and it is not yet clear how well they apply to past conditions in the montane and desert regions.

Fluvial cycles comprise the rise and fall of alluvial water tables and associated trends in the deposition and erosion of alluvial deposits.[7] Alluvial groundwater variability has a period of approximately 550 years with primary maxima centered at ca. A.D. 50, 600, 1150, and 1750, and first order minima at ca. A.D. 350, 850, 1450, and (projected) 2000. Extrapolating this curve back into the B.C.period would produce projected maxima at 500, 1050, 1600, and 2150 BC and projected minima at 200, 750, 1300, and 1850 B.C.. Due to lagged relationships, the relatively smooth hydrologic curve is associated with an asymmetrical sequence of alluvial aggradation and degradation with primary depositional periods occurring at approximately 200 B.C.–A.D. 250, A.D. 400–750, 925–1275, and 1475–1875, and major erosional intervals at A.D. 250–400, 750–925, 1275–1475, and 1875 to the present. During the A.D. 300–1300 interval, LFP changes in effective moisture, as reconstructed by palynology, parallel the hydrologic curve to a remarkable degree.[8]

The relevance of the above LFP fluctuations to this paper lies in the fact that in a wide variety of circumstances, at least on the Colorado Plateaus, groundwater rather than rainfall is the primary limiting factor in farming alluvial lowlands. For this reason, during periods of aggradation and rising alluvial water tables, farming conditions in alluvial settings are optimal, and precipitation falling directly on fields is of secondary importance to crop success. Conversely, during intervals of falling water tables and alluvial erosion, rainfall becomes highly important to floodplain cultivation. The parallel behavior of effective moisture indicates that inimical fluvial situations were rarely offset by increased effective moisture and, therefore, that environmentally induced changes in the fitness landscapes of populations practicing bottomland farming are expectable during transitions between hydrologic states.[23,24]

LFP environmental conditions provide a major set of limiting conditions on human adaptive strategies, but the effects of LFP variability can be impacted by HFP variability. For example, during periods of rising water tables, high or low precipitation has less effect on alluvial farming than during periods of declining water levels. Likewise, above-average rainfall and low-temporal variability in climate during periods of groundwater decline might compensate for less soil moisture. Or, low rainfall and high-temporal variability during groundwater minima could exacerbate risky situations. Consequently, LFP and HFP interact in complicated ways that require detailed regional and local analyses to assess their relationship to sociocultural change.

[7] Dean,[25] Figure 1; Euler et al.[34]; Gumerman and Dean,[55] Figure 15; Plog et al.,[100] Figure 3.3.
[8] Dean et al.,[25] Figure 1b.

EVOLUTION OF SOUTHWESTERN AGRICULTURAL SYSTEMS

The agricultural economies of prehistoric southwestern people were created by adapting nonindigenous domesticated plants to local environments. Maize, squash, and beans were domesticated in Mesoamerica and first adopted by southwestern hunter-gatherers around 1200 B.C., or perhaps several centuries earlier. These plants spread rapidly throughout the Southwest, probably within a few human generations, and are found in a wide variety of ecological contexts.[41,61,62,86,136,152]

Despite the initial wide distribution of these plants, there is little evidence for diversity in the plants or the economic systems in which they were used for more than a thousand years following their adoption. It is particularly interesting that maize exhibits no phenotypic trends for a millennium that are attributable to human manipulation, indicating that selective pressure for increased yields through cultivation must have been quite low.[155,154] One possible explanation is that cultigens moved through regional hunter-gatherer populations so quickly that the opportunity for specialized adaptations to local environments did not occur. After these resources had essentially flooded available economic niches, the relatively high mobility of recipient hunter-gatherer settlement systems may have militated against the kind of intensive management required to produce genetic changes.

While the adoption of domesticates apparently did not produce immediate alterations of the basic economic strategies of Southwestern people, the archaeological record suggests that recipient groups used these new resources to increase the range of economic options within existing subsistence organizations. For example, the cultivation of maize and squash by people utilizing the upland mountain zones of west-central New Mexico appears to represent a shift toward longer periods of occupation in that region. Preagricultural settlement systems are interpreted as based on seasonal movement from low-elevation areas occupied during winter and spring months, such as the Rio Grande Valley, to upland resource zones during summer and fall months.[120,132,152,155] The appearance of cultigens at 1200 to 1000 B.C. at Bat Cave in the Mogollon Highlands indicates occupation of this upland region during late spring or early summer, sooner than expected in the preagricultural settlement system.[152] In order to grow domesticates at the higher elevations near Bat Cave (2020 m), cultivators had to forego available natural food resources at lower elevations to be in the uplands where there was much less natural resource abundance during the spring. One reason for a shift to longer seasonal use of the uplands may have been to monitor future resource patterning by sending portions of a population into the highlands before the arrival of the main body. If this is an accurate interpretation, early agriculture in the central mountains seems to have expanded the ability of essentially foraging economies to effectively exploit a wider range of resource zones simultaneously, a result that is more closely linked to benefits accruing from increased information about the environment than from intensification of production per se.

A change toward greater information exchange among cultivating groups as a result of economic specialization in subsistence activities is consistent with Arthur's[9] argument that portions of an economy that are knowledge-based may be characterized by increasing returns, so that large initial investments (such as moving part of a population into a new seasonal stance) may have very high payoffs. Likewise, Moore's[89] studies of information exchange among hunter-gatherer groups demonstrate that reorganization of information flow among independent economic units can greatly enhance the carrying capacity of a local environment. Agriculture is fundamentally about predictability, about the increased probability that particular plants will be available in specific places. The capacity for high yields that is often considered the hallmark of food production is in fact a byproduct of the control afforded by the information component of plant cultivation. Hence, the first millennium of agriculture in the Southwest represents a significant economic development, one that created new subsistence options for groups electing to use these new resources.

In southeastern Arizona, extensive investigations of preceramic agricultural sites by Huckell[61,62] and Roth[112] have outlined one of those subsistence options that contrasts with the first adoption of domesticates in the Mogollon Highlands. Along streams that drain into the Tucson Basin from surrounding mountain ranges are numerous sites dating between ca. 1000 and 800 B.C.containing abundant evidence for maize cultivation. These sites are often characterized by large quantities of fire-cracked rock, high densities of lithic artifacts, bell-shaped pits, and small, circular structures interpreted as dwellings. Huckell[10] argues that many of these sites represent relatively large aggregations of people that resided for several months at a time in those places. The positioning of sites in alluvial bottomlands provided access to high water tables for cultivation and were close enough to mountain ranges that foraging parties could have operated out of these farming settlements.[38] Although preagricultural people apparently had similar settlement systems,[78,141] Huckell[11] suggests that maize cultivation allowed for more intensive occupation of certain key resource areas. Thus, while the incorporation of domesticates does not seem to have promoted a sudden shift to a *new* economic adaptation in the Sonoran Desert, it did apparently open up the range of subsistence tactics within ongoing forager strategies.

To some degree these expanded variations on basic economic strategies may be viewed as evidence for increased specialization. Storage, whether for seed crops alone or for delayed consumption, is a tactic that allows for greater control of future resource availability. The extent of preagricultural storage strategies is uncertain, probably due in part to highly mobile settlement systems, but the introduction of domesticates is clearly associated with the use of subsurface storage facilities.[62,152] These types of storage features probably indicate continued high group mobility

[9] Arthur,[6] p. 95.

[10] Huckell,[62] p. 363.

[11] Huckell,[62] p. 351.

with pits used as caches for future site reoccupation.[27] The numbers of documented storage features increases dramatically in the Southwest between ca. A.D. 200 and 500.[122]

A less obvious but very important aspect of economic specialization during the first millennium of domesticate use in the Southwest is in the organization of production, particularly the division of labor within local groups. In the Mogollon Highlands during the Late Archaic, local groups may have split into smaller production units at certain times of the year. During late spring, advance groups (probably young males) left the main winter aggregate for the highland regions. They were followed by most of the main group, with older or infirm individuals left behind at cultivation localities. This reconstruction implies that local groups allocated labor in a complex and sophisticated organizational structure in which the income from cooperation exceeded that which small, independent groups could obtain.[158] It is not economic specialization in the sense that an entire system is geared toward a small range of activities (sensu Leonard[75]), rather it appears that individuals within local groups may have taken on new and specific subsistence roles that increased the flexibility of overall economies.

EARLY CERAMIC PERIOD (CA. A.D. 200–750)

Throughout the Southwest, the end of the Archaic and the beginning of the ceramic period is widely perceived by archaeologists as a time of increasing dependence on agricultural production.[12] A fairly dramatic increase in the construction of substantial dwellings (pithouses), the first use of ceramic vessels, and extensive midden formation at some sites between ca. A.D. 200 and 500 is taken to indicate a qualitative shift toward increasingly sedentary adaptations. Because the Southwest is dominated by arid environments with low human carrying capacity, most researchers feel that sedentism was achieved through agricultural intensification leading to food surpluses capable of sustaining groups throughout large portions of a year.[42,64,130] Although the causes of intensification a thousand years or more after the introduction of cultigens are unclear, there is solid evidence that during the interval from A.D. 200 to 700 rapid transformations in settlement systems did take place in many areas of the Southwest.

The locations and internal structure of archaeological sites dating between ca. A.D. 200 and 750 indicate a wide range of economic strategies. A good example is the culture historical classification called "Basketmaker III," the earliest ceramic period in the Colorado Plateau. For archaeologists, the term Basketmaker III refers to a period between about A.D. 500 to 750, with ceramic assemblages dominated by undecorated graywares and, typically, pithouse architecture. Basketmaker III sites are found throughout the northern Southwest, from the northern Canadian River

[12] Cordell,[16] p. 214; Cordell and Gumerman,[17] p. 7.

drainage in northeastern New Mexico to south-central Utah, to the Zuni Plateau of western New Mexico and eastern Arizona.

Formal variation among Basketmaker III sites with pithouse architecture is low compared to later puebloan periods. While some clinal stylistic differences seem to occur over large areas,[139] Basketmaker III dwellings are similar in size and construction throughout the Colorado Plateaus.[85] Coupled with a settlement distribution of sites in nearly every part of the Plateaus region,[50] the relative lack of differentiation among Basketmaker III sites suggests a basic shared organizational structure. With the exception of some large settlements that may be the result of periodic population aggregations, most Basketmaker III sites indicate a small group, probably consisting of an extended family.[8] Hence, the Basketmaker III period archaeological record suggests a widely dispersed population of small, economically redundant groups. By redundancy we mean that each group probably duplicated any other local group in its ability to meet basic subsistence requirements. These local groups cannot have been economically autonomous for extended periods,[10,101] but there are no indications that any Basketmaker III settlements specialized in the production or procurement of particular resources.

It seems more likely from the spatial dispersion of Basketmaker III sites that individual settlements experienced proportional differences in the range of dietary items as a function of local resource variation. Basketmaker III groups occupying grassland ecotones of the upper Canadian River drainage in northeastern New Mexico probably had greater access to ungulates such as antelope than contemporaneous populations in the forested mountains of southwestern Colorado. Similarly, local groups in upland regions undoubtedly enjoyed greater access to pinon nuts than those in other areas, such as the Rio Grande Valley near Albuquerque, who conversely had more opportunity to obtain riparian resources. In most respects, the Basketmaker III period seems to represent a continuation of extensive land-use patterns established during the Late Archaic.

The primary difference between the Basketmaker III period and the Late Archaic, and one that implicates agriculture as a driving force, is an increase in occupational intensity at residential locales. Pithouses in the Basketmaker III period represent major labor investments and generally occur with large, well-constructed storage features. There are pithouses and storage pits in the Late Archaic of the Colorado Plateaus,[90,122] but Basketmaker III houses are typically much more substantial and are assumed to reflect longer anticipated occupation.

Associated with increased investment in settlement architecture is the first obvious evidence for genetic variation in maize attributable to human manipulation.[43,44,86] Therefore, it appears that the mobility characteristic of Basketmaker III settlements was spatially constrained to a degree that localized adaptations of maize became possible. Phenotypic selection also suggests that local groups were able to produce successful crops on a regular basis from which particular seeds could be set aside, a process that may have necessitated increased cultivation efforts. The storage emphasis apparent in Basketmaker III sites is consistent with efforts to extend the temporal availability of cultivated plants, although storage features seem to indicate caching tactics among groups that periodically left habitation sites rather than facilities for storaging large, regularly accessible food supplies.[13]

Curiously, despite evidence for increased investment in cultivation during the Basketmaker III period, archaeologists have not yet recorded formal field systems or water control devices for this time period. The lack of such features points to extensive systems of land-use rather than intensive localized production. Extensive systems are successful by pooling production in order to reduce the average variance of all the plots.[14]

The risk reduction that comes from pooling or averaging productivity among several cultivation plots is characteristic of economic systems that achieve constant or diminishing returns. Increasing returns systems, where increased input leads to increases in marginal returns, require a more risk-prone strategy in which potential catastrophic losses are exchanged for large surpluses in successful years.[123] The land-extensive systems that seem to be implied by the spatial distributions and small sizes of Basketmaker III sites point to strategies of risk-avoidance, or ones expected to achieve some minimum level of production while avoiding the risk of large losses.

If Basketmaker III people were more concerned with risk-avoidance than surpluses, it is possible that the goals of agricultural production between A.D. 200 and 750 may have remained essentially the same as during the Late Archaic. Some support for this conclusion can be found in a recent study of settlement organization at Shabik'eschee Village, a Basketmaker III site in Chaco Canyon, New Mexico.[108] Shabik'eschee Village consists of over 60 pit structures, 21 of which have been excavated, and at least 50 storage features distributed in extramural areas between the pithouses. One of the structures is exceptionally large (94 m^2) and is interpreted as communal architecture.[15] Wills and Windes[155] believe that Shabik'eschee Village was occupied only periodically by large population aggregates, brought together by a local combination of conditions favorable to both agricultural and natural resource productivity. They argue that this pattern of aggregation is the same as that found among most mobile hunter-gatherer societies, where resource abundance stimulates large group formations. However, the potential for agricultural surplus made extended occupations possible at Shabik'eschee Village. The small size of individual dwellings (ca. 17 m^2) argues against a correlation between families and pithouses, while the public placement of storage facilities argues against private ownership of stored foods.

In Flannery's[39] terms, Shabik'eschee Village represents a communally organized group in which food is owned and shared among all members of the settlement. Communal economies provide little incentive for individuals to produce beyond their own needs, because surplus is likely to be dissipated quickly to the demands of generalized reciprocity.[113] Consequently, the organization inferred for Shabik'schee

[13] Wills and Windes,[155] p. 357.

[14] Winterhalder,[158] p. 69.

[15] Roberts,[108] p. 80.

Village is similar to the risk-averse economies of the Late Archaic, except that larger corporate groups appear to have established greater residential commitments to particular places, even if on an irregular schedule.

Several Basketmaker III sites equivalent in size to Shabik'eschee Village have recently been reported west of Chaco Canyon in the Chuska and Lukachukai mountains,[2,3] and these may represent a similar pattern of seasonal aggregation, although investigators have proposed varying interpretations of site histories. Nonetheless, survey and excavations throughout the Colorado Plateau continue to report small, dispersed settlements as the predominant Basketmaker III settlement pattern.[54,58,133,140] If larger sites are the remains of periodic aggregations, it is possible that the aggregating groups also maintained a number of other settlements between which they moved opportunistically.[16] Basketmaker III settlement systems would therefore have been similar to those characteristic of historical Navajo groups, with families operating on seasonal and annual schedules between residences scattered over a large area and a variety of environmental settings.[109]

These reconstructions assume that extended families were the common economic unit and that communal systems of reciprocal sharing were typical among Basketmaker III populations. However, in some Basketmaker III settlements, the appearance of private storage located within dwellings signals the beginnings of a shift toward household economies.[17] This is not necessarily a progressive evolutionary trend; internal storage facilities occur in earlier Late Archaic structures in the Durango, Colorado, area,[90] are absent in Basketmaker III sites in the Prayer Rock area of northeastern Arizona between A.D. 600 and 800,[91] but may be present as antechambers in sites dating from A.D. 500 to 800 in many parts of the San Juan River drainage. [Evidence for storage in baskets and ceramic vessels is not considered in this discussion.] In other words, there is good evidence that Basketmaker III groups utilized a range of subsistence and social tactics in a variety of configurations, probably in response to local demographic and environmental conditions.[18]

In sum, Basketmaker III populations flourished on the principles of flexibility and mobility that required large geographical areas in which people could move between stable residential locations. Individual groups probably experimented with a number of approaches to food production, involving different mixes of cultivated and natural resources, as well as varying forms of economic organization.

In the Mogollon region of the central and southeastern Southwest, economic strategies during the transition from the Late Archaic to the early ceramic period between approximately A.D. 200 and 500 suggest a range of economic tactics similar to those among Basketmaker III groups, with the exception that some settlements indicate intensive local cultivation tactics. Unfortunately, data from the initial ceramic period in the Mogollon area are far fewer than those available for

[16] Wills and Windes,[155] p. 365.

[17] See S. Plog,[103] p. 185.

[18] See Cordell,[16] p. 214.

the Colorado Plateau, and thus interpretations of this time frame must be based on investigations at a limited number of sites.[19]

One of the few Mogollon sites where the nature of organizational patterns can be evaluated is the SU site, an early pithouse site located in the Mogollon Mountains. First investigated by the Field Museum,[81,82] and later by the University of New Mexico,[155,154] the SU site consists of at least 35 pit structures, 28 of which were excavated. In distinct contrast to Basketmaker III sites like Shabik'eschee Village or Broken Flute Cave, storage within dwelling structures was a major feature at the SU site, with houses averaging about 2.8 m^3 of potential pit storage volume.[154] SU houses were much larger than contemporaneous Basketmaker III structures, averaging almost 40 m^2 in floor area and ranging to more than 70 m^2. The size and corresponding investment in construction labor and building material costs for SU houses indicates a residential unit greater than a single household.

Private storage and the probable affiliation of extended families with single pithouses at the SU site point to an economy based on household production. Household systems of production involving agriculture require restricted patterns of resource sharing so that an incentive exists for surplus production.[39] In household economies, the group gives up the advantages of pooling risk among numerous individuals in exchange for the surplus yields available from increased labor investment; the risk reduction from averaging among a large group is made up by dependence on surplus. However, the household system is therefore more risk-prone,[123] and thus, while early Anasazi and Mogollon settlements are alike in many respects, there is evidence for different approaches to food production.

Those differences encompass the Hohokam region of southern Arizona as well. Information about the early ceramic period (pre-A.D. 200–500) among the Hohokam is scant relative to later time periods. Red Mountain Phase (pre-A.D. 200) sites contain only small structures, but Vahki Phase sites contain both small and extremely large structures (averaging 85 m^2). Wilcox, McQuire, and Sternberg[20] suggest that several of these structures surrounded an open plaza area at Snaketown. Haury[21] argues that these large structures were occupied by extended families, while Sayles[22] and Gladwin[23] believed that they were communal or ceremonial structures. If Haury is correct, the presence of large storage pits within the structures, together with the presence of two entryways and multiple hearths in some of the structures, suggests a possible analog with the SU site pithouses in the presence of a household-based economy.

Data concerning the degree of dependence on agricultural produce and level of farming technology are scant for the early Hohokam occupation. While floodwater farming was probably practiced, agricultural features are lacking for the Vahki

[19] E.g., LeBlanc,[73] p. 45.

[20] Wilcox, McQuire, and Sternberg,[149] p. 143.

[21] Haury,[57] p. 68.

[22] Gladwin et al.,[49] p. 82.

[23] Gladwin,[48] p. 118.

and Red Mountain phases. A study of manos from Pueblo Patricio suggests low dependence on corn for the Vahki phase, although macrobotanical remains from the site include abundant corn remains.[24] Doyel[25] interprets the data as indicating a broad-based economy, and suggests that populations may have moved between larger villages and smaller summer hamlets on a yearly basis.

To summarize, we believe the early ceramic period throughout the Southwest was characterized by an increase in socioeconomic complexity over the Archaic Period and that agriculture had a role in that change. In some cases, local groups appear to have selected risk-prone strategies that permitted large surpluses, but these experiments with restricted sharing were situational and not a progressive evolutionary trend throughout the region. By and large, most people lived in small, highly dispersed settlements with extensive communication and exchange of mates among these settlements. Economic variation was largely a matter of local resource possibilities and not specialization at the settlement level. Depending on a variety of factors, including the development cycle of an individual household, the economic status of household, or the specific crop mixes, small settlements probably employed a variety of methods for meeting production needs.[147] The ability to move between settlements and to respond quickly to local environmental and demographic variation might be described as "ecological fine-tuning."[26] Such a strategy indicates greater concern with meeting minimal subsistence thresholds than with over production for surplus.

LATER SEDENTARY VILLAGES (CA. A.D. 750–1300)

The importance of residential mobility apparently continued for Anasazi populations after A.D. 750. For example, the archaeological record in southwestern Colorado shows a steady increase in population between A.D. 600 and 1250, despite intervals of population decline.[115] Most of the sites dating to this period were small pueblos assumed to have been occupied for a generation or two. Schlanger[27] views evidence for frequent site abandonment and short occupation spans as a result of periodic shifts in an hypothesized "farmbelt," a zone of adequate moisture and growing season length for maize cultivation. As climatic changes affected the location of this farmbelt, horticultural communities moved residence either up or down slope in response.

The farmbelt model of residential mobility outlined by Schlanger[115] assumes that climatic effects on plant cultivation were the primary factors producing settlement shifts. Kohler and Matthews[72] also argue for the "higher order priority" of

[24]Doyel,[30] p. 240.

[25]Doyel,[30] pp. 40–241.

[26]See Netting,[96] p. 40.

[27]Schlanger,[115] p. 775.

climate impact on food production as the stimulus for mobility among Dolores populations, but suggest that climate alone cannot explain all of the empirical record for mobility. They argue that local deforestation around settlements resulting from fuel and construction use also contributed to high levels of residential mobility.[28]

The issue of fuel depletion is one aspect of the larger problem of resource depression and increased procurement costs associated with increased sedentism.[131] The longer a group resides in a location, the greater the costs of obtaining mobile resources such as game and the more likely that negative impacts on local resources will occur. For example, soil depletion is still evident in agricultural terraces in the Mimbres area that are over 900 years old.[114]

Many nonagricultural resource depression effects associated with sedentism would be felt before be the onset of delerious climatic factors relative to agriculture. Production intervals for food production are so long (annually in temperate environments) that it may take years to distinguish trends in agricultural production.[158] Natural resource productivity will be apparent in a much shorter temporal span. Consequently, the factors contributing to residential mobility among agriculturalists with land extensive production systems are numerous, and as Stephen Plog[102] argues for Black Mesa, in northern Arizona, population declines and local abandonments precede the beginnings of deficient climate regimes for agriculture. Dendrochronological evidence in several areas of the Southwest after A.D. 500 for wood stockpiling prior to building construction provides further support for the conclusion that agricultural production was part of economic systems in which population movement was anticipated and planned in advance.[22,53]

In some cases, population shifts may have resulted from population increase; the archaeological sequence on northern Black Mesa is one example. According to S. Plog,[29] between A.D. 850 and 1100, "the mobility of groups living on Black Mesa declined as population density increased, competition for resources increased, and there was more intense exploitation of localized resources." A key variable in Plog's reconstruction is the catchment size of potential field locations. As population grew, field catchments declined and "the smaller the catchment areas the lower the probability a field in that catchment will receive sufficient moisture."[30] People apparently left Black Mesa by A.D. 1150 as it became difficult to buffer annual variance in rainfall through dispersed fields.

It is interesting that on Black Mesa, as in most parts of the Southwest, abandonment was relatively total—everyone seems to have left. What makes this interesting is the apparent connectivity among the settlements in the region. If the problem were simply one of alleviating land pressure, only some elements of the population would need to move. Instead, the entire population left, suggesting that the linkage between local social units was a critical part of their adaptive systems, meaning that

[28] Kohler and Matthews,[72] p. 559.

[29] S. Plog,[102] p. 197.

[30] S. Plog,[102] p. 200.

the exchange relationships among these communities was central to their adaptive systems.

Southwestern abandonments have conventionally been seen as economic failures. In the case of Black Mesa, should we think of the system as having failed, or did it simply reach a threshold that set off a predictable process inherent to the strategy? We suggest instead of economic "collapse," the significance of long-term persistence of mobility as a tactical option among mixed farming/forger economies in most parts of the prehistoric Southwest is that movement was a successful adaptation to the spatial variation in factors affecting agricultural production.[94] The importance is not that agricultural productivity sometimes failed to meet a farmer's expectations, but that farmers were able to transfer their agricultural systems to new places with apparent ease as long as unpopulated areas were available to receive them.

The ability to shift subsistence strategies to new locales is therefore related to the emergence of large aggregated and relatively sedentary communities in various parts of the Southwest between A.D. 800 and 1000. What role did agriculture have in the establishment of multigenerational communities of hundreds of individuals in Chaco Canyon, Mesa Verde, the Mimbres Valley, or the Phoenix and Tucson basins? Most data seem to indicate that nucleated population centers evolved in conjunction with a greater diversity of cultivation techniques, and especially with a greater division of labor in agricultural production.

The large size of aggregated communities and the obvious requirements for cooperative labor in building and cultivation point to the evolution of relatively specialized economic roles among members of these communities. By specialized we do not mean that individuals contributed to communities in only one way, but that some individuals must have spent a disproportionate amount of time and energy in certain economic activities, such as hunting or farming, and that stable organizations with fluctuating membership emerged to accomplish specific tasks.

The elaboration of specialized social and economic roles within large aggregated communities does not mean that communities were specialized at the settlement level. We believe that available studies of intersite subsistence variation indicate that every community probably produced the same basic set of domesticated resources, primarily corn and turkeys, and that the proportions of these domesticates in the diet was probably not much different than would have been found hundreds of years earlier.[11,28,88,127] Households were likely the basic unit of production regardless of community size and the "economy" of a community should probably be seen as a byproduct of aggregated but economically independent households, rather than an entity itself.[59,60] Thus the significant economic relationships were those that integrated these households.

One such relationship existed in the shift from settlement movement among dispersed cultivation localities to greater reliance on intensive localized production. By concentrating on more investment in smaller areas, corporate groups increased the risk of catastrophic crop loss.[31] The source of subsistence problems for these communities lay not so much in dependence on a limited set of food stuffs, although this surely was not the most helpful feature of localized agriculture, but in the spatial constraints on production.

Chaco Canyon provides evidence for one of the earliest intensive plant cultivation systems in the Colorado plateaus. Vivian[137,139] reported at least three methods of farming in Chaco Canyon during the eleventh-century period of major architectural building. These include sand dune fields, small walls and terraces for capturing runoff, and large canyon bottom field systems comprised of diversion dams, canals, ditches, and headgates. Small canals took rainfall runoff from major tributary drainages and effectively directed it to prepared fields averaging about 100 garden plots per hectare.[32] Because the Chaco fields were dependent on rapid response by farmers to localized runoff events, the effective use of these systems may have required some level of decision-making above the individual farmer.[33] A similarly complex diversion system is also documented for Kin Bineola, a Chacoan outlier. Consequently, Chacoan agriculture seems to have been based on two strategies, an intensive water diversion system founded and operated at some cooperative level (Vivian suggests communities) combined with smaller, diverse gardening tactics that could have been maintained by much smaller units, perhaps on the order of families.

Although several field systems are known in Chaco Canyon, estimates of available arable land are less than is thought to have been needed to support probable population densities.[34] If the land base in Chaco was indeed inadequate, the people living in the canyon either acquired additional food stuffs from other areas or periodically moved out. Sebastian[35] suggests that at least two agricultural strategies were operating in the San Juan Basin around Chaco Canyon during the eleventh and twelfth centuries, one involving "land extensive" systems of dispersed fields and the other based on "labor intensive" management of spatially concentrated fields. As yet the complicated nature of multiple land-use strategies among the San Juan Basin Anasazi is not clear, but the emergence of field systems built and used by large social units is an apparent break with the past. In one sense, corporate food production strategies, reflected in formal field systems, were a shift from averaging the productive variance of individual fields to averaging the labor of individual farmers. This change likely produced a gain in productive efficiency through greater division of labor (by reducing redundancy among producing units).

The development of intensive agricultural production strategies seems to characterize many episodes of population aggregation in the Southwest. For example, localized river bottom agriculture is posited for the Mimbres Valley in the Mogollon

[31]S. Plog,[102] p. 200.

[32]Vivian,[139] p. 312.

[33]Sebastian,[117] p. 58.

[34]Judge,[66] p. 227.

[35]Sebastian,[117] pp. 134–137.

area at about the same time that formal field systems appear at Chaco Canyon. Minnis[87] argued that successful farming strategies in the well-watered Mimbres Valley led to rapid population growth (.03% annually) that ultimately could not be sustained, resulting in widespread abandonment of the area after A.D. 1150. Elsewhere in the Mogollon region, small groups of mobile farmers were more common than the intensive agriculturalists of the Mimbres Valley.[94,105,107]

In contrast to the Anasazi and Mogollon regions, maize production in the Hohokam region was based on intensive canal irrigation by A.D. 800.[20,21] Hohokam agricultural systems associated with canals had a much more reliable water source than contemporaneous systems in the Anasazi or Mogollon regions, and, when coupled with the longer growing season in the Sonoran Desert, provided a relatively stable level of productivity. The Hohokam have traditionally been viewed as master farmers whose skills included cultivation of nonfood plants such as cotton.[57] Recent studies show that the Hohokam had even more complex agricultural strategies than once imagined. For the Salt-Gila Basin, Crown[18] described a wide range of agricultural features found in addition to canal systems, including terraces, check dams, grid gardens, catchment pools, and rock piles. She argued that this diversity reflected a corresponding diversity in tactics within single economic systems, possibly utilized in order to increase the range of foodstuffs available or to buffer the risk of crop loss from irrigated field systems, as should be seen as integral to agricultural success.[36]

In the Marana area, Suzanne and Paul Fish[36] describe a complex system of sites and agricultural features. One of the primary Hohokam crops apparently grown in nonirrigated contexts was agave (*Agave parryi* and *Agave murphyi*). Rock pile complexes are consistently associated with roasting pits containing charred agave remains, and therefore the rock piles are assumed to have served to facilitate agave production.[37] The rock piles probably acted as a mulch and as protection for small plants. Some agave roasting pits date prior to A.D. 1000, but these become extensive during the Classic period (ca. A.D. 1150-1600). An important aspect of agave cultivation is the requirement for an intensive focus on plants that take many years of maturation before they become useful economically. Agave cultivation is therefore one more indication of the very localized productive systems that characterized Hohokam agriculture.

The preceding examples suggest that by about A.D. 1000, two basic approaches to agricultural production had emerged from earlier periods of economic diversity. These were (1) large-scale, intensive field systems consisting of permanent features managed at levels above the household, and (2) land-extensive cultivation facilitated by movement of small residential groups between agricultural localities. Labor-intensive strategies did not replace other food production tactics, but seem related to corporate or community-level decisions about agriculture. In areas where people had some room in which to move, dispersed settlement systems of small sites with corresponding distributions of agricultural fields were common.

[36] Crown,[18] p. 24.

Nucleated settlements dating after A.D. 1100 in many, but not all, parts of the Southwest tend to be associated with large and complex arrays of agricultural features.[16,20,99,138,160] Large sites in the Mesa Verde region,[111] the Mogollon Rim,[76,135,159] the Zuni region,[69,142] the Rio Grande Valley,[1,13,84] and southern Arizona[83] occur with extensive systems of water control features utilized for agriculture.

In contrast, areas without large nucleated population centers of the twelfth and thirteenth centuries A.D. exhibit settlement patterns consisting of small, highly dispersed settlements. In some areas pithouse architecture continued, such as the Gallina region and portions of the Mogollon region, while small pueblos were common throughout the Kayenta area and the Rio Grande Valley.[80,129,144] Pueblos of the Coalition Period (ca. A.D. 1200–1325) in the northern Rio Grande Valley are often associated with nearby agricultural features such as cobble mulch beds or parallel ridges.[56,97,104] Coalition sites are highly standardized in size and site plan, which suggests very little functional differentiation among them. In many ways, these later puebloan settlements indicate economic systems that might have been organizationally like earlier Basketmaker economies. A major difference is that Pueblo adaptive systems indicate more distinctive stylistic boundaries and consequently perhaps less mobility than during the Basketmaker period.

In the Rio Grande Valley there is no evidence currently available that would indicate significant differences between large communities in terms of dietary reliance on agricultural production (see especially Spielmann et al.[125]) although presumably settlements varied in *productivity* relative to environmental constraints. However, excavations throughout the Rio Grande Valley suggest that every settlement cultivated maize, beans, and squash and kept turkeys.[15,143] The major economic difference between aggregated groups was probably in access to naturally occurring resources, not domesticated ones.

The eastern pueblos, extending from Taos in the north, through Pecos, the Galisteo Basin, and Tijeras Pueblo to the Salinas Pueblos in the south, were gateway communities to the High Plains where hunters could acquire antelope and bison. To some extent these communities probably specialized in game procurement and as interaction nodes for Pueblo and Plains people.[124] This kind of division of labor among communities does not mean that individual settlements were economically dependent on other settlements for survival. It does mean that some communities allocated a portion of their labor to procuring or producing products for exchange with other communities.

This sort of intersettlement exchange, which may well have been common hundreds of years earlier,[101] probably became increasingly complex when Athapaskan-speaking people arrived in the fifteenth or sixteenth centuries A.D. These Plains foragers appear to have moved into the hinterlands around sedentary Pueblos and developed mutualistic ties to farmers, providing Pueblo people with game and other natural resources while receiving agricultural products and practicing their own small-scale gardening.[148] While one may separate these ethnic groups for the purposes of discussion, in a practical sense the economic adaptations of interacting groups should be seen as components of a larger economic system. If Puebloan people were producing crop surpluses for the purpose of exchange with non-Pueblo people, the agricultural part of the Pueblo system was surely susceptible to pressures created by individuals outside the community. So, too, if other economic activities, such as pottery manufacture, were tied to exchange.

SUMMARY

Temporal patterns in prehistoric southwestern agricultural strategies appear to meet expectations for the evolution of complexity outlined in the beginning of this chapter. The initial adoption of domesticates gave hunter-gatherers the ability to expand tactical subsistence options without significantly altering their economies. Sometime between about A.D. 200 and 500, people began explore a wide range of social and subsistence strategies, but by A.D. 1000, two basic agricultural strategies had evolved.

One of these strategies was simply the continuation of economic systems established during Archaic; that is, a high degree of residential mobility that achieved a secure but low-yield productivity by averaging subsistence returns from geographically dispersed locations. In contrast, a newer strategy based on averaging surplus production from high-yield, intensive cultivation systems developed in contexts of population aggregation. This newer, risk-prone strategy involved greater division of labor among community members and hence greater complexity than the older approach to food production. Hohokam irrigation communities and the large villages of Chaco Canyon appear to have been among the first examples of intensive, corporately managed agricultural systems.

In the Southwest, as in many other parts of the world, the appearance of large-scale farming systems was apparently associated with demographic thresholds and resource depression. Worldwide historic and ethnographic data consistently associate agricultural intensification with land pressure stemming from population density in excess of perceived tolerance levels.[96] In the Southwest, intensive and diverse agricultural features occur with large nucleated settlements, indicating a connection between aggregation and increased complexity in food production. However, aggregation per se need not be the cause of agricultural complexity.

Population aggregation quickly leads to local natural resource depletion in an arid environment like the Southwest, provided the aggregate stays in one place for any length of time. Consequently, the costs of acquiring natural resources increases positively with length of occupation as resources close to the settlement are exhausted. Szuter and Bayham[37] argue that sedentary farmers will increasingly specialize in procuring high-return resources as the travel costs for obtaining these items increase. Therefore an expectable relationship in the aggregation process is

[37]Szuter and Bayham,[131] p. 88.

greater specialization through time in nonagricultural components of a community's economy.

But in agricultural production, large, nucleated villages could pursue the same strategy of spatial averaging as smaller communities. Historically, many Pueblos did just that; they dispersed to field houses or agricultural villages during the summer and returned to aggregated villages in the winter. Consequently, intensification of agricultural production involving specialization implies one or more factors besides population nucleation, including (1) social or natural barriers to seasonal dispersal and (2) economic advantage in concentrating producers.

In agrarian societies, labor is the force behind surplus production.[65,113] Among modern Pueblos, individual farmers depend on help from work parties during critical periods, especially for field preparation and harvesting.[40] The same condition exists among maize farmers throughout Latin America.[70,151] While households are relatively independent in consumption of food stuffs, they rely heavily on exchange of labor for production.

The role of labor exchange among subsistence farmers is strongly affected by the location and nature of agricultural fields. Stone's[128] analysis of Kofyar field systems revealed that the Kofyar will only travel a limited distance to work in another farmer's field. Obviously, the closer together fields are, the easier for a field owner to assemble a work party to help him. Therefore, there is a strong motivation to concentrate fields in one place, provided that production depends on labor exchange.[38] In this sense, fields attract farmers. Once established as a strategy, concentrating field locations has a strong positive feedback aspect, since if fields attract labor, and the availability of labor then improves production, this should in turn attract more farmers.

An analysis of labor allocation among the Shan (Thailand) also indicates the positive feedback aspect of formal field systems. Durrenberger[32] demonstrated that among the Shan, harvest productivity was highly correlated with the amount of area and labor available to individual farmers who cultivate both irrigated and swidden fields. Because the area of irrigated fields is fixed, area determines the necessary labor to meet production expectations. In contrast, swidden plots are not fixed, and thus labor determines the area cultivated. One might generalize from Durrenberger's study that field systems comprised of fixed features and structures, such as those characteristic of many aggregated southwestern settlements, require a stable labor pool, while less formal fields operate on a more flexible and less predictable source of labor.

Since the availability of labor is a matter of demography, a shift toward highly productive but risk-prone agricultural strategies seems strongly density dependent. It is not critical to our discussion whether changes in regional population density are driven by changes in agricultural production, as some researchers have suggested, or vice versa; the reality is certainly that various factors promote demographic change in different contexts. We simply think that available data indicate that the

[38] Stone,[128] p. 350.

evolution of two basic strategies for food production was related to the demographic structure of regional populations.

The competitive advantage in one or the other agricultural strategy need not have involved direct conflict between communities. For example, if a community adopted an intensive production strategy that achieved higher production levels through labor exchange within the group, it may not have needed labor from other communities, thereby denying the other communities access to labor and diminishing the effectiveness of their economies. Such linkage is analogous to the fitness landscape concept described earlier in this chapter and similar to concepts proposed by Redding[106] for understanding the origins of domesticates. Synchroneity in regional Southwestern abandonments is a strong argument that socioeconomic connectivity among communities was probably more important to long-term survival than the particular agricultural strategies employed by any single community. In other words, exchange relationships (i.e., specialization) among settlements may have had as much economic significance as individual community production strategies.

The development of agricultural strategies based on long-term investments in formal field systems may have resulted in new cultural relationships with environmental variables. Low-return, risk-averse agricultural systems would probably be more responsive to high-frequency processes but might well be oblivious to low-frequency phenomena. Likewise, large field systems that relied on labor to maintain productivity likely were resilient to high-frequency processes but more susceptible to long-term changes in environmental periodicity. If this were the case prehistorically in the Southwest, adaptive shifts in cultural systems should have been more closely related to low-frequency cycles after A.D. 1000.

We cannot evaluate this complex relationship between environmental change and human adaptive systems in this paper, but we note that during the interval from ca. A.D. 800 to 1300, the Colorado Plateaus experienced a period of rising alluvial water tables and effective moisture (with brief episodes of departure) and from A.D. 1000 to 1300, a period of low temporal variability in rainfall.[39] In *general*, these conditions should have favored stable agricultural field systems and this is when the first formal fields appear in this region with aggregated settlements.[40] Additionally, a precipitous drop in water levels and an increase in temporal variability in precipitation just before A.D. 1300 corresponds to widespread demographic restructuring through regional abandonments. We do not suggest that environmental conditions caused the evolution of different agricultural strategies, but we do believe that variable climatic conditions affected different production strategies in expectable ways.

In sum, we think that in modeling the evolutionary significance of agriculture, it is critical not to detach food production from its demographic and economic

[39] Dean et al.,[25] p. 541.

[40] Dean et al.,[25] p. 549.

context. Agriculture was never an independent economic system, but rather a component of a larger economic realm. In order to appreciate fully the emergence and persistence of at least two primary approaches to food production, we need to think of the communities utilizing these strategies as interacting entities. And with that observation we return to the idea that mobility was fundamental to the adaptive systems of Southwestern farmers, regardless of community size. The episodes of regional and local abandonment that occur throughout the agricultural period among communities with complex food production systems are perhaps simply larger scale versions of the same process characteristic of smaller settlements with less complex economies. Those abandonments are not failures of agriculture to support a community, but rather the strength of the overall economic system to sustain a population regardless of environmental variation.

EVOLUTION OF POTTERY PRODUCTION

As with domesticated plants, ceramic technology was familiar to Southwestern populations long before it became incorporated into the economy in a "significant" way. Thus, fired clay figurines are found in Archaic sites by 800 B.C., perhaps as much as a thousand years before ceramic vessels are manufactured. Pottery vessels appear first in the Hohokam area prior to A.D. 200,[30] in the Mogollon area about A.D. 200,[41] and in the Anasazi area by A.D. 500.[157] In all three areas, the earliest ceramic vessels were plain brownware. There are several possible reasons for the lag between clear knowledge of ceramic technology and actual manufacture of ceramic vessels. First, as often noted, pottery is more fragile than basketry and would thus be less easy to transport among mobile populations. The fact that relatively few hunting and gathering populations make pottery is often cited in support of this argument. The period of the year in which pottery manufacture may have been possible in most portions of the Southwest (roughly March through October due to problems with firing during cold months) is roughly the same period in which wild resources would be available for harvesting. This might present a scheduling conflict for individuals responsible for both activities. The advantages of pottery include the greater storage security provided by solid vessels and the greater efficiency of manufacture. Pottery was probably incorporated into the material culture of Southwestern peoples when greater sedentism made it easier to possess these heavier, breakable vessels, made greater containment security (particularly from pests) a higher priority, and when farming lessened the scheduling conflict of pottery manufacture.

[41]LeBlanc,[74] p. 30.

Once adopted, the variety of pottery produced among different groups in the Southwest appears to undergo a slow evolution toward increasing diversity, followed by the disappearance of some types and apparently less variation. For instance, in the Hohokam area, the earliest pottery (prior to A.D. 300) is brownware. By the Vahki Phase (approximately A.D. 300-500), pottery types included both a plain brownware and a redware.[42] Probably the greatest variability in Hohokam pottery technology occurs during the succeeding Estrella (A.D. 500s) and Sweetwater Phases (A.D. 600s) (dating from Dean[43]). Techniques of vessel thinning include both paddle-and-anvil and scraping; techniques of vessel finishing include red-slipping, painting, and incising. Vessel types produced include a plainware, redware, red-on-gray, and polychrome (red and yellow paint).[44] The polychrome was short-lived, the redware disappeared after A.D. 800, and the red-on-gray was transformed into a red-on-buff through the addition of a wash and changes in the firing technology.[57] As with pottery technology, decorative styles undergo a shift from greater variability to greater homogeneity during this time period. In his discussions of Pioneer Period pottery designs from Snaketown, Haury[57] often refers to the "experimental" and "innovative" nature of the early pottery,[45] while the pottery after A.D. 700 shows greater "rigidity" of expression.[46] He notes that, "The adoption of strict standards by the potters in the Snaketown Phase must somehow be related to the maturing of an art style."[47] Even vessel forms appear to become less diverse during the Snaketown Phase.[48]During the early Pioneer Period, ceramic technology was used to produce a variety of forms, in addition to ceramic vessels, including human and animal effigies, miniatures, and miscellaneous objects.

In general, the adoption of pottery vessels among the Hohokam seems to have been followed by a long period (300 to 500 years) of relative homogeneity in the manufacturing tradition. Once decorated ceramics were introduced, there is strong evidence for increased "experimentation" with both designs and technology for approximately two centuries, followed by the adoption of a fairly uniform technology and decorative style after A.D. 700. Innovations, such as polychrome paint and patterned grooving, apparently disappeared from the repertoire after this time period. Change and innovation obviously occurred after A.D. 700, but most of these involved "fine tuning" of the existing process, so that the range of techniques used to produce pottery did not reach the early Pioneer levels probably until after A.D. 1150/1300.

[42]Doyel,[30] p. 236.

[43]Dean,[26] Table 3.3.

[44]Haury,[57] pp. 217–222.

[45]Haury,[57] pp. 217, 219.

[46]Haury,[57] p. 214.

[47]Haury,[57] p. 216.

[48]Compare Haury,[57] Figures 12.42 and 12.48.

Evidence for specialization in pottery production is ambiguous.[93] At La Cuidad, the even distribution of evidence for ceramic production among courtyards during the Colonial Period was interpreted as indicating production by all households.[71] However, Neitzel[49] suggests that a decrease in the production steps needed to produce Santa Cruz Phase red-on-buff ceramics might be associated with a shift to specialized production. There is also evidence that a limited number of households participated in ceramic production (producer specialization) during the Sedentary Period at Snaketown.[119] Some researchers have also argued that only a few Hohokam sites produced red-on-buff ceramics (site specialization).[29,57] Wilson and Blinman[157] argue that the presence of dedicated pottery production or firing areas within sites is probably evidence for increased specialization because higher production is associated with greater investment in facilities. At Snaketown, a dedicated production and firing area is associated with Sacaton Phase structures in one portion of the site,[50] minimally suggesting producer specialization at this time period. This question obviously deserves more attention, for the distinction between producer specialization and site specialization is important in evaluating the degree of economic dependence within and between communities.

Specialized production is documented early in the Anasazi sequence as well, with producer specialization in white wares by A.D. 725, and production of San Juan Red Ware vessels on a regional basis in southeastern Utah by A.D. 750 with widespread exchange from there.[157]

The general principles outlined at the Santa Fe Institute thus describe the process of the adoption of pottery in a general fashion. The adoption of basic pottery technology was indeed followed over a period of centuries by experimentation with the technology and the development of many variants. Eventually however, some variants were dropped in favor of others, resulting in a decline in diversity. Specialization within sites at the level of the household probably developed fairly early in the sequence, with site specialization developing later. While these basic principles describe the process, they do not explain it. All of the variants were viable containers, and the question remains as to why some forms disappeared and others became the standard. As an analog of the bicycle model presented above, it could be argued that experimentation led to the recognition that some ways of making pottery are inherently better than others, so that inferior forms are dropped in favor of superior ones. In this instance, "superiority" probably relates to efficiency in production, accessibility of the basic resources needed to produce the variants, and achievement of desirable vessel properties (thermal shock resistance, porosity).

In terms of craft specialization, early specialization in pottery may well have been associated with differential access to agricultural land, with pottery production by poorer households providing a means for supplementing subsistence products.[4] However, it is possible that later site specialization in particular wares developed

[49] Neitzel,[93] p. 185.

[50] Haury,[57] p. 194.

due to access to or control of an essential material used in the production of these wares (Shepard[121] and Rice[[51]]) rather than location in a poor farming area.

Viewed over the entire sweep of Southwestern prehistory, patterning is apparent that matches the expectations of the general principles as outlined. The temporal parameters are quite broad however. It is clear that technological innovations were familiar to Southwestern populations long before they were incorporated into their lives in any significant fashion. The adoption of these innovations took many centuries in some cases. The period of experimentation and greater diversity apparently took centuries as well. Here, however, it is possible that we view archaeological manifestations at too gross a level to recognize experimentation and improvement in innovations that actually occurred. Our methods may not be sufficiently improved to elucidate all of the steps in the process. Alternatively, it may be that such processes only occurred over long periods of time in prehistory, so that they are of more general utility for describing that past.

CONCLUSIONS: ARCHAEOLOGICAL PERSPECTIVES ON COMPLEX SYSTEMS

Archaeology takes as a major goal the scientific understanding of human behavior, especially as patterned behavior changed and evolved over the time spanned by the archaeological record. At present, the discipline seems to have reached something of an impasse in its pursuit of this objective. Established approaches appear to have exhausted much of their explanatory vigor, while newer approaches, or recycled older ones, have yet to evince much explanatory power.[134] Joe Tainter, in his introductory remarks to this conference, identified the proliferation of theory as one symptom of this crisis, and he and Linda Cordell enumerate a plethora of contenders for the next unifying body of theory. This conceptual stagnation demands new ways of looking at important general issues and their local manifestations. Projecting general theories of complex systems into the sociocultural domain may help propel archaeology onto the next explanatory plateau.

Sociocultural complexity has been an important interest of both anthropology and archaeology since the inception of these disciplines more than a century ago. An immense amount of effort has been devoted to characterizing and explaining the nature and evolution of complex social systems. Numerous cultural classifications are based implicitly or explicitly on increasing degrees of complexity as measured by a wide range of variables. From rudimentary savagery-barbarism-civilization continual[79,92] to more sophisticated, multivariate, band-tribe-chiefdom-state sociopolitical progressions,[118,113] social anthropologists have sought to capture the

[[51]]Rice,[107] p. 262.

complexity inherent in human cultures. Archaeologists have created similar typologies ranging from nineteenth-century stone-bronze-iron age sequences to hamlet-village-city-state progressions.

Another common theme of both anthropological and archaeological attempts to understand sociocultural change and evolution has focused on the changing relationships between human social units and behavior on the one hand and environmental stability and variability on the other. Different relationships between human behavior and the physical environment are thought to characterize the different stages of sociocultural complexity. Changes in these interactions are often thought to stimulate the transition from one level to the next. Thus, increasing complexity often is explained in terms of cultural evolution impelled by human behavioral adaptation to stasis and variability in the social and physical environments.[118,126,146]

The foregoing considerations leave no doubt that sociocultural systems change in response to internal and external stimuli and thereby evolve into different, more complex systems. Human societies therefore conform to the definition of "complex adaptive systems."[47] Because complex adaptive systems result from fundamental natural laws, their behavior can be characterized mathematically. Recent advances in the modeling of such systems should aid in the understanding of sociocultural systems.[5,12,63,67] These studies represent the initial stages in the development of a mathematics capable of dealing with the manifest complexities of human behavior and its derivative, the archaeological record. Until very recently this capability simply did not exist, but in the next decade it is likely that there will be considerable refinement of such mathematics, and that the archaeological record will be a prime medium for the evaluation and further development of these analytical techniques.

Models of complex systems are designed to represent their fundamental structure and to specify the rules that govern their behavior, development, and evolution. Even the most complex behavior can be generated from a finite, relatively simple set of entities, properties, boundaries, initial conditions, and rules. In addition, interactions between entities and between entities and their environments are built into the models, which are used to simulate possible spatial and temporal outcomes of the involved processes. To the extent that cultural systems conform to basic natural laws, the application of these models should elucidate the dynamics of cultural processes. To the extent that tested and reformulated models fail to account for cultural dynamics, fundamental differences between natural and cultural systems will have been revealed and new model classes, perhaps even a new mathematics, will have to be created to account for these differences. At the very least, new concepts and perspectives for viewing sociocultural complexity will emerge. Any of these eventualities would represent a significant scientific advance.

Southwestern archaeology provides a suitable context for the application of complex adaptive system modeling to the human domain. The prehistoric behavior involved is sufficiently complex to be relevant but not too complex to mask important features. The archaeological record is long and detailed enough to evaluate models that encompass extensive time intervals and a broad range of behavioral variability. A clear trajectory of increasing sociocultural complexity exists in the region and the paleoenvironmental record is sufficiently refined to allow accurate description of the external variables that impinge on adaptive systems. Combining the archaeological and paleoenvironmental data allows detailed examination of the interactions among human behavioral and environmental stability and change. Thus, simulations generated from the parameters and rules of the models can be adequately tested against hard data.

This paper has discussed in very preliminary terms some of the exciting explanatory possibilities inherent in new theoretical and methodological approaches to the study of complex adaptive systems. Obviously, we had neither the time nor space to cover the entire range of possibilities, to construct a family of mathematical models of the evolution of cultural complexity, or to discover which these constructs best captures the dynamics of Southwest prehistory. Instead, we confined ourselves to a few key concepts that we believe to be relevant to understanding complex cultural systems. We see particular potential in three classes of concepts that should generate expectations of the archaeological data. One suite of concepts pertains to the co-evolution of complex entities on ever-changing (rugged) fitness landscapes. Another set of concepts is used in generating complex entities and interactional networks through the operation of random grammars and other schemata on symbol strings that define the entities involved. Another potentially productive set of concepts and procedures are concerned with the analysis of phase boundaries, phase transitions, and other phenomena. Two kinds of models and their implications for the evolution of Southwestern cultural complexity should be useful: (1) global or top-down constructs, and (2) bottom-up formulations. We believe that models of this type will eventually be used to create simulations of complex adaptive systems that can be empirically evaluated against archaeological data.

REFERENCES

1. Acklen, John C., Mark E. Harlan, Stephen C. Lent, and James L. Moore. *BA Station to Norton Station 345 kV Transmission Project Archaeological Testing Phase Report.* Albuquerque: Public Service Company of New Mexico, 1984.
2. Altschul, Jeffrey H. "Social Complexity and Residential Stability During the Basketmaker III Period." Paper presented at the 56th Annual Meeting for the Society for American Archaeology, New Orleans, 1991.
3. Anyon, Roger, and Michael McFaul. "Tempo, Duration and Sequence: Microstratigraphic Excavation and Basketmaker III Settlement Systems." Paper presented at the 56th Annual Meeting of the Society for American Archaeology, New Orleans, 1991.
4. Arnold, Dean E. *Chemical Theory and Culture Process.* Cambridge: Cambridge University Press, 1985.
5. Arthur, W. Brian. "Self-Reinforcing Mechanisms in Economics." In *The Economy as an Evolving System*, edited by P. W. Anderson, K. J. Arrow, and D. Pines, 9–31. Santa Fe Institute Studies in the Sciences of Complexity, Proc. Vol. V. Reading, MA: Addison-Wesley, 1988.
6. Arthur, W. Brian. "Positive Feedbacks in the Economy." *Sci. Amer.* **262** (1990): 92–99.
7. Baumol, William J., and Alan S. Binder. *Microeconomics: Principles and Policy.* San Diego: Harcourt Brace Jovanovich, 1991.
8. Birkedal, Terje G. *Basketmaker III Residential Units: A Study of Prehistoric Social Organization in the Mesa Verde Archaeological District.* Unpublished Ph.D. dissertation, Department of Anthropology, University of Colorado, 1976.
9. Boyd, Robert, and Peter J. Richerson. *Culture and the Evolutionary Process.* Chicago: University of Chicago Press, 1985.
10. Braun, David P., and Stephen Plog. "Evolution of 'Tribal' Social Networks: Theory and Prehistoric North American Evidence." *Amer. Antiquity* **47** (1982): 504–525.
11. Breternitz, Cory D., and David E. Doyel. "Methodological Issues for the Identification of Chacoan Community Structure: Lessons From the Bis Sa'ani Community Study." *Amer. Archaeology* **6** (1987): 183–189.
12. Brown, Joel S. "Habitat Selection as an Evolutionary Game." *Evolution* **44** (1990): 732–746.
13. Buge, David. "Prehistoric Subsistence Strategies in the Ojo Caliente Valley, New Mexico." In *Prehistoric Agricultural Strategies in the Southwest*, edited by S. Fish and P. Fish, 27–34. Anthropological Research Papers No. 33. Tempe: Arizona State University, 1984.

14. Cavalli-Sforza, L. L., and Marcus W. Feldman. *Cultural Transmission and Evolution: A Quantitative Approach.* Monographs in Population Biology, Vol. 16. Princeton, NJ: Princeton University Press, 1981.
15. Cordell, Linda S. "Late Anasazi Farming and Hunting Strategies: One Example of a Problem in Congruence." *Amer. Antiquity* **42** (1977): 449–461.
16. Cordell, Linda S. *Prehistory of the Southwest.* New York: Academic Press, 1984.
17. Cordell, Linda S., and George J. Gumerman. "Cultural Interaction in the Prehistoric Southwest." In *Dynamics of Southwest Prehistory*, edited by L. S. Cordell and G. J. Gumerman, 1–18. Washington, DC: Smithsonian Institution Press, 1989.
18. Crown, Patricia L. "Adaptation Through Diversity: An Examination of Population Pressure and Agricultural Technology in the Salt-Gila Basin." In *Prehistoric Agricultural Strategies in the Southwest*, edited by S. K. Fish and P. R. Fish, 5–25. Anthropological Research Papers No. 33. Tempe: Arizona State University, 1984.
19. Crown, Patricia L. "Classic Period Hohokam Settlement and Land Use in the Casa Grande Ruins Area." *J. Field Archaeology* **14** (1987): 147–162.
20. Crown, Patricia L. "Water Storage in the Prehistoric Southwest." *The Kiva* **52** (1987): 209–228.
21. Crown, Patricia L. "The Hohokam of the American Southwest." *J. World Prehistory* **4** (1990): 223–255.
22. Dean, Jeffrey S. *Chronological Analysis of Tsegi Phase Sites in Northeastern Arizona.* Tempe: University of Arizona Press, 1969.
23. Dean, Jeffrey S. "A Model of Anasazi Behavioral Adaptation." In *The Anasazi in a Changing Environment*, edited by G. J. Gumerman, 25–44. Cambridge: Cambridge University Press, 1988.
24. Dean, Jeffrey S. "Dendrochronology and Paleoenvironmental Reconstruction on the Colorado Plateaus." In *The Anaszazi in a Changing Environment*, edited by G. J. Gumerman, 119–167. Cambridge: Cambridge University Press, 1988.
25. Dean, Jeffrey S., Robert C. Euler, George J. Gumerman, Fred Plog, Richard H. Hevly, and Thor N. V. Karlstrom. "Human Behavior, Demography, and Paleoenvironment on the Colorado Plateaus." *Amer. Antiquity* **50** (1985): 537–554.
26. Dean, Jeffrey S. "Thoughts on Hohokam Chronology." In *The Hohokam: Prehistoric Desert Peoples of the American Southwest*, edited by George J. Gumerman, 61–149. Albuququerque, NM: University of New Mexico Press, 1991.
27. DeBoer, Warren R. "Subterranean Storage and the Organization of Surplus: The View from Eastern North America." *Southeastern Archaeology* **7** (1988): 1–20.

28. Decker, Kenneth W., and Larry L. Tieszen. "Isotopic Reconstruction of Mesa Verde Diet from Basketmaker III to Pueblo III." *The Kiva* **55** (1989): 33–48.
29. Doyel, David E. "Hohokam Social Organization and the Sedentary to Classic Transition." In *Current Issues in Hohokam Prehistory*, edited by D. E. Doyel and F. Plog, 23–40. Arizona State University Anthropological Research Papers 23. Tempe, AZ: Arizona State University, 1980.
30. Doyel, David E. "Hohokam Cultural Evolution in the Phoenix Basin." In *Exploring the Hohokam*, edited by G. J. Gumerman, 231–278. Albuquerque: University of New Mexico Press, 1991.
31. Dunnell, Robert C. "Aspects of the Application of Evolutionary Theory in Archaeology." In *Archaeological Thought in America*, edited by C. C. Lamberg-Karlovsky, 35–49. Cambridge: Cambridge University Press, 1989.
32. Durrenberger, E. Paul. "An Analysis of Shan Household Production Decisions." *J. Anthropological Resh.* **35** (1979): 447–458.
33. Eldredge, Niles. "Information, Economics, and Evolution." *Ann. Rev. Ecology* **17** (1986): 351–369.
34. Euler, Robert C., George J. Gumerman, Thor N. V. Karlstrom, Jeffrey S. Dean, and Richard Hevly. "The Colorado Plateaus: Cultural Dynamics and Paleoenvironment." *Science* **205** (1979): 1089–1101.
35. Farmer, J. Doyne, and Alletta d'A. Belin. *Artificial Life: The Coming Revolution.* Cambridge: Cambridge University Press, 1990.
36. Fish, Paul R., and Suzanne K. Fish. "Hohokam Political and Social Organization." In *Exploring the Hohokam*, edited by G. Gumerman, 151–176. Albuquerque: University of New Mexico Press, 1991.
37. Fish, Suzanne, K., Paul R. Fish, Charles Miksicek, and John Madsen. "Prehistoric Agave Cultivation in Southern Arizona." *Desert Plants* **7** (1985): 107.
38. Fish, Suzanne K., Paul R. Fish, and John Madsen. "Sedentism and Settlement Mobility in the Tucson Basin Prior to A.D. 1000." In *Perspectives on Southwestern Prehistory*, edited by P. E. Minnis and C. L. Redman, 76–91. Boulder: Westview Press, 1990.
39. Flannery, Kent V. "The Origins of the Village as a Settlement Type in Mesoamerica and the Near East: A Comparative Study." In *Man, Settlement and Urbanism*, edited by P. Ucko, R. Tringham, and G. Dimbleby, 1–31. Cambridge, MA: Duckworth, 1972.
40. Ford, Richard I. "An Ecological Analysis Involving the Population of San Juan Pueblo, New Mexico." Ph.D. dissertation, University of Michigan. University Microfilms International, Ann Arbor, 1968.
41. Ford, Richard I. "Gardening and Farming Before A.D. 1000: Patterns of Prehistoric Cultivation North of Mexico." *J. Ethnobiology* **1** (1981): 6–27.
42. Ford, Richard I. "Ecological Consequences of Early Agriculture in the Southwest." In *Selected Papers on the Archaeology of Black Mesa*, edited

by S. Plog and S. Powell, 127–138. Carbondale: Southern Illinois University Press, 1984.

43. Ford, Richard I. "The Processes of Plant Food Production in Prehistoric North America." In *Prehistoric Food Production in North America*, edited by R. I. Ford, 1–18. Anthropological Papers of the University of Michigan, No. 75. Ann Arbor, 1985.
44. Galinat, Walton C. "Domestication and Diffusion of Maize." In *Prehistoric Food Production in North America*, edited by R. I. Ford, 245–278. Anthropology Papers No. 75, Museum of Anthropology, University of Michigan, Ann Arbor, 1985.
45. Gatewood, John B. "Cooperation, Competition, and Synergy: Information-Sharing Groups Among Southeast Alaskan Salmon Seiners." *Amer. Ethnologist* (1984): 350.
46. Gell-Mann, Murray. "Simplicity and Complexity in the Description of Nature." *Engineering and Science* (1988): 3–9.
47. Gell-Mann, Murray. "The Santa Fe Institute. Presentation given in Santa Fe, New Mexico." January, 1990.
48. Gladwin, Harold S. *Excavations at Snaketown IV: Review and Conclusions.* Medallion Papers No. 38. Gila Pueblo, Globe, Arizona, 1948.
49. Gladwin, Harold S., Emil W. Haury, E.B. Sayles, and Nora Gladwin. *Excavations at Snaketown Material Culture.* Medallion Papers No. 25. Gila Pueblo, Globe, Arizona, 1937.
50. Glassow, Michael A. "Changes in the Adaptations of Southwestern Basketmakers: A Systems Perspective." In *Contemporary Archaeology: A Guide to Theory and Contributions*, edited by M. P. Leone, 289–302. Carbondale: Southern Illinois University Press, 1972.
51. Gooding, John D. ed. *The Durango South Project: Archaeological Salvage of Two Late Basketmaker III Sites in the Durango District.* Anthropological Papers of the University of Arizona No. 34. Tucson: University of Arizona, 1980.
52. Graham, Martha, and Alexandra Roberts. "Residentially Constrained Mobility: A Preliminary Investigation of Variability in Settlement Organization." *Haliksa'i* **5** (1986): 104–115.
53. Graves, Michael W. "Growth and Aggregation at Canyon Creek Ruin: Implications for Evolutionary Change in East-Central Arizona." *Amer. Antiquity* **48** (1983): 290–315.
54. Gumerman, George J. *The Archaeology of the Hopi Buttes District, Arizona.* Center for Anthropological Investigations Research Paper No. 49. Carbondale: Southern Illinois University, 1988.
55. Gumerman, George J., and Jeffrey S. Dean. "Prehistoric Cooperation and Competition in the Western Anasazi Area." In *Dynamics of Southwest Prehistory*, edited by L. S. Cordell and George J. Gumerman, 99–148. Washington, DC: Smithsonian Institution Press, 1989.

56. Haecker, Charles M. "Puebloan Limited Activity Sites Within the Western Uplands of the Ortiz Mountains." In *Secrets of a City: Papers on Albuquerque Area Archaeology*, edited by Anne V. Poore and John Montgomery, 99–107. The Archaeological Society of New Mexico, 1987.
57. Haury, Emil. *The Hohokam: Desert Farmers and Craftsmen.* Tucson: University of Arizona Press, 1976.
58. Hayes, Alden. "A Survey of Chaco Canyon Archeology." In *Archeological Surveys of Chaco Canyon. New Mexico*, edited by A. C. Hayes, D. M. Brugge, and W. J. Judge, 1–68. Publications in Archaeology 18A. Washington, DC: National Park Service, 1980.
59. Hegmon, Michelle. "Risk Reduction and Variation in Agricultural Economies: A Computer Simulation of Hopi Agriculture." *Research in Economic Anthropology* **11** (1989): 89–121.
60. Hegmon, Michelle. "The Risks of Sharing and Sharing as Risk Reduction: Interhousehold Food Sharing in Egalitarian Societies." In *Between Bands and States*, edited by Susan A. Gregg, 309–332. Center for Archaeological Investigations, Occasional Paper No. 9, Southern Illinois University, Carbondale, 1991.
61. Huckell, Bruce B. "Late Archaic Archaeology of the Tucson Basin: A Status Report." In *Recent Research on Tucson Basin Prehistory: Proceedings of the Second Tucson Basin Conference*, edited by W. Doelle and P. Fish, 57–80. Institute for American Research Papers No. 10. Tucson, AZ:IAR, 1988.
62. Huckell, Bruce B. "Late Preceramic Farmer-Foragers in Southeastern Arizona: A Cultural and Ecological Consideration of the Spread of Agriculture into the Arid Southwestern United States." Ph.D. dissertation, Arid Lands Resource Sciences. University of Arizona, Tucson, 1990.
63. Huges, D. J., and J. B. C. Jackson. "Do Constant Environments Promote Complexity of Form?: The Distribution of Bryoson Polymorphism as a Test of Hypotheses." *Evolution* **44** (1990): 889–905.
64. Hunter-Anderson, Rosalind L. *Prehistoric Adaptation in the American Southwest.* Cambridge: Cambridge University Press, 1986.
65. Johnson, Allen W., and Timothy Earle. *The Evolution of Human Societies from Foraging Group to Agrarian State.* Stanford: Stanford University Press, 1987.
66. Judge, W. James. "Chaco Canyon—San Juan Basin." In *Dynamics of Southwest Prehistory*, edited by L. S. Cordell and G. J. Gumerman, 209–262. Washington, DC: Smithsonian Institution Press, 1989.
67. Kauffman, Stuart A. "Random Grammars: A New Class of Models for Functional Integration and Transformation in the Biological, Neural, and Social Sciences." Santa Fe Institute, Santa Fe, New Mexico, 1990.
68. Kauffman, Stuart A., and Sonke Johnsen. "Coevolution to the Edge of Chaos: Coupled Fitness Landscapes, Poised States and Coevolutionary Avalances." Santa Fe Institute Paper 90-013. Santa Fe, New Mexico, 1990.

69. Kintigh, Keith W. *Settlement Subsistence and Society in Late Zuni Prehistory.* Anthropological Papers of the University of Arizona No. 44. Tucson: University of Arizona, 1985.
70. Kirkby, Anne V. T. *The Use of Land and Water Resources in the Past and Present Valley of Oaxaca, Mexico.* Memoirs of the Museum of Anthropology, No. 5. Ann Arbor: University of Michigan, 1973.
71. Kisselburg, J. A. "Specialization and Differentiation: Nonsubsistence Economic Pursuits in Courtyard Systems at La Ciudad." In *The Hohokam Village: Site Structure and Organization*, edited by D. E. Doyel, 159–170. American Association for The Advancement of Science, Glenwood Springs, Colorado, 1987.
72. Kohler, Timothy A., and Meredith H. Matthews. "Long-Term Anasazi Land Use and Forest Reduction: A Case Study From Southwest Colorado." *Amer. Antiquity* **53** (1988): 537–564.
73. LeBlanc, Steven A. "The Advent of Pottery in the Southwest." In *Southwestern Ceramics: A Comparative Review*, edited by A. H. Schroeder, 27–52. The Arizona Archaeologist 15. Phoenix: Arizona Archaeological Society, 1982.
74. LeBlanc, Steven A. *The Mimbres People.* London: Thames and Hudson, 1982.
75. Leonard, Robert D. "Resource Specialization, Population Growth, and Agricultural Production in the American Southwest." *Amer. Antiquity* **54** (1989): 491–503.
76. Longacre, William A., Sally J. Holbrook, and Michael W. Graves, eds. *Multidisciplinary Research at Grasshopper Pueblo Arizona.* Anthropological Papers of the University of Arizona No. 40. Tucson: University of Arizona, 1982.
77. Lumsden, Charles J. "Gene-Culture Coevolution: A Test of the Steady-State Hypothesis for Gene-Culture Translation." *J. Theor. Biol.* **130** (1990): 391–406.
78. Mabry, Jonathan B. *A Late Archaic Occupation at AZ AA:12:105 (ASM).* Technical Report No. 90-6, Center for Desert Archaeology. Tucson, 1990.
79. Maine, Henry. *Ancient Law.* London: J. Murray, 1861.
80. Marshall, Michael P., and Henry J. Walt. *Rio Abaio: Prehistory and History of a Rio Grande Province.* New Mexico Historic Preservation Division, Santa Fe, 1984.
81. Martin, Paul S. *The SU Site: Excavations at a Mogollon Village Western New Mexico: Second Season, 1941.* Anthropological Series, Field Museum of Natural History, Vol. 32, No. 2. Chicago: Field Museum of Natural History, 1943.
82. Martin, Paul S. "Early Development in Mogollon Research." In *Archaeological Researches in Retrospect*, edited by G. Willey, 3–29, 1974.
83. Masse, Bruce W. "The Quest for Subsistence Sufficiency and Civilization in the Sonoran Desert." In *Chaco & Hohokam: Prehistoric Regional Systems*

in the American Desert, edited by P. L. Crown and W. J. Judge. Santa Fe: School of American Research Press, 1991.

84. Maxwell and Anschuetz. "Variability in Garden Plot Locations and Structure in the Lower Rio Chama Valley, New Mexico." Paper presented at the 52nd Annual Meeting of the Society for American Archaeology, Toronto, 1987.
85. McKenna, Peter J., and Marcia L. Truell. *Small Site Architecture of Chaco Canyon, New Mexico.* Publication in Archeology, 18D. Santa Fe: National Park Service, 1986.
86. Minnis, Paul E. "Domesticating Plants and People in the Greater American Southwest." In *Prehistoric Food Production in North America*, edited by R. I. Ford, 309–340. Anthropological Papers No. 75. Ann Arbor: Museum of Anthropology, University of Michigan, 1985.
87. Minnis, Paul E. *Social Adaptation to Food Stress: A Prehistoric Southwestern Example.* Chicago: University of Chicago Press, 1985.
88. Minnis, Paul E. "Prehistoric Diet in the Northern Southwest: Macroplant Remains From Four Corners Feces." *Amer. Antiquity* **54** (1989): 543–563.
89. Moore, James A. "The Effects of Information Networks in Hunter-Gatherer Societies." In *Hunter-Gatherer Foraging Strategies: Ethnographic and Archeological Analyses*, edited by B. Winterhalder and E. A. Smith, 194–217. Chicago: University of Chicago Press, 1981.
90. Morris, Earl H., and Robert F. Burgh. *Basketmaker III Sites near Durango, Colorado.* Publication 604, Carnegie Institution of Washington, 1954.
91. Morris, Elizabeth A. *Basketmaker Caves in the Prayer Rock District, Northeastern Arizona.* Anthropological Papers of the University of Arizona No. 35. Tucson: University of Arizona, 1980.
92. Morgan, Lewis H. *Ancient Society.* New York: Holt, 1877.
93. Neitzel, Jill. "Hohokam Material Culture and Behavior: The Dimensions of Organization Change." In *Exploring the Hohokam*, edited by G. J. Gumerman, 177–230. Albuquerque: University of New Mexico Press, 1991.
94. Nelson, Ben A., and Steven A. LeBlanc. *Short-Term Sedentism in the American Southwest: The Mimbres Valley Salado.* Albuquerque: University of New Mexico Press, 1986.
95. Nelson, Margaret C. "Comments: Sedentism, Mobility and Regional Assemblages: Problems Posed in the Analysis of Southwestern Prehistory." In *Perspectives on Southwestern Prehistory*, edited by P. E. Minnis and C. L. Redman, 150–156. Boulder: Westview Press, 1990.
96. Netting, Robert McC. "Population, Permanent Agriculture and Polities: Unpacking the Evolutionary Portmanteau." In *The Evolution of Political Systems*, edited by S. Upham, 21–61. Cambridge: Cambridge University Press, 1990.
97. Orcutt, Janet D. "Environmental Variability and Settlement Changes on the Pajarito Plateau, New Mexico." *Amer. Antiquity* **56** (1991): 315–332.
98. Pianka, Eric R. *Evolutionary Ecology.* New York: Harper & Row, 1978.

99. Plog, Fred, and Cheryl K. Garrett. "Explaining Variability in Prehistoric Southwestern Water Control Systems." In *Contemporary Archaeology: A Guide to Theory and Contributions*, edited by M. P. Leone, 280–288. Carbondale: Southern Illinois University Press, 1972.
100. Plog, Fred, George J. Gumerman, Robert C. Euler, Jeffrey S. Dean, Richard H. Hevly, and Thor N.V. Karlstrom. "Anasazi Adaptive Strategies: The Model, Predictions, and Results." In *The Anasazi in a Changing Environment*, edited by George J. Gumerman, 230–276. Cambridge: Cambridge University Press, 1988.
101. Plog, Stephen. "Village Autonomy in the American Southwest: An Evaluation of the Evidence." In *Models and Methods in Regional Exchange*, edited by R. E. Fry, 135–146. Society for American Archaeology Papers No. 1. Washington, DC: SAA, 1980.
102. Plog, Stephen. "Group Mobility and Locational Strategies: Tests of Some Settlement Hypotheses." In *Spatial Organization and Exchange: Archaeological Survey on Northern Black Mesa*, edited by S. Plog, 187–223. Carbondale: Center for Archaeological Investigations, Southern Illinois University, 1986.
103. Plog, Stephen. "Agriculture, Sedentism, and Environment in the Evolution of Political Systems." In *The Evolution of Political Systems: Sociopolitics in Small-Scale Sedentary Societies*, edited by S. Upham, 177–202. Cambridge: Cambridge University Press, 1990.
104. Pruecel, R. W. "Settlement Succession on the Pajarito Plateau, New Mexico." *The Kiva* **53** (1987): 3–33.
105. Rautman, Alison E. "The Environmental Context of Decision-Making: Coping Strategies Among Prehistoric Cultivators in Central New Mexico." Ph.D. dissertation, University of Michigan, University Microfilms International, 1990.
106. Redding, Richard W. "A General Explanation of Subsistence Change: From Hunting and Gathering to Food Production." *J. Anthropological Archaeology* **7** (1986): 56-97.
107. Rice, Prudence. "Specialization, Standardization, and Diversity." In *The Ceramic Legacy of Anna O. Shepard*, edited by R. L. Bishop and F. W. Lange, 257–279. Boulder: University of Colorado Press, 1991.
108. Roberts, Frank H. H. "Shabik'eschee Village: A Late Basketmaker Site in the Chaco Canyon New Mexico." Bulletin 92. Bureau of American Ethnology, Smithsonian Institution, Washington, DC, 1929.
109. Roberts, and Graham. "Residentially Constrained Mobility: A Preliminary Investigation of Variability in Settlement Organization." *Haliksa'I* **5** (1986): 104–115.
110. Rocek, Thomas R. "Sedentism and Agricultural Dependence: Perspectives from the Pithouse to Pueblo Transition in the American Southwest." Paper presented at the 56th Annual Meeting of the Society for American Archaeology, New Orleans, 1991.

111. Rohn, Arthur H. "Northern San Juan Prehistory." In *Dynamics of Southwest Prehistory*, edited by L. S. Cordell and G. J. Gumerman. Washington, DC: Smithsonian Institution Press, 1989.
112. Roth, Barbara. "Late Archaic Settlement and Subsistence in the Tucson Basin." Ph.D. dissertation, Department of Anthropology, University of Arizona, University Microfilms International, Ann Arbor, 1989.
113. Sahlins, Marshall. *Stone Age Economics.* Chicago: Aldine, 1972.
114. Sandor, J. A., P. L. Gersper, and J. W. Hawley. "Soils at Prehistoric Agricultural Terracing Sites in New Mexico: I. Site Placement, Soil Morphology, and Classification." *Soil Science Society of America Journal* **50** (1986): 166–180.
115. Schlanger, Sarah H. "Patterns of Population Movement and Long-Term Population Growth in Southwestern Colorado." *Amer. Antiquity* **53** (1988): 773–793.
116. Schlanger, Sarah H. "Artifact Assemblage Composition and Site Occupation Duration." In *Perspectives on Southwestern Prehistory*, edited by P. E. Minnis and C. L. Redman, 103–121. Boulder: Westview Press, 1990.
117. Sebastian, Lynne. "Leadership, Power and Productive Potential: A Political Model of the Chaco System." Ph.D. dissertation, Department of Anthropology, University of New Mexico, Albuquerque, 1987.
118. Service, Elman R. *Primitive Social Organization.* New York: Random House, 1962.
119. Seymour, Deni, and Michael B. Schiffer. "A Preliminary Analysis of Pithouse Assemblages from Snaketown, Arizona." In *Method and Theory for Activity Area Research: An Ethnoarchaeological Approach*, edited by S. Kent, 549–602. New York: Columbia University Press, 1987.
120. Shackley, M. Steven. "Lithic Raw Material Procurement and Hunter-Gatherer Mobility in the Southwest Archaic." Paper presented at the 50th Annual Meeting of the Society for American Archaeology, Denver, 1985.
121. Shepard, Anna O. "Rio Grande Glaze-Paint Pottery: A Test of Petrographic Analysis." In *Ceramics and Man*, edited by Frederick R. Matson, 62–87. Viking Fund Publications in Anthropology, vol. 41. New York: Wenner-Gren Foundation for Anthropological Research, 1965.
122. Smiley, F. E. "The Chronometrics of Early Agricultural Sites in the American Southwest." Ph.D. dissertation, University of Michigan, University Microfilms, Ann Arbor, 1985.
123. Smith, Eric A., and Robert Boyd. "Risk and Reciprocity: Hunter-Gatherer Sociecology and the Problem of Collective Action." In *Risk and Uncertainty in Tribal and Peasant Economies*, edited by E. Cashdan, 167–192. Boulder: Westview Press, 1990.
124. Spielmann, Katherine A. "Inter-Societal Food Acquisition among Egalitarian Societies: An Ecological Analysis of Plains-Pueblo Interaction in the American Southwest." Unpublished Ph.D. dissertation. Department of Anthropology, University of Michigan, Ann Arbor, 1982.

125. Spielmann, Katherine A., Margaret J. Schoeninger, and Katherine Moore. "Plains-Pueblo Interdependence and Human Diet at Pecos Pueblo New Mexico." *Amer. Antiquity* **55** (1990): 745–765.
126. Steward, Julian H. *Theory of Culture Change: The Methodology of Multilinear Evolution.* Urbana: University of Illinois Press, 1955.
127. Stiger, Mark. "Mesa Verde Subsistence Patterns from Basketmaker to Pueblo III." *The Kiva* **44** (1979): 133–144.
128. Stone, Glenn D. "Agricultural Territories in a Dispersed Settlement System." *Current Anthropology* **32** (1991): 343–353.
129. Stuart, David E., and Farwell. "Late Pithouse Occupations in the Highlands of New Mexico." In *High Altitude Adaptations in the Southwest*, edited by J. Winter, 115–158. Albuquerque, NM: U.S.Forest Service, 1983.
130. Stuart, David E. "The Rise of Agriculture: An Essay on Science, the Rule of Unintended Consequences, and Hunter-Gatherer Behaviour." New Mexico Humanities Council, University of New Mexico, Albuquerque, 1986.
131. Szuter, Christine R., and Frank E. Bayham. "Sedentism and Prehistoric Animal Procurement Among Desert Horticulturalists of the North American Southwest." In *Farmers as Hunters: The Implications of Sedentism*, edited by S. Kent, 80–95. Cambridge: Cambridge University Press, 1989.
132. Toll, M. S., and A. C. Cully. "Archaic Subsistence in the Four Corners Area: Evidence for a Hypothetical Seasonal Round." In *Economy and Interaction Along the Lower Chaco River*, edited by P. Hogan and J. C. Winter, 385–392. Albuquerque: Office of Contract Archaeology, University of New Mexico, 1983.
133. Toll, H. Wolcott. "Patterns of Basketmaker III Occupation in the La Plata Valley, New Mexico." Paper presented at the 56th Annual Meeting of the Society for American Archaeology, New Orleans, 1991.
134. Trigger, Bruce. *A History of Archaeological Thought.* Cambridge: Cambridge University Press, 1989.
135. Upham, Steadman. *Polities and Power: An Economic and Political History of the Western Pueblo.* New York: Academic Press,1982.
136. Upham, Steadman, Richard S. MacNeish, Walton C. Galinat, and Christopher M. Stevenson." Evidence Concerning the Origin of Maiz de Ocho." *Amer. Anthropologist* **89** (1987): 410–419.
137. Vivian, R. Gwinn. "An Inquiry into Prehistoric Social Organization in Chaco Canyon, New Mexico." In *Reconstructing Prehistoric Pueblo Societies*, edited by W. A. Longacre, 59–83. Albuquerque: University of New Mexico Press, 1970.
138. Vivian, R. Gwinn. "Conservation and Diversion: Water-Control Systems in the Anasazi Southwest." In *Irrigation's Impact on Society*, edited by T. E. Downing and M. Gibson, 95–112. Tucson, AZ: University of Arizona Press, 1974.
139. Vivian, R. Gwinn. *The Chacoan Prehistory of the San Juan Basin.* New York: Academic Press, 1990.

140. Vytlacil, Natalie, and J. J. Brody. "Two Pit Houses Near Zia Pueblo." *El Palacio* **65** (1958): 174–184.
141. Waters, Michael R. *The Geoarchaeology of Whitewater Draw, Arizona.* Anthropological Papers No. 45, University of Arizona, Tucson, 1986.
142. Watson, Patty Jo, Steven A. LeBlanc, and Charles L. Redman. "Aspects of Zuni Prehistory: Preliminary Report on Excavations and Survey in the El Morro Valley of New Mexico." *J. Field Archaeology* **7** (1980): 201–218.
143. Wetterstrom, Wilma. *Food, Diet, and Population at Prehistoric Arroyo Hondo Pueblo, New Mexico.* Santa Fe: School of American Research Press, 1986.
144. Whalen, Michael E. "Cultural-Ecological Aspects of the Pithouse-to-Pueblo Transition in a Portion of the Southwest." *Amer. Antiquity* **46** (1981): 75–92.
145. White, Leslie A. *The Science of Culture: A Study of Man and Civilization.* New York: Farrar, Straus and Giroux, 1949.
146. White, Leslie A. *The Evolution of Culture: The Development of Civilization to the Fall of Rome.* New York: McGraw-Hill,1959.
147. Wiber, Melanie G. "Dynamics of the Peasant Household Economy: Labor Recruitment and Allocation in an Upland Philippine Community." *J. Anthropological Resh.* **41** (1985): 427–441.
148. Wilcox, David R. "The Entry of Athapaskans into the American Southwest: The Problem Today." In *The Protohistoric Period in the North American Southwest,* A.D. *1450–1700,* edited by D. R. Wilcox and Bruce Masse. Anthropolical Research Papers No. 24. Tempe, AZ: Arizona State University, 1981.
149. Wilcox, David R., T. R. McGuire, and C. Sternberg. *Snaketown Revisited.* Arizona State Museum Archaeological Series 115. Tucson: ASM, 1981.
150. Wilcox, David R. "The Entry of Athapaskans into the Southwest: The Problem Today." In *The Protohistoric Period in the North American Southwest,* A.D. *1450-1700,* edited by D. Wilcox and B. Masse, 213–256. Arizona State University Anthropological Papers No. 24. Tempe: ASU, 1981.
151. Wilk, Richard R. "Households in Process: Agricultural Change and Domestic Transformation among the Kekchi Maya of Belize." In *Households: Comparative and Historical Studies of the Domestic Group,* edited by R. Netting, R. Wilk, and E. Arnould, 217–244. Los Angeles: University of California Press, 984.
152. Wills, W. H. *Early Prehistoric Agriculture in the American Southwest.* Santa Fe: School of American Research, 1988.
153. Wills, W. H. "Patterns of Prehistoric Food Production in West-Central New Mexico." *J. Anthropological Resh.* **45** (1989): 139–157.
154. Wills, W. H. "Organizational Strategies and the Emergence of Prehistoric Villages in the American Southwest." In *Between Bands and States,* edited by S. A. Gregg, 161–180. Occasional Paper No. 9. Center for Archaeological Investigations, Southern Illinois University, Carbondale, IL, 1991.

155. Wills, W. H., and Thomas C. Windes. "Evidence for Population Aggregation and Dispersal During the Basketmaker III Period in Chaco Canyon, New Mexico." *Amer. Antiquity* **54** (1989): 347–369.
156. Wills, W. H., Bruce B. Huckell, and F. E. Smiley. "Economic Implications of Changing Land-Use Patterns in the Late Archaic." In *The Evolution of Prehistoric Southwestern Society*, edited by G. J. Gumerman. Santa Fe: School of American Research Press, in press.
157. Wilson, C. Dean, and Eric Blinman. "Changing Specialization of White Ware Manufacture in the Northern San Juan Region." Paper presented at the 56th Annual Meeting of the Society for American Archaeology, New Orleans, 1991.
158. Winterhalder, Bruce. "Open Field, Common Pot: Harvest Variability and Risk Avoidance in Agricultural and Foraging Societies." In *Risk and Uncertainty in Tribal and Peasant Economies*, edited by E. Cashdan, 67–88. Boulder: Westview Press, 1990.
159. Woodbury, Richard B. *Prehistoric Agriculture at Point of Pines, Arizona.* Memoirs of the Society for American Archaeology, vol. 17, 1961.
160. Woosely, Anne I. "Agricultural Diversity in the Prehistoric Southwest." *The Kiva* **45** (1980): 317–336.

Norman Yoffee
Museum of Anthropology, University of Michigan, Ann Arbor, MI 48103

Memorandum to Murray Gell-Mann Concerning: The Complications of Complexity in the Prehistoric Southwest

INTRODUCTION: PERCEPTIONS OF COMPLEXITY

Dear Murray:

You'll remember that at the Santa Fe Institute (SFI) conference that George Gumerman and you organized, I was a member of the group commissioned to write an essay on "historical processes in Southwest prehistory." This mandate was, I fear, confusing. Archaeologists have not always been clear about what historians do and, as a student of Mesopotamian historic civilization, the maxim "no writing, no history" has seemed unproblematic. Some archaeologists call themselves "prehistorians" precisely because they work in periods and in places in which there are no written documents. This distinction doesn't always hold, however, because many prehistorians frame inferences about the archaeological record through colonial texts (e.g., *visitas* and *relaciones* in the New World) and especially through ethnographic analogies, namely the texts created by social anthropologists or ethnoarchaeologists. Archaeologists also read "texts" created by other archaeologists

Understanding Complexity in the Prehistoric Southwest,
Eds. G. Gumerman and M. Gell-Mann, SFI Studies in the Sciences of
Complexity, Proc. Vol. XVI, Addison-Wesley, 1994

and the past is thus inscribed, revised, and "textualized" in an internal discourse that in history is called historiography. Perhaps archaeologists, increasingly self-conscious about how they know the past, will soon be teaching courses on "prehistoriography." (Not.)

Our group's conception of the "history" of nonliterate peoples was largely a reaction to the reliance by so-called "new archaeologists" on adaptation as both a definition for culture (following Leslie White and Lewis Binford) and as an explanation of cultural change. We distanced ourselves from that view by letting Randy McGuire recite the word "contingency" as much he wanted. By this we wanted to imply that ancient people were "agents," making choices among various behaviors, and that these choices had to be negotiated in the context of previous choices which themselves were the products the cultural tensions and struggle for power. (All this is sounding more poststructuralist than our discussions were. In fact, we behaved as contractees, charged—we felt—to show that local archaeological sequences were "history" and that meaning as well as subsistence could be read from the ancient data.) I can't claim our chapter is ultimately successful: Haas and McGuire wrote all the substantive parts and Levy and I tried to fit our introduction and conclusion into the extremely rapid review of data in the chapter. Sometimes it's good to be the victim of discrimination, in my case of implacable alphabetism: I hope our chapter will be cited (if it ever is) as Haas et al.

Since I attended the conference as a minor member of a panel, I was quite surprised when George asked me to write a "commentary" chapter for the volume. Of course, I recognized the syndrome: find an outsider to keep the Southwest experts honest; that is, make them speak to a larger audience than their own brethren and sistren. Outsiders are also thought to be able to identify areal or disciplinary biases that are less visible to the initiated assembly. Sometimes the procedure works well as, for example, in Greg Johnson's[15] outsider's commentary on Southwest archaeology, which has stirred up some useful controversy among Southwest archaeologists. However, the trade-off can be that the discussant is so distant from the action that he/she never quite gets the point or writes something so hopelessly mired in his/her own bailiwick that one regrets the entire enterprise.

I'm writing this memo to you, Murray, because, having been assigned my present role, I've got more than the usual discussant's problems and I think you're just the person to help me. After all, you're an outsider yourself to Southwest archaeology (though you know a lot about the subject, according to the insiders) and you're a *real* complexity humanoid in a conference that purports to consider what is complex about the Southwest and what that designation of complexity might mean to future research. Also, you are known as one not happy with cant or the academic posturing designed to make reputations rather than advance knowledge. My problems (as discussant for the papers of this conference) concern what I think might be some misunderstandings between the SFI complexity folk (mainly Stu Kauffman and Chris Langton) and the collected Southwest archaeologists; on a more pessimistic note, I'm worried about one of the assumptions that has justified the conference: if Southwest archaeologists can only learn about "complexity" from the SFI experts, they will become better archaeologists. I write this memo, then, presuming that you'll tell me where I'm off base on how the SFI investigates "complexity" and also whether my own notions of complexity in archaeology—which incorporate a consideration of Southwest prehistoric cultures—are of any interest to the mission of the SFI.

I do not presume to say anything significant about Southwest archaeological data in this memo. As the participants in this conference demonstrated, the control of data and method in the Southwest is exemplary. It is particularly impressive how the dimensions of geology, geography, climate, bioanthropology, and other specialities are regularly and fruitfully brought into Southwest investigations. Furthermore, Southwest archaeologists like to work together, to talk together, and corporately to advance knowledge in their field. (They also give good parties, if you like cheap beer and tortilla chips.)

It is the openness of Southwest archaeologists to new interpretations, however, that has led to their recent susceptibility to what has become a pandemic archaeological disease. They think good theory lies "out there" in some field other than their own and, if they just read enough and talk to the right people, they'll get an idea that will transform their understanding of their data. This is getting truly weird! When I came to the Southwest in 1972 (well, I was a student at the Vernon field school in 1965, when Paul Martin, Jim Hill, John Fritz, and Mark Leone were my mentors; Fred Plog alluded to this at the conference, but I don't think anyone picked up on it and I didn't want to jeopardize my outsider status by admitting actually to have worked at a Southwest site), my Tucson archaeology colleagues were adamant that the Southwest was where the real action in archaeology theory was. Now in the 1990s, theory seems to be anywhere but in the Southwest! Indeed, the very reason for the conference (for many of the participants) is that an appreciation of what are "complex adaptive systems," especially as they are understood by certified scientists, will turn out to be the key for interpreting ancient social organization and social change in the prehistoric Southwest.

I think some SFIers never really perceived why some Southwest archaeologists are so keen on "complexity." "Complexity" is one of those "now" words being hyped in archaeology: we've got "complexity" in the Paleolithic in Central/Eastern Europe,[25] we've got "complex" hunter-gatherers,[23] the Natufian in the Levant is complex enough to be called a "matrilineal chiefdom,"[12] preceramic, pre-(maize) agricultural Peruvian coastal cultures are or are nearly "states,"[1,24] some chiefdoms are "complex" while others aren't,[30] prehistoric Hawaii is a state,[14] and so is Cahokia[21]; thus, it's no surprise that Dave Wilcox[27] thinks Chaco deserves to be a state. Naturally, not everyone agrees with these classifications, but it's clear that many archaeologists want their data (and their people) to be "complex."

In light of previous classifications of prehistoric societies, what we see in archaeology is a kind of "complexity inflation" (like "grade inflation" in universities). The term "complexity" also fits well with normative opinions of social/cultural anthropologists (the dominant species of anthropologist in departments in which archaeologists work) who believe that *all* human social and cultural systems are *complex.* Hunter-gatherers, for example, think rich and complex thoughts about their lives and their environment, speak the most complex of languages, create an incredibly complex array of kinship categories, and perform complex ceremonials.

Whereas all this is right and proper, complexity becomes more complicated when the term "evolution" is introduced into the discussion. Sociocultural anthropologists are not often concerned with social evolution for the good reasons that (1) the term has conjured up notions of progress, changes in cognitive abilities, and the like, all of which are redolent of social Darwinist doctrines that find our (Western) society at the apex; (2) social anthropologists aren't worried much about long-term patterns of change, say, from the Neolithic (e.g., at ca. 8000 B.C. in Mesopotamia) in which small villages flourished, to the development of city-states (ca. 3000 B.C. in Mesopotamia). For archaeologists, however, such changes require social evolutionary theory and social evolution, by any other name (development, change) is just as evolutionary.

Unfortunately, the term "complex society"/"complexity" came to be used in social evolutionary theory (starting in the 1960s, I think) to describe those socially and economically driven societies in which there is a politically centralized governmental system, namely states, or to those societies that were nearly states (usually assumed to be "chiefdoms"). One thus spoke of the "evolution of complex societies" or the "evolution of social complexity" while the study of hunter-gatherers and early agricultural societies was excluded from such evolutionary considerations (or relegated to an earlier stage of precomplexity).

In the perspective of sociocultural anthropological insistence that hunter-gatherers were not living in a Rousseauian, egalitarian state-of-nature, archaeologists began to consider some Paleolithic and Neolithic societies as "complex." Furthermore, as archaeological skills improved and archaeological knowledge grew and, most important, I think, archaeological data accumulated massively, it was far from simple to interpret archaeological remains. Big pithouse villages in the Southwest, for example, presented complex problems of digging, identifying formation processes, classifying data, and understanding ancient inhabitants. Thus, *complexity of modern recovery and analysis became a shorthand for complexity of ancient social organization.*

Furthermore, in the Southwest, the classic investigations of modern puebloan societies that were used to model ancient pueblos began to be reevaluated. The understanding in these earlier studies (see Levy[17]) was that the nature of the kinship system more-or-less explained the (limited) kinds of economic and political stratification in pueblos. Thus, Southwest societies (modern and ancient) were not "complex," i.e., states or on the way to becoming states. That model was attacked, initially by Cordell and Plog,[5] and now it is generally accepted that ancient Southwest societies (especially Chaco, Hohokam, and Casas Grandes) were more "complex" than modern pueblos. Elsewhere in the Southwest, the introduction of European diseases and the incorporation of native peoples into a colonial system led to a reduction of "complexity" from late prehistoric to historic times. However condescending (among other things) to modern puebloan peoples this revisionist view is, when it is coupled with the general anthropological objections to the term "complexity," the Southwest has clearly joined in the archaeological trend toward "complexity inflation." I think you can see from this brief discussion, that Southwest archaeologists brought a lot of baggage with them to the conference on "complexity in the prehistoric Southwest" at the SFI. If some SFI types were only dimly aware of why some Southwest archaeologists are into "complex systems," it was equally apparent that the Southwest archaeologists had only dim notions of what you (SFI folk) mean by "complexity."

IS THIS WHAT SFI MEANS BY COMPLEXITY?

While I have relied on some recent articles (Holland,[13] Kelly,[16] Casti,[3] Gell-Mann,[8,9] and abstracts from the SFI workshop "Common Themes in Complex Adaptive Systems," from conversations with Stu Kauffman and Chris Langton while I was up the road at the School of American Research (SAR) in 1991–1992, and the new book by Roger Lewin[18]), I thought I would make sure that I have some of the basic principles of "complex adaptive systems" right, or at least not totally wrong.

A complex system is a network of interacting parts that exhibits a dynamic, aggregate behavior. This system behavior cannot be reduced to the "sum of its parts" because the action of some parts is always affecting the action of other parts so that equilibrium of the entire system is never reached, or maintained for very long. There is no optimum state of system performance and the system can always surprise, as when a small initial perturbation can result in a large outcome. Per Bak's principles of avalanches, Chris Langton's "phase transitions," and Stu Kauffman's (and others') "edge of chaos" ideas all account for instances of rapid change, the disproportional output from small inputs, and the nonreducibility of such systems. The system can always perform better or can unravel into new parts that were not the constituents of the old system.

Such complex systems cannot be calculated or predicted through standard mathematical descriptions because their operations are not linear and fixed within a few variables and so controlled by some central process or mechanism. Rather, aggregate—or emergent—behavior derives from interactions, which themselves continuously change internally, and because many systems are in contact with one another and so affecting the landscape of them all. Indeed, it is often difficult to specify the boundaries of a system (or between systems) and the environment that embeds systems is hardly "out there," providing a standard through which performance of systems can be measured. The natural property of complex adaptive systems is that of self-organization ("self-organized criticality") and it is the internal dynamics of such systems, not only a response to something external, that can occasion rapid and profound change.

Indeed, complex systems have the capacity to learn from interactions with other systems and the environment (however defined). Complex systems tend to anticipate the future by seeking new interactive ways while also applying old ways of interaction that have already "worked." Thus, complex systems are often (or maybe always) inefficient, sacrificing efficiency for the ability to remain flexible.

Complex adaptive systems include prebiotic chemical evolution, central nervous systems, vertebrate immune systems, political systems, and the global economy. It is one of SFI's main goals to bring together specialists in these domains to learn if there is some common kernel to all complex adaptive systems.

I hope that the above description, consisting in the main of transcriptions from the papers I've read, has not missed the point of what some poeple at SFI are doing. From having heard a number of lectures at SFI by mathematicians, chemists, biologists, and others last year, I can attest that there is a definite feeling that "complexity" signals a new way in which scientists are thinking about things. Of course, I couldn't understand most of what was being said at the lectures, but I did have a lot of sympathy with those proteins (as complex adaptive systems) that were described as "periodic, disordered, and frustrated"!

What several of the Southwest archaeologists at the conference seem to have missed, is that those of us who investigate ancient states (see Paynter[22]) have been talking SFI language (but with a different lexicon) in the last years. We're well passed the idea that there were "control mechanisms" in ancient states or that we can draw flow diagrams of neatly boxed social institutions or that ancient states are homeostatic systems in the way archaeologists of the 1970s and early 1980s talked about ancient state systems (see Brumfiel[2]). Rather, ancient states are messes, filled with institutional struggle among classes and ethnic groups and there is conflict within such social groups, too. Leaders half understand what's happening in their societies. Bureaucracies consist of a mixture of patrimonial retainers and those recruited by ability. Finally, in ancient states, long-term stability is sacrificed regularly for short-term gain; against an old image of timeless rule by beneficent kings who are toppled only by victorious barbarians or more militaristic neighbors, we now investigate internecine strife, the creation of wealth and misery, factional solidarities, and environmental degradation (in no particular order and sometimes all at once).

However, not only are ancient states and civilizations "complex systems" in your terms, so are *all* human societies playgrounds for social negotiation, for the empowerment of the few, in which the parts remain far from some equilibrium with each other and their environment. In SFI terms, it is perfectly obvious that all human social systems, from Paleolithic hunter-gatherers to states are complex systems.

TOWARDS A THEORY OF RELATIVITY OF COMPLEX SOCIAL SYSTEMS

I hope you and I agree that Southwest prehistoric societies are (1) complex systems and (2) by that designation we know absolutely nothing more than we did before so declaring them "complex." Now we need to get to the real work of social complexity theory. Are some societies more complex than others? In what ways can we measure complexity? Why are there different kinds of socially complex societies? And is there a theory of the relativity of social complexity that can fall within the kinds of complex phenomena SFI is interest in?

Actually, I know from your paper on "the evolution of human languages" that you've already said some things about this, for example, that "simplicity" and "complexity" can be evaluated according to "schemata." A schema is the compression of regularities identified by a complex adaptive system from its surroundings and about itself, including importantly regularities about the behavior of the system. A schema evaluates other schemata so as to anticipate future situations in a fitness landscape. If I understand correctly, a schema is the shortest message that will describe the system at some appropriate level of significance: in human societies such schemata are social institutions, and traditions, and myths. A schema then can unfold to yield predictions of behavior. If the schemata of one system are more numerous than in another system, then system one, by definition, is more complex. The more specialization, the more ranking, the more elaborate patterns of change in a system, the more complex is the system.

If I've got all your ideas of schematic relativity right, then I think we can start talking about social complexity and social evolution: for example, we can proceed to "rank" societies (see next section) according to amounts of social differentiation, economic specialization, and nature of political institutions. But we must make sure that we compare societies only on this set of schemata rather than another possible set of social schemata. For example, I reckon that hunter-gatherers in Australia have a more elaborate set of ritual and ceremonial schemata than scientists at the SFI! Your own example of languages is instructive since, as far as I'm aware, all linguists talk about the evolution of language (and can do so in a variety of ways) but few or no linguists would say that modern English is more or less "complex" as a whole than Proto-Indo-European (PIE). (For example, complex word-order principles in English may replace complex conjugational patterns in PIE.) If complexity, then, is in the idea of the beholder, to some extent, it is the different kinds of complexity that are of common interest to us both. And in this perspective the research into the kinds of complexity that flourished in the prehistoric Southwest might be most productive.

A BIRD'S-EYE VIEW OF NEW SOCIAL EVOLUTIONARY THEORY

At the SFI conference I distributed the draft of a paper (now Yoffee[32]) entitled "Too Many Chiefs?" from which I now wish to extract some points that relate directly to a central question about (the kind of) complexity in the prehistoric Southwest. Let me begin this brief discussion by turning my paper inside-out and presenting one of the concluding diagrams on "possible evolutionary trajectories."

As the full paper[32] makes clear (I hope), the main purpose of this diagram (Figure 1) is to offer an alternative to the well-known stage-level, stepladder model used by archaeologists in the 1960s–1980s (Figure 2). For present purposes, Figure 1 implies that not all human societies fall along one continuum, in which lower rung examples (like modern pueblo societies) have failed to progress to the top of the ladder (as Figure 2 implies). Although I cannot pause here to discuss other implications of the diagram, I do wish to note that the diagram does *not* intend to claim that there are only four ideal (and epigenetic) "types" of societies, that social, economic, and political relations are fixed within a type, and that change within a type must occur in all relations at the same time and in the same direction.

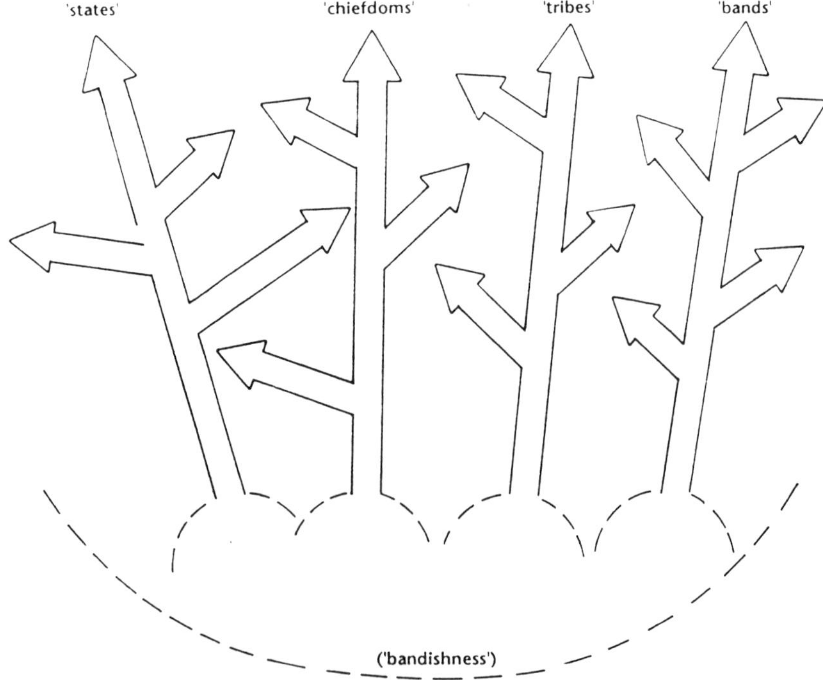

FIGURE 1 Possible Evolutionary Trajectories.

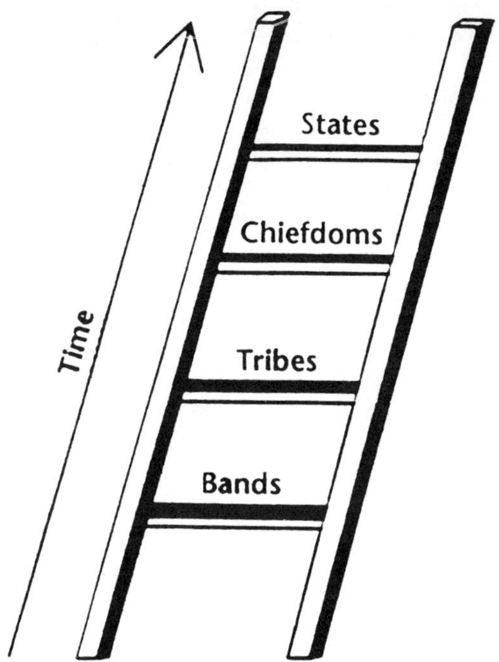

FIGURE 2 Evolutionary, Stepladder Mode.

What I hope the diagram shows is that modern (ethnographic and ethnohistoric) "chiefdoms" may not be good models for those societies that preceded states in the areas of the world in which states evolved. (Of course, chiefdoms could become states through contact with states, as the arrows in the diagram indicate; contact with large states might also reduce small states to chiefdoms.) From our conversations in Santa Fe, I know you can see exactly why Figure 1 should be of interest to SFI theorists of complexity: it implies that there are different *trajectories* (or pathways) in social evolution and that, by identifying specific and differing *initial conditions* (in the time of "bandishness"), we might predict (or, more modestly, begin to understand) what distinguishing qualities there are among the various trajectories.

While I tentatively proffer some possible variables that I have culled from the papers of this conference (in the next section of the memo), I want to spend a moment longer on two (related) issues that are significant to the investigation of Southwest complexity. First, in the perspective of "new social evolutionary theory," since there are different kinds of complexity, one should be chary of ascribing the Chaco "phenomenon" or Hohokam "complicatedness" (Hallpike,[11] see below) as precursors to states. If these Southwest examples of complexity are indeed complex (and they are), one must ask whether they look like those documented prestate societies like Mesopotamian Ubaid, Naqada II Egypt, early Harappan, pre-Shang, preclassic Maya, formative Teotihuacan, or Chavin Peru societies. (I don't think they do, and I'll come back to the implications of that.)

Second, as Jerry Levy argues in his chapter on analogy,[17] the rejection of modern puebloan society as a direct historical model for ancient pueblos itself raises problems. Since all archaeologists must model the residues of the past upon some other (usually more recent) societies which they reckon are like ancient ones, the abandonment of ethnographic-historic pueblos as models of prehistoric Southwest pueblos *necessarily* requires the substitution of other societies as models. Levy notes that archaeologists (who criticize the direct historic approach) either have not been explicit about what analogies they put in place of the former one or they have facilely equated Southwestern complexity with near-state level complexity. Levy's paper makes the case for kinship-based complexity, as it were, as a model that would explain the nature of prehistoric alliances and conflicts and population aggregations and dispersals in the Southwest—exactly the situations that some prehistorians feel merits rejection of modern puebloan analogues.

THE MAJOR THEME OF THE SFI CONFERENCE ON PREHISTORIC COMPLEXITY IN THE SOUTHWEST: BAD WEATHER

As I sit in my study today (Jan. 8, 1993) in the sunny "Old Pueblo," watching the third day of steady rain, listening to radio reports of street closures (streets that are paved-over arroyos) and noting that the Tanque Verde Wash is only one foot lower than its maximum in 1983 (the year of the 500/1000-year flood that washed away houses nearly a mile from its banks, including the house of Sue Philips, a linguistic anthropologist colleague of mine in Tucson), it is hard to miss the major theme of the papers on prehistoric complexity in the Southwest: bad weather.

The position paper by Alan Swedlund[26] and the chapter by Nelson et al.[19] on demography ("sick-o's in the Southwest" according to D. L. M. Martin) repeatedly stress the "clear limiting factors to population growth in a semi-arid environment": "things were good and they got bad"; "on Black Mesa. . .one sees strong evidence for a population that was consistently under marginal nutritional stress, in a marginal environment, and one in which stress markers and disease incidence tend to increase through time"; "agriculture is probably always marginal in the Southwest-somewhere"; and "undernutrition [at Arroyo Hondo and some other sites]. . .became malnutrition and even starvation." Swedlund concludes his essay with a meditation on how "wood limitations present. . .a plausible model for a very serious resource limitation."

Nelson et al.[19] discuss areas that are "environmentally marginal" and suggest that there may be "some sort of absolute threshold of community size beyond which health maintenance becomes increasingly problematic" and that "politically autonomous groups" were able "to foresee the limitations of their subsistence systems." Thus, the logic of abandonment of one region and the aggregation in another is one of managing scarcity by population movement. Having cited directly from these papers, it only requires comment that the population numbers of sites and regions, in the perspective of the earliest states, are small (see the various estimates of Nelson, Kohler, and Kintigh[20]): 5,500 people in Chaco or a maximum of 2,000 in a "humongous" pueblo.

The demography group paper (Nelson et al.[19]) represents, I think, something quite new in Southwestern studies and perhaps in archaeology more broadly. Interrelations among diet, resources, population size, health, economy, and politics are evaluated and—even if the data sets are imperfect—hypotheses are invented and tested and health is not simplistically understood as just biology. Deb Martin informs me that on a global scale of agricultural societies, the Southwest seems different because "disease seems so *endemic* to the area from the earliest skeletal remains from Basketmaker times through to historic times." Deb's (e-mailed) comments are worth quoting further:

> There seems to be a shared profile of health that goes with agricultural economies, but there are certainly key things that put a further strain on health, like food shortages, marginal environment, and lack of good water (like you have in the Southwest) which makes nutrition more of a variable in explaining poor health. In other places in the world, it is the increased density [of population] and increased transmissibility of infectious disease that are the most important issues.

In the context of environmental marginality, it's no wonder that Joe Tainter finds it "irrational to become complex" (he means politically and economically stratified) in the Southwest. From my outsider's view, of course, it is amusing that Tainter considers Chaco an "800-pound gorilla" that affects core-periphery relations in his study area. But scale is everything: in a lecture last semester in Tucson on pithouse villages that Pat Gilman excavated in a remote corner of Southeast Arizona, she described the largest set of such pithouses as "incredibly complex." Right.

Matters of scale and a comparative method seem to me especially curious in Stead Upham's paper on "system modeling and evolution." He lists some of his "biases" about social complexity:

1. "Political systems fundamentally reflect the systems organization of society."
2. "Political and social systems are really nothing more than solutions to specific problems of management and production."
3. "Overall population size [is] strongly correlated with...centralized political control."
4. "Economies and production strategies...are merely solutions to problems of management."

As nearly as I can tell, however, most analyses of ancient states ought to cause him to adopt some new "biases." In the Mesopotamian case there is certainly no overall correlation between population size and political centralization; in neighboring Egypt, on the other hand, which has a small areal population (ca. 2 million in New Kingdom times), there is a vast amount of centralization. Over the populous Maya city-states, there is likewise no overall centralization. In fact, in Mesopotamia (and I think this generally holds in the earliest states), there is no area-wide systemic organization of societies; indeed, within and among various city-states and social segments, individuals contest for power, honor, and privilege. "Problems of management and production" are certainly not reducible to the political or any other single system. If there are no pan-regional politically managed regions among the earliest states (with the exception of Egypt), it is difficult to agree with Stead that there are such "regional and pan-regional organizations" in the Southwest. (I'll take a couple of paragraphs in the last section of this memo, Murray, to delineate the nature of complexity in ancient states and civilizations and contrast that with the kind of complexity that some argue characterizes the prehistoric Southwest.)

Linda Cordell's paper,[6] the panel on explanation (Cordell et al.[7]), Chip Wills' paper,[28] and his group's paper[29] get at the heart of the complications of prehistoric Southwest complexity. These papers ask what does the trajectory toward social complexity in the Southwest tell us about social evolutionary theory in general and then how does social evolutionary theory fit into studies of complexity in general? I conclude this section with a comment on the first topic and I'll devote part of the conclusion to the second.

The conceit of Southwest archaeology, as is depicted especially in Wills' paper, is that the Southwest is a laboratory for "develop[ing] general methods that are adequate for identifying their relative effects on human systems, or incorporating them in interpretations." Cordell et al.'s assessments about developing a comparative archaeology, the necessity of integrating CRM data into larger investigative projects, and the need to know and the need to check facts in the Southwest, however, betrays (I think) a certain wariness about celebrating the achievements of Southwest archaeologists. This cautionary note needs to be juxtaposed with Chip Wills' comment that "the archaeological record of the prehistoric Southwest is one of the most extensively studied anywhere." Cordell et al. remind us that it is still to be determined whether these lovely data and analyses are typical of anything outside their own "boundary conditions." Perhaps the real value of using the Southwest as a case study in building social evolutionary theory is that it may be an example of a trajectory that has been little studied (on a global scale). In this context, the Southwest is truly exceptional and far more interesting than just an example of a lower rung in the great chain of becoming—just like us—a state.

Wills' statement that "a major trend in the prehistoric Southwest is toward increasing complexity" is both inadequate and incomplete—although it is correct enough in itself. It is inadequate because, as I've already boringly repeated, there are all sorts and kinds of complexity. Yes, in the Southwest, population grows, there are degrees of differentiation and stratification, and there is regional interaction.

All of that simply begs the questions, however, of whether population growth had limits, what was the nature of stratification, and what were the kinds of regional interaction. His statement is further incomplete because an equally major trend in the Southwest is "collapse"—abandonment of sites and regions, disaggregation, transportable societies.

The Cambrian phenomenon—an initial explosion of diversity followed by a thinning out of species—is discussed by Wills' group as a useful model of prehistoric Southwest activities. (Lewin does it better in his book on complexity.) In the actual analytical heart of the paper, however, Wills rehearses some of his previous, excellent work on the nature and meaning of agriculture in the Southwest: Agriculture is a risk-avoidance strategy, households are basic units of production, nucleated settlements have more agricultural features than non-nucleated ones, and various patterns of exchange between regions are established because cooperation is a good thing given the bad Southwest weather. "Population aggregation quickly leads to local natural resource depletion in an arid environment like the Southwest."

Levy's paper presupposes much of this and builds upon it. Hopi clans manage the agricultural landscape in ways to ensure that some part of the crop might survive an environmental disaster and an egalitarian ideology promotes areal cooperation. Political confederacies ("alliances" in Southwest-speak) are managed among Iroquois through kinship principles. One is tempted to see the Hopi as wise successors of the cycle of aggregation and dispersal: by rejecting trends towards political and economic power and dominion, they achieve their own richly complex ceremonial and intellectual behavior within a kinship system in which contests over status and leadership are resolved in the Hopi way.

ALTERNATE COMPLEXITIES IN THE PREHISTORIC SOUTHWEST

Judith Habicht-Mauche[10] describes prehistoric Southwest (actually just eastern pueblos) as neither economically and politically stratified societies nor as deeply conservative, egalitarian organizations, but as "complex tribes." In these societies "the fundamental unit of economic and social interaction remains the extended family or lineage...The integration of these units into larger social systems is achieved through the existence of pan-residential corporate social and religious institutions which crosscut these kin groups."

For Habicht-Mauche the expansion of these local social and religious institutions can integrate relatively large geographic regions (which in other views are "provinces"). Interarea exchange and regional specialization are largely decentralized while decisions about agricultural production and craft specialization occur at the level of households and lineages. "Political leadership is vested in a 'council' which represents the various pan-residential social and religious societies. Leaders have respect and status in their individual communities and the regional system at large, but they do not control access to large quantities of subsistence resources and elite 'status' goods." I quote further, with Habicht-Mauche's permission, her next two paragraphs:

> The duties of these tribal leaders can include: (1) the organization and scheduling of religious ceremonies; (2) the supervision of labor for communal projects; (3) the coordination of subsistence activities such as communal hunts, planting and harvesting; and (4) the negotiation of alliances with external groups. Although these leaders may control some coercive force (i.e., war captains and warrior societies might act as tribal police), the real basis for their authority lies in their respect for group consensus and cultural sanctions against uncooperative behavior. Complex tribes, thus, perform many of the same integrative functions of more centralized hierarchical systems, such as chiefdoms. Both can be characterized by regional specialization, inter-regional trade, intensification of subsistence practices, organization of labor for public projects, and the existence of regionally shared religious and social institutions. The two types of socio-political systems differ primarily in the way group decisions are made and enforced, and the degree of control that leaders have over access to natural and social resources.

In her essay, Habicht-Mauche discusses the archaeological evidence for regional clusters of villages, the absence of centralized workshops and storage facilities, and how standardization of crafts is managed as cottage industries by individual craft workers and families, and she states that the majority of exchanged goods are relatively utilitarian objects not limited to any elite segment of society. There is, of course, good evidence for kivas and plazas, religious venues at which inter-pueblo exchange and the distribution of food are reviewed. "Collapse" results in the breakdown of ties of economic interdependency and the (retro-)development of self-sufficient villages. In other words, Habicht-Mauche has written the very article Jerry Levy is calling for: while modern pueblo social organization may not supply a 1:1 model for the past, it does offer a relevant set of continuities and discontinuities that can be explained: in a word, puebloan history.

If, as I suspect, there are trends towards social complexity in the prehistoric Southwest that differ from evolutionary trajectories of ancient states and civilizations (and thus, I repeat, the Chaco phenomenon and the Hohokam phenomenon are NOT stages of pre-state societies), I should devote at least a few words on what kinds of complexity are demonstrated in trajectories to states. (I've already written about this in 1979[31] and 1993[32] and in a few places in-between and I'm trying to finish a book tentatively titled *New Rules of the Game: The Evolution of Ancient States and Civilizations*, so I'll only glide over some of the more vexing issues here.)

All over the world in post-Pleistocene times, there evolved increasingly differentiated social groups, organized in relatively specific and autonomous frameworks, with their own symbolic and occupationally specialized orientations, that interacted within the confines of an overarching institutional system. These differentiated and specialized groups became internally hierarchized and were themselves hierarchically related to one another. New groups, specialized for the purpose, emerged to supply integration among the various forces of production and of interaction in order to ensure the continued existence of the system. In trajectories to states, then, progressive differentiation is resolved in a moment of near-explosive change (in every instance of the evolution of the first states) as a specialized decision-making sub-system began to regulate the flow of goods, people, and information among the various social components. This political sub-system, the state, further maintained itself by use of its power to mobilize resources that were not totally embedded with the various social components.

However, with the exception of ancient Egypt, the earliest states were politically independent city-states, while the regional interaction among them, including the creation of boundary-establishing codes—standardized languages and set of symbols that transcended political borders—we denominate as "civilizations." (Thus, there are Mesopotamian, Maya, Indus, and pre-Ch'in China city-states/civilizations; pre-Inka Andean city states: Wari, Tiwanaku, Chan Chan; Teotihuacan is a city-state gone pituitary.) This city-state/civilization distinction is not the only thing that can be said about the earliest states, of course, but it is a salient point in the kind of complexity we can contrast with evolutionary developments in the Southwest.

Within the earliest city-states/civilizations, there is a "ubiquity of conflict," since dominant sociopolitical goals are never accepted by all the differentiated components of the system. Ubiquity of conflict results in a partial, "consensual" resolution in which a legitimation of inequality is effected (that is, inequality is justified as a state of nature). Modes of conflict are partly managed and political struggle is channeled as civilizational values are contested and reproduced. While "collapses" of states (political systems) are normal, so is the ability to reformulate states beyond collapse, since the "idea" of the state to a certain extent, is, independent from the existence of the institution itself (see the papers in Yoffee and Cowgill[33]).

In social evolutionary theory and in the empirical archaeological investigations of the development of states and civilizations, it is the process towards socioeconomic differentiation and political integration that is to be traced and explained. Whence come the economically, ideologically, and politically defined social orientations, the arrangements of people not related to one another through rules of kinship, but through the unequal distribution of resources and power? In the evolution of states one sees dramatic changes in the content of people's roles such that the moral economy of kinship is supplanted, partly by the political economy of power relations: in states the relations between governors and the governed is no longer explained by kinship but through wealth, status, and force.

Hallpike's distinction[11] between "complex" and "complicated" provides a good beginning to describe the differences between trajectories to states and other evolutionary trajectories. "Complex societies" are those with various parts that are interconnected with significant degrees of economic inequality and subordination;

social groups maintain separate identities and political rules and full-time political specialists are developed to manage their interaction. "Complicated societies" are those in which social relations are little differentiated, groups look much like one another, and consequently the societies are "mixed up" and "involved." Naturally, in the SFI and anthropological views of "complexity," we must emend Hallpike: all human social interactions are complex and different evolutionary trajectories simply exhibit different kinds of complexity.

THE FUTURE OF COMPLEXITY AT SFI: FINALE

While it's obvious, Murray, that the term "complexity" as a category that has been used in studies of the prehistoric Southwest, must be unpacked, I think we agree that the scientists at SFI can benefit from SFI's involvement with Southwest archaeology. Indeed, if I have one policy recommendation, it is that SFI needs to increase its programs in archaeology and social evolutionary theory—if you truly are interested in the subject of "complexity."

Archaeologists—trained to identify connectedness in social institutions, to measure patterns in social organizations and how these change over periods of time, to delineate the feedback between "natural" and social environments, and to oppose reductionist formulas that ignore sources of variation—have already learned some useful things from the SFI conference, but, as I've argued, archaeologists also have some things to contribute to the study of complexity.

Reading the papers in this volume from my distant perch as a Mesopotamianist, it seems most apparent that a social theory of relativity is critical and that we are poised at the edge of a rapid transition in our ability to construct it. The dynamic system of self-organized archaeologists exhibits the emergent property that arises from locally interactive schemata (conferences convened by George Gumerman) in which we can disarticulate kinds of complexity, speak clearly about the need to study initial conditions, and so explain the data in its different outcomes—outcomes that are not themselves ends, but part of historically contingent and continuous processes. Of course, some ideas are fitter than others. Over to you, Murray.

REFERENCES

1. Bawden, G. "The Andean State as as State of Mind." *J. Anthro. Res.* **45** (1989): 327–332.
2. Brumfiel, E. "Breaking and Entering the Ecosystem—Gender, Class, and Faction Steal the Show." Distinguished Lecture in Archaeology. *Amer. Anthro.* **94** (1992): 551–567.
3. Casti, J. "The Simply Complex: Trendy Buzzword or Emerging New Science?" *The Bulletin of the Santa Fe Institute* **7(1)** (1992): 10–13.
4. Cordell, L., and G. Gumerman, eds. *Dynamics of Southwest Prehistory.* A School of American Research Book. Washington: Smithsonian Institution Press, 1989.
5. Cordell, L., and F. Plog "Escaping the Confines of Normative Thought: A Reevaluation of Puebloan Prehistory." *Amer. Antiquity* **44** (1979): 405–429.
6. Cordell, L. "The Nature of Explanation in Archaeology: A Position Statement." This volume.
7. Cordell, L. S., J. H. Kelley, K. W. Kintigh, S. H. Lekson, and R. M. Sinclair. "Toward Increasing our Knowledge of the Past: A Discussion." This volume.
8. Gell-Mann, M. "Complexity and Complex Adaptive Systems: Fundamental Concepts and Questions." Notes to paper delivered at Workshop Integrative Themes of Complex Adaptive Systems, Santa Fe, New Mexico, July 8, 1992.
9. Gell-Mann, M. "Complexity and Complex Adaptive Systems." This volume.
10. Habicht-Mauche, J. "Town and Province: Regional Integration and Economic Interaction Among the Classic Period Rio Grande Pueblos." Manuscript, (n-d).
11. Hallpike, C. W. *The Principles of Social Evolution.* Oxford: Clarendon Press, 1988.
12. Henry, D. *From Foraging to Agriculture: The Levant at the End of the Ice Age.* Philadelphia: University of Pennsylvania Press, 1989.
13. Holland, J. "Complex Adaptive Systems." *Daedalus* **(Winter)** (1992): 17–29.
14. Hommon, R. "Social Evolution in Ancient Hawaii." In *Island Societies: Archaeological Approaches to Evolution and Transformation*, edited by Patrick Kirch, 55–68. New Directions in Archaeology. Cambridge: Cambridge University Press, 1986.
15. Johnson, G. "Dynamics of Southwestern Prehistory: Far Outside–Looking In." In *Dynamics of Southwest Prehistory*, edited by Linda Cordell and George Gumerman, 371–389. Washington: Smithsonian Institution Press, 1989.
16. Kelly, K. "A Distributed Santa Fe System." *The Bulletin of the Santa Fe Institute* **7(1)** (1992): 4–6.
17. Levy, J. E. "Ethnographic Analogs: Strategies for Reconstructing Archaeological Cultures." This volume.

18. Lewin, R. *Complexity: Life at the Edge of Chaos.* New York: Macmillan, 1992.
19. Nelson, B. A., D. L. Martin, A. C. Swedlund, P. R. Fish, and F. J. Armelagos. "Studies in Disruption: Demography and Health in the Prehistoric American Southwest." This volume.
20. Nelson, B. A., D. Kohler, and K. W. Kintigh. "Demographic Alternatives: Consequences for Current Models of Southwestern Prehistory." This volume.
21. O'Brien, P. "Cahokia: The Political Capital of the 'Ramey' State?" *North Amer. Archaeol.* **10** (1989): 275–292.
22. Paynter, R. "The Archaeology of Inequality and Inequality." *Ann. Rev. Anthropology* **18** (1989): 369–399.
23. Price, T. D., and J. Brown, eds. *Prehistoric Hunter-Gatherers: The Emergence of Cultural Complexity.* New York: Academic Press, 1985.
24. Raymond, J. S. "The Maritime Foundations of Andean Civilization: A Reconsideration of the Evidence." *Amer. Antiquity* **46** (1981): 806– 821.
25. Soffer, O. "Patterns of Intensification as Seen From the Upper Paleolithic of the Central Russain Plain." In *Prehistoric Hunter-Gatherers: The Emergence of Cultural Complexity*, edited by T. Douglas Price and James Brown, 235–270. New York: Academic Press, 1985.
26. Swedlund, A. C. "Issues in Demography and Health." This volume.
27. Wilcox, D. "The Changing Structure of Macroregional Organization in the North American Southwest: A New World Perspective." Manuscript, 1991.
28. Will, W. H. "Evolutionary and Ecological Modeling in Southwestern Archaeology." This volume.
29. Will, W. H., P. L. Crown, J. S. Dean, and C. G. Langton. "Complex Adaptive Systems and Southwestern Prehistory." This volume.
30. Wright, H. "Prestate Political Formations." In *On the Evolution of Complex Societies: Essays in Honor of Harry Hoijer*, edited by Timothy Earle, 41–77. Malibu: Undena Publications, 1984.
31. Yoffee, N. "The Decline and Rise of Mesopotamian Civilization." *Amer. Antiquity* **44** (1979): 5–35
32. Yoffee, N. "Too Many Chiefs? or Safe Texts for the 90s." In *Archaeological Theory–Who Sets the Agenda?*, edited by Norman Yoffee and Andrew Sherratt, 60–78. New Directions in Archaeology. Cambridge: Cambridge University Press, 1993.
33. Yoffee, N., and G. L. Cowgill, eds. *The Collapse of Ancient States and Civilizations.* Tucson: University of Arizona Press, 1988.

Index

A

abandonment, 49, 315, 351
accelerator dating, 159
Acoma, 31, 276, 279
    region, 223
actualistic studies, 171
adaptation, 174
Adler, M. A., 117
agave, 317
age-at-death values, 98
aggregation, 49, 61
agriculture, 212, 216-217
    adoption of, 121-122
    Chacoan, 179, 316
    innovation, 41
    production, 117, 304, 308, 310, 314-315, 317
    surplus production, 320
    with canal irrigation, 219
Akins, N., 168
alliances, 124, 269-270
alluvial water tables, 304
alternative histories, 155
Alvarez, L., 153, 157
Alvarez, W., 157
ambiguities, 198
Anasazi, 27, 32, 44, 54, 70, 132, 210-214, 277, 291, 313, 316, 322, 324
antelope, 318
archaeology, 151
    as art, 150
    comparative, 175-178
    explanation, 151, 154-155, 159
    as humanity, 150
    inference, 151, 153, 155
    method and theory, 150
    and paleontology compared, 151, 155
    perspective, 18
    record, 151-153, 155
    school of thought, 151
    as science, 150, 155
    types of, 25
Archaic sites, 322
architecture, 212
Arroyo Hondo, 61, 75-76, 78-79, 84, 97-99
artifacts, 287, 289, 292
    accumulation rates of, 128, 130
Athapaskan, 318
Atrisco, 31, 276, 278-279
authority, 199

B

ball court, 215
Basketmaker III, 308-311
Basketmaker period, 46
Bat Cave, 306
Bayham, F. E., 319
behavior, individual vs. group, 300
bias, 155
Binford, L. R., 172
biological disruption, 60, 65-66, 88, 100
bison, 318
Black Mesa, 52, 61, 70-73, 75, 96-97, 99, 237, 314-315
Blake, M., 118
Blinman, E., 135, 324
boundary conditions, 153
boundedness, 196
Boyd, R., 273
Breternitz, D., 180
British functionalist school, 203

C

Cambrian Explosion, 301
Canonical Ensemble, 17-18
Casas Grandes, 61, 91-92, 94-95, 97, 99, 178, 180, 185, 221, 224
catastrophic explanations, 157
Cebolleta Mesa, 29, 32
centrality, 20, 216
ceramics
    see pottery
ceramic breakage rates, 133-134
Chaco Canyon, 18-20, 30, 54, 61, 86, 88-89, 92-93, 97, 151-152, 158, 168, 177, 179, 198, 205, 214-215, 217-218, 310-311, 316, 349
city-state/civilization distinction, 355
clans, 237, 241
Clarke, D., 25

Classic period, 317
climatic variation, 303
clowns, 195, 199
co-authorship in archaeology, 183
Coalition Period, 318
coherence theory, 150
collapse, 61, 65
Colorado Plateau, 49, 215, 222, 311, 321
Colton, H. S., 269
communal ceremonial structures, 211
communities, 81
community organization, 123
comparative archaeology, 175-178
competing hypotheses, 172
complex adaptive systems, 3, 16-17, 297, 326, 343, 345
complexity, 5, 26-27, 170, 347
    complex vs. complicated, 356
    cost of, 32
    evolutionary, 344
    inflation, 343
    process of, 26
    social, 293
    sociocultural, 325
    sociopolitical, 123
conditioned action, 199
Conkey, M. W., 155, 252
contemporaneity, 124
context of structure, 196
context of symbols, 196
contingency, 155, 157, 199-200, 342
contingent explanation, 207
continuous area comparison, 235
cooperation-competition, 49
Cordell, L., 25, 170, 325, 352
core-periphery relations, 25, 28
corn dancers, 199
Coronado entrada, 200
cotton, 317
covering law, 152, 154
Cowgill, G., 166
creative thinking, 5
Creek Confederacy, 240-241
Cretaceous-Tertiary (K-T) extinctions, 157
critical tests, 169
cultural resources management (CRM), 182-183
Crown, P. L., 167, 317
crucial tests, 168-169
Crumley, C. L., 198
Culbert, T. P., 278
culture, 195-196, 288-289, 298
    area, 291
    behavior, 152-154
    environment, 153
    evolution, 152, 173-174
    generalities, 235
    history, 204, 293
    institutions, 121
    of science, 155
    process, 152, 204
    stability, 40
    system, 247, 298

D

Darwinian competition, 17
data, accumulation of, 159
Dean, Jeffrey, 43
deductive nomological (DN) model, 165
demography, 39, 41, 59, 113, 320, 351
    intensity of, 309
    variability in, 248
    see also paleodemography
demographic anthropology, 40
dendrochronology, 42, 70
density dependency, 47, 66, 101
deposition, 128, 135
Desert Culture, 184
determination, 199
development, uneven, 197-199
Di Peso, C., 185
diet, 78, 81, 83
differentiation, 213
diffusion, 195, 301
dimensions of analysis, 43
discovery and justification, 167
dispersal, 61
diverse perspectives, need for, 163
division of labor, 308
Dolores Archaeological Project (DAP), 115, 130
Dolores area, 114, 116, 132, 134, 314

domesticated plants, 303
 see also agriculture
Dray, W., 173
Duhem thesis, 167
Dunnell, R. C., 173
Durrenberger, E. P., 320
dynamic equilibrium, 47

E

Eastern Keresans, 241
ecological
 functionalism, 121
 modeling, 287
economic
 imperatives, 206
 specialization, 315
 ties, 31
 variability, 250-251
edge of chaos, 345
effective scale, 198
Eggan, F., 233, 237
Elena Gallegos Project, 276
elites, 238
emblemic behavior, 252
emergence of order, 16
empirical generalizations, 155-156
endogamy, 250
Engels, F., 251
environmental
 limitations, 205
 marginality, 67, 100
equifinality, 167, 170, 186, 266
equilibrium, 40
Estrella phase, 323
ethnogenesis, 221
ethnographic
 analogs, 233-234
 cases, 153
evaluation
 procedures, 183
 and testing in archaeology, 157
evolution, 156, 173-174, 200, 344
 general evolution, 205
 modeling, 287
 specific, 205
 theory, 173
 trajectory, 175
exchange systems, 31
explaining change, 246
explanation
 ahistorical, 151
 complete, 154
 in archaeology, 165, 178
 nomological, 150, 154
 probabilistic, 154
extinctions, 157

F

family level of social integration, 209
farmbelt model, 313
farming methods, 316
 see also agriculture
Feinman, G., 249
fertility, 68, 98-99
finite mixture distributions, 135
Fish, P., 317
Fish, S., 317
fitness landscapes, 301, 321
Flannery, K. V., 122, 175
footnotes, use of, 172, 183
formation processes, 155
Foster, G. M., 133
founder houses, 219
Four Corners region, 223
frameworks, 165, 167
fuel depletion, 314
fuel wood, 54

G

Galisteo Basin, 318
Gallina region, 318
Galton's problem, 236
Geertz, C., 196
general evolution, 205
general laws, 152, 155-156
general theory, 156
generalization, 199
Genetic Algorithm, 11, 272, 274
genetic tree, 301
goals of archaeology, 163, 170
Gould, S. J., 156, 200
Grand Canonical Ensemble, 299
Grasshopper burial data, 170
gray literature, 183

Great Basin Shoshoneans, 237
Great Houses, 215
growth
    and change models, 46
    disruption, 73
    rate, 116
    see also demography
Guadalupe Ruin, 31
Guatemala, 64
Gumerman, G., 269

H

Haas, J., 342
Habicht-Mauche, J., 353
Haisla, 238
Hallpike, C. W., 356
Hanen, M. P., 172
Hano, 236
Hanson, F. A., 128
Harappan Civilization, 19
Haury, E., 312, 323
health, 39, 54, 59, 72, 96
Hegman, M., 238
Hempel, C. G., 165, 167
Hempelian program, 164
hermeneutical archaeology, 25-26
high-frequency processes (HFP), 304
Hill, J. N., 166, 171
Hispano, 193-194, 196
historical
    explanation, 150-151, 155-156
    process, 157, 195, 199-200, 203-204, 227, 299
    reconstruction, 203
historiography, 342
history, 195
    as cultural difference, 195-196
Hodder, I., 155
Hohokam, 21, 29, 210-215, 217, 219, 241, 291, 312, 317, 322-323, 349
Hohokam Classic Period, 198
Hohokam regional system, 178-179
homeostasis, 40, 47
    models of, 47
Hopi, 60, 222-223, 236-238
Hosta Butte, 179
"how possibly" statements, 155, 173
household elements, 130
household-size measurement, 136
Huckell, B. B., 307
Huichol Indians, 194-196
human
    behavior, 151-152, 154, 156, 172
Human Relations Area File (HRAF), 178
Hunter-Anderson, R. D. L., 122
hunter-gatherers, 153
hydrological cycles, 52
hypothesis testing, 165

I

ideology, 165, 167
Indus Valley, 19
inference to the best explanation, 172
innovations, role of, 303
introspection as a method of inference, 153
Iroquois League, 240
isochrestic variation, 252

J

Jemez, 239
Johnson, G., 342
Joint Casas Grandes Project, 185
Jorgensen, G. P., 239
Jorgensen, J. G., 238
Judge, W. J., 168

K

Kane, A. E., 116
Katsina religion, 223
Kauffman, S. A., 300
Kayenta area, 318
Kelly, J. H., 172
Keresan, 239
Kin Bineola, 316
Kincaid settlement system, 125
King, Thomas, 26
Knowledge-Base Controlled Genetic Algorithm (KBGA), 274, 280-282
Kofyar field systems, 320
Kohler, T. A., 135, 313
Kuhn, T., 167
Kwakiutl, 238

L

Ladd, E., 225
land use, 279
large population aggregate, 124
Latin America, 320
laws
  of cultural behavior, 152
  of human behavior, 154, 156, 172
  of nature, 152
  of physics, 158
Leacock, E. B., 251, 254
LeBlanc, S., 118
Lee, R. B., 254, 48
Leonard, R. D., 302
Levy, J. E., 342, 350, 353
life tables, 45, 68, 79, 86, 89, 94, 98
Lightfoot, K. G., 134
Linton, R., 195
lithics, 270
Little Colorado network, 222, 224
lived human experience, 195
local cultivation, 311, 315
locational analysis, 46
Lofgren, L., 136
logical positivism, 167
lone scholar approach, 8
Longacre, W. A., 136, 166, 171-172, 233-234
low-frequency processes (LFP), 304

M

maize, 309, 313, 317, 320
Malthusian assumption, 41, 47, 54
Marana area, 317
Mark Twain, 199
Marquardt, W., 198
Martin, D., 52, 351
Marxism, 173
Matson, R. G., 123
Matthews, M. H., 313
McGuire, R., 342, 28
McPhee Village, 132
Mesa Verde, 19-20, 61, 80-81, 83-84, 97, 99, 318
Mesoamerica, 175, 195, 198
method, theory and explanation, 150
microeconomic theory, 17
Middle Little Colorado River, 20
Mimbres, 29, 214, 216
  Classic Mimbres, 179
  Mimbres Valley, 118-119, 316-317
Minnis, P. E., 317
models
  building, 181
  deductive nomological model, 165
  explaining, 168-169
  farmbelt model, 313
  homeostasis, 40, 47
  paleoepidemiological, 46
  redistribution, 168
  statistical relevance, 173
  time series, 42
  tribute, 168
modes of explanation, 171
Mogollon, 210-211, 213-214, 291, 311-312, 316-318, 322
  Mogollon Highlands, 306-307
  Mogollon Rim, 318
Mohawks, 240
moiety, 241
Moore, J. A., 307
mortality, 68, 95, 98
mortuary variability, 92
Moundville, 64
multiethnic sociopolitical interaction spheres, 221
multiple lines of evidence, 166, 172, 183
multiple perspectives, 172, 183
multiple working hypotheses, 183
Murdock, G. P., 235, 237, 241

N

Naroll, R., 249
narratives of the past, 155
natural science, 152
Navajo, 311
Near East, 175
Neher, H. V., 150
Neitzel, J., 249, 324
Nelson, B. A., 133, 136, 350
Netting, R. M., 251

new archaeology, 164, 195, 204, 342
New Mexico, 195
nomothetic, 204
non-cultural process, 151
novelty and originality, 167
NP-hard problems, 280
Nuvaqeotaka, 123-124

O

O'otam network, 222, 224
Ohio Hopewell region, 64
Onandagas, 240
oral traditions, 225
organizational state, 61, 65-66, 101
outliers, 124, 168
    see also Chaco Canyon

P

paleodemography, 44-45, 68-69
paleoepidemiological models, 46
paleontology, 151, 155-156
Palmer Drought Severity Indices, 117
Papagos, 241
Paquime, 91
pathologies, 72, 79, 83, 88, 93, 96
patterned grooving, 323
Pecos, 318
Pecos Classification, 29
phase transitions, 345
philosophy of science, 167
Phoenix Basin, 198, 215, 219, 221, 224
physical laws, 151
physical science, 152, 155
Picuris Pueblo, 193-194, 196-197, 199-200
pilgrimage model, 168
Pimas, 241
Pioneer Period, 323
pit structures, 114
pithouse architecture, 309
Plog, F., 52
Plog, S., 252-253, 314
political
    centralization, 61
    evolution, 245
    structures, 239-240
    systems, 246
    variability, 255
polychrome paint, 323
Popper, K., 167
population
    aggregation, 81
    density, 248-249
    estimates, 115-116, 118
    and agricultural productivity, 117, 122
    size, 248-249
    relative, 137
positivism, 186
Positivist Paradigm, 46
Post-Pleistocene Adaptations, 172
post-processualist, 164, 167
    critiques, 164-165
pottery, 212, 270, 308
    distribution of types, 238
    production, 303, 322, 323-324
    specializing in, 324
    see also ceramic breakage rates
    see also regional specific wares
Powell, M. L., 65
Prayer Rock, 311
pre-state farming communities, 175
principles of avalanches, 345
processual approach, 46
prognosis, 199
province, 269
proxy measures, 267
Pueblo Bonito, 89, 97
Pueblo III, 29
Pueblo IV, 29
Pueblo Patricio, 313

R

Raish, C., 175
reconstructed contingencies, 155
reconstructing culture history, 235
Red Mountain Phase, 312
Redding, R. W., 321
redistribution model, 168
redundancy, 309
regional
    alliance systems, 234
    systems, 221, 268-269
    traditions development of, 209
    specific wares, 210

residence rule, 171
Reynolds, R. G., 272, 274-275
Richerson, P. J., 273
Rio Grande Valley, 31-32, 222-223, 306, 318
Rio Puerco, 30-31, 276
Rio Salado, 31
River Yumans, 239
Roth, B., 307

S

Sackett, J. R., 252
Sahlins, M. D., 196, 205
Salado network, 221, 224
Salado polychromes, 167
Salinas Pueblos, 318
Salmon, W. C., 173
Salt-Gila Basin, 317
San Juan Basin, 27, 29-32, 80, 86, 200
San Juan Pueblo, 60
San Juan River, 311
San Lorenzo, 193-194, 196-198, 200
San Mateo Ixtatán, 136
Sand Canyon, 117
Santa Clara Pueblo, 60
Santa Cruz Phase, 324
SARG, 271, 292
scale, 171, 198
Schelberg, J., 168
schema, 3
schemata, 347
Schiffer, M. B., 151
Schlanger, S. H., 115-116, 313
science, 157
  and discovery, 157
  vs. humanity problem, 151
  philosophy of, 167
scientific, 154
  beliefs, 154
  confirmation, 155
  evaluation, 152, 156, 164
  explanation, 165
  method, 156, 164-165
seasonal transhumance, 31
Sebastian, L., 179, 316
second-level testing, 172
Sedentary Period, 324
sedentism, 125, 308, 314
self-organization, 345
Senecas, 240
Services, E. R., 205
Shabik'eschee Village, 310
Shan, 320
Snaketown, 323-324
social
  complexity theory, 347
  evolutionary theory, 344, 349, 352, 355
  network, 269
  organization, 251
  variability, 252
socio-political alliances, 180
socio-political factors, 181
sociological considerations in professional archaeology, 164, 167, 181
sociological issues, 182
sociopolitical inequality, 123
socipolitical complexity, 81
Sonoran Desert, 307
Southwest Anthropological Research Group (SARG), 8
Spanish, 195-196
Spaulding, A. C., 154
specialization, 302, 321
specific evolution, 205
stability, 304
statistical explanation, 154
statistical relevance model, 173
stepladder model, 348
Stewart, J. H., 205, 233, 291, 293
Stodder, A. W., 83
Stone, G. D., 320
storage strategies, 307, 310, 312
stratification, 49
stress, 43, 47, 53
Strong, W. D., 233
structure, 114, 195-196
  arrangement of, 175
  for determining population size, 116, 118-119, 115
  of history, 196
SU site, 312
supra-tribal political aggregates, 29
survivorship curves, 45

Swedlund, A., 350
Sweetwater phase, 323
symbols, 195-196
synchroneity, 321
synergistic mutualism, 17
synergy, 194
systems approach, 234
systems modeling, 245, 265
systems of group dependence, 256
Szuter, C. R., 319

T

Tainter, J. A., 168, 276, 325, 351
Tanoan, 239
Taos, 318
taphonomic studies, 151, 155
Tarahumara, 224
Tewa, 236, 239, 241
Textor, R. B., 236
Theories
  coherence, 150
  microeconomic, 17
  social complexity theory, 347
  social evolutionary theory, 344, 349
Theories of human behavior, 174
Tijeras Pueblo, 318
Tikal, 64
time dependency, 66, 101
time-series models, 42
Tiwa, 197
Toltec, 92
tools, 288
transformation process, 151, 155
tribute model, 168
Trigger, B. G., 150, 154, 163, 173
Tucson Basin, 307
turkeys, 315
Turner II, C. G., 136
Twain, M., 199

U

uneven development, 197-198
uniformitarian theory, 45
universals of human behavior, 152
Upham, S., 123, 234, 255, 272, 351

V

Vahki Phase, 312
Valley Nisenan, 239
Valley of Oaxaca, 125
Van West, C., 117, 179
VGA, 275
Virgin River, 20
Vivian, R. G., 158, 179, 316

W

Wallersteinian world-system, 28
water clowns, 194, 197
White, L. A., 173, 205
Wiessner, P. W., 252-253
Wilcox, D. R., 219
Wilk, R. R., 138, 251
Willey, G. R., 163
Wills, C., 352-353
Wills, W. H., 122, 310
Wilson, C. D., 324
Windes, T. C., 310
Wobst, M., 153
Wolf, 199
World Systems approach, 167
Wylie, A., 155

Y

Yoffee, N., 175
Yuman, 224

Z

Zuni, 222-223
  area, 120
  origin story, 225
  region, 318